Supramolecular Structure and Function 8

Supramolecular Structure and Function 8

Edited by

Greta Pifat-Mrzljak
Ruđer Bošković Institute
Zagreb, Croatia

Kluwer Academic / Plenum Publishers
New York, Boston, Dordrecht, London, Moscow

Library of Congress Cataloging-in-Publication Data

Supramolecular structure and function 8 / edited by Greta Pifat-Mrzljak.
p. cm.
"Proceedings of the 8th International Summer School in Biophysics: Supramolecular Structure and Function, held September 14-26, 2003 in Rovinj, Croatia"
Includes bibliographical references and index.
ISBN 0-306-48661-X -- ISBN 0-306-48662-8 (ebook)
1. Proteins--Structure--Congresses. 2. Proteins--Conformation--Congresses. I. Pifat-Mrzljak, Greta. II. International School on Biophysics "Supramolecular Structure and Function" (8th : 2003 : Rovinj, Croatia)
QP551.S89 2004
572'.633--dc22

2004051662

Proceedings of the 8th International Summer School on Biophysics: Supramolecular Structure and Function, held September 14–26, 2003, at Red Island, Rovinj, Croatia.

ISBN 0-306-48661-X
eISBN: 0-306-48662-8

233 Spring Street, New York, New York 10013

http://www.wkap.nl/

10 9 8 7 6 5 4 3 2 1

A C.I.P. record for this book is available from the Library of Congress

Printed in the United States of America

Preface

An enormous amount of new knowledge on the molecular basis of various biological phenomena has emerged in the rapidly expanding field of bioscience. Since the frontiers in scientific research are difficult to define, the creation of new knowledge depends not only on new methods and concepts but also on interaction with other fields of research. The principles and methods of biophysics should be a rational language for discussion not only between scientists of the different disciplines of natural sciences, such as physics, mathematics, biochemistry, molecular biology and biotechnology, but also for medicine and social sciences as well. This is the general philosophy behind the organization of the Summer Schools organized by Rudjer Bošković Institute, Zagreb, Croatia and the Croatian Biophysical Society.

The International Summer Schools on Biophysics have a very broad scope. This is in contrast to the other workshops or schools which are centred mainly on one topic or technique. The intention was to organize courses which provided advanced training at doctoral or postdoctoral level in biosciences. Therefore, the Schools essentially have a catalytic role and are complementary to, rather than competing with, activities of parallel national or international programmes.

Internationally recognized and successfully established these eight international summer schools have been organized under the title «Supramolecular Structure and Function». The Schools were devoted to the structure-function relationship of biological macromolecules and to mayor biophysical techniques. The Biophysics Schools provide a significant contribution to the capacity building of science in Europe. This is in line with UNESCO's declaration that, "it should be increased, support to regional and international programmes of higher education and to networking of graduate and postgraduate institutions". The organizers are deeply committed to the idea that science, scientific and educational collaborations help to create the European framework we all want to be a part of. These summer schools, as Master Classes of UNESCO, supported by IUPAB and UNESCO, could be a part of the mosaic forming the European Research Area (ERA) and the European Higher Education Area (EHEA). The organizers of the International Summer School on Biophysics think, therefore, that the publication of this volume and its distribution within the scientific community will serve towards the objectives of expanding, sharing and providing easy access to scientific knowledge.

The publication of the volume is due to the substantial financial support provided by the UNESCO Venice Office - Regional Office for Science and Technology for Europe (UVO-ROSTE) as well as by the intellectual efforts of Prof. Greta Pifat-Mrzljak from the Ruđer Bošković Institute - initiator of the series of the International Schools on Biophysics and Chairperson of the 8th School held in Rovinj in 2003.

Organizing Committee

Contents

Structural Basis of Protein Synthesis

CHRISTIANE SCHAFFITZEL AND NENAD BAN
Institute for Molecular Biology and Biophysics, Swiss Federal Institute of Technology, ETH Hönggerberg HPK, CH - 8093 Zürich, Switzerland

In the cell, proteins are synthesized from amino acids by the ribosome using messenger RNA as the template. The ribosome is a large macromolecular assembly consisting of RNA and proteins. In bacteria, the translating 70S ribosome is formed from two unequally sized subunits: the large 50S and the small 30S subunit. Structures of both ribosomal subunits have been determined at near-atomic resolution. The large ribosomal subunits are from the halophilic archaebacterium *Haloarcula marismortui*[1] and from the eubacterium *Deinococcus radiodurans*[2]. The small subunit has been solved from the thermophilic eubacterium *Thermus thermophilus*[3,4]. The structure of the 70S ribosome was determined from *Thermus thermophilus* with mRNA and tRNAs at a resolution of 5.5 Å[5]. More recently, the 70S ribosome from *Escherichia coli* was published at a resolution of 9 Å[6]. Furthermore, several ribosomal structures in complex with tRNA analogues[7-10], small translation factors[11] and antibiotics[8,9,12-15] were solved giving further insight in the molecular mechanism of protein synthesis.

The small subunit of the ribosome is responsible for binding mRNA and selection of the cognate aminoacyl-tRNA. Based on early electron microscopic studies, the 30S subunit is classically divided into head, body, neck, shoulder and platform (Fig. 1). It consists of one 16S rRNA chain and 20 ribosomal proteins (S1-S20). The shape of the small subunit is mostly determined by ribosomal RNA. The proteins are distributed over the top, sides and back of the 30S subunit, while the interface with the 50S subunit is

Supramolecular Structure and Function 8, Edited by Pifat-Mrzljak
Kluwer Academic/Plenum Publishers, New York 2004

mostly formed by rRNA[3,4]. The Shine Dalgarno (SD) helix at the 3' terminal tail of 16S rRNA, which interacts with the ribosome binding site (SD sequence) on the mRNA is located between the head and the platform of the small subunit.

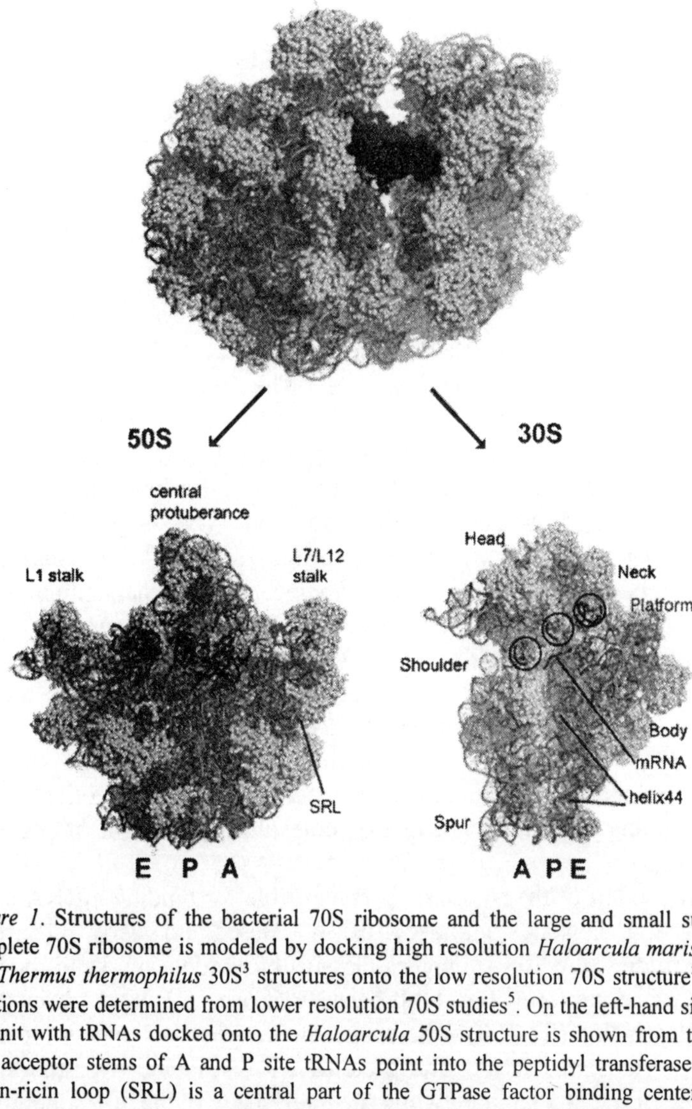

Figure 1. Structures of the bacterial 70S ribosome and the large and small subunits. The complete 70S ribosome is modeled by docking high resolution *Haloarcula marismortui* 50S[1] and *Thermus thermophilus* 30S[3] structures onto the low resolution 70S structure[5]. The tRNA positions were determined from lower resolution 70S studies[5]. On the left-hand side, the large subunit with tRNAs docked onto the *Haloarcula* 50S structure is shown from the interface. The acceptor stems of A and P site tRNAs point into the peptidyl transferase center. The sarcin-ricin loop (SRL) is a central part of the GTPase factor binding center. The three characteristic protuberances of the large subunit are labeled. On the right-hand side, the 30S subunit is shown from the interface with the A, P and E site tRNA anticodon stem-loops. The anticodon loops of the A and P tRNAs contact the mRNA. Architectural characteristics and important features of 30S are labeled.

The large ribosomal subunit is mostly composed of 23S and 5S rRNA and contains approximately 31 to 35 proteins (L1 – L35). The 50S structure shows three characteristic protuberances: the L1 stalk, the central protuberance and the L7/L12 stalk (Fig. 1). The stalks are involved in interactions with the small subunit or interact with tRNAs or GTP-binding translation factors and thus are likely to be dynamic elements of the 50S. In the large subunit, the ribosomal proteins are also not distributed equally over the ribosome, but cluster at the solvent exposed surface[1,2]. They have extensions, which play an important role in the stabilization of the RNA tertiary structure[1]. Besides a globular domain, many ribosomal proteins have extensions or tails that penetrate far into the rRNA core and closely interact with it, thereby stabilizing the RNA domains.

Peptide synthesis occurs in cycles of distinct steps (Fig. 2). The two most important steps in elongation - decoding and peptidyl transfer - are allotted to the two ribosomal subunits. Thus, the crystal structures of the separate subunits revealed many mechanistic details of these processes. The atomic structures of the large ribosomal subunit in complex with substrates and inhibitors demonstrate that the peptidyl transferase center (PTC) is exclusively composed of rRNA[7]. The structure of the small subunit also unequivocally shows that ribosomal RNA is responsible for all functional interactions in the decoding center[13]. Therefore, these structures substantiate that ribosomal RNA is in fact responsible for all aspects of peptide synthesis. In other words, the ribosome is a ribozyme. This finding strongly supports the pre-protein RNA world hypothesis.

Except during initiation, the first step in an elongation cycle is the selection of the correct aminoacyl-tRNA in the A site of the small subunit, i.e. in the decoding center. This selection step determines the 10^{-3} to 10^{-4} accuracy of translation, which corresponds to one error in 3000 amino acids. The ribosome reaches this high specificity mainly by monitoring the Watson-Crick base pair geometry of the mRNA codon and tRNA anticodon duplex, i.e. the positioning of the phosphate-sugar backbone and the major and minor groove. 16S rRNA plays a critical role in this process. In case of a correct codon-anticodon interaction, two adenines of the 16S rRNA (A1492 and A1493) flip out of an internal loop of helix44 and form A-minor motif tertiary interactions with the first two base pairs of the codon-anticodon helix [8,16] (Fig.3). The third wobble position base pair is less stringently checked. Here, when a cognate aminoacyl-tRNA is bound, a guanine (G530) flips towards the minor groove of the codon anticodon helix by switching from a syn- to an anti-conformation. G530 interacts with the second and the third base pair occupying the minor groove of the third base pair only partially. The third base pair also interacts with C1054 and contacts proline48 of ribosomal protein S12 in a metal-mediated interaction (Fig.3).

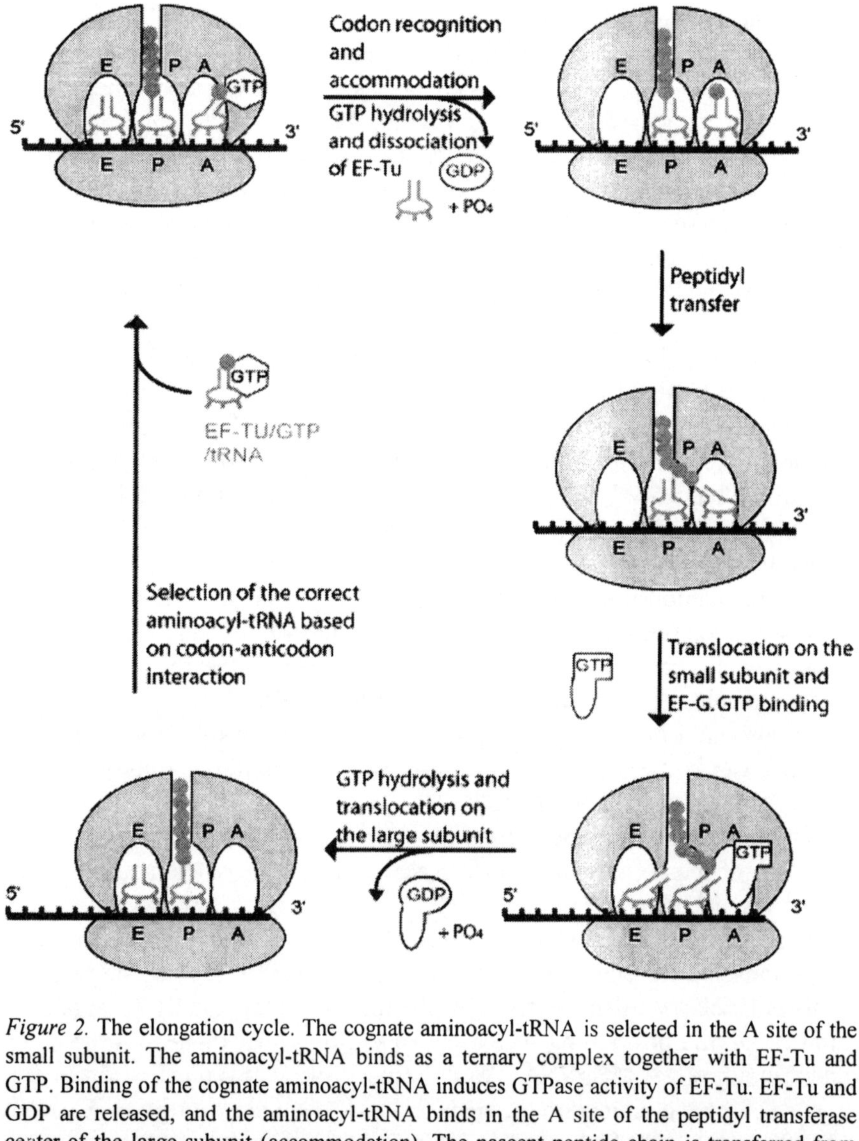

Figure 2. The elongation cycle. The cognate aminoacyl-tRNA is selected in the A site of the small subunit. The aminoacyl-tRNA binds as a ternary complex together with EF-Tu and GTP. Binding of the cognate aminoacyl-tRNA induces GTPase activity of EF-Tu. EF-Tu and GDP are released, and the aminoacyl-tRNA binds in the A site of the peptidyl transferase center of the large subunit (accommodation). The nascent peptide chain is transferred from the P site peptidyl-tRNA to the amino group of the A site aminoacyl-tRNA. The peptide chain leaves the peptidyl transferase center through the ribosomal tunnel, which spans the large subunit. According to the hybrid state model, translocation of mRNA and tRNA on the small subunit occurs spontaneously after peptidyl transfer. EF-G catalyzed GTP hydrolysis drives translocation of tRNAs from the A to the P site and from the P to the E site on the large subunit.

This type of interactions is not possible if the codon-anticodon interaction is non- or near-cognate[9]. Consistent with their important role in decoding, A1492, A1493 and G530 are universally conserved and essential for cell viability suggesting that the mechanism of decoding is similar in all kingdoms of life.

Aminoacyl-tRNA binds to the ribosome as a tertiary complex with EF-Tu and GTP (Fig. 2). The initial, low affinity interaction of this complex with the ribosome takes place only in the decoding center[17-19]. The formation of the cognate codon-anticodon interactions rapidly triggers GTP hydrolysis by EF-Tu. This is accompanied by a transition of the small subunit from an open to a closed conformation where the aminoacyl-tRNA is tighter bound. This induced fit is a further source of selectivity during decoding, providing a "proofreading mechanism", since near-cognate tRNA is frequently rejected at this step. EF-Tu and GDP dissociate from the aminoacyl-tRNA, and the acceptor stem of the aminoacyl-tRNA can enter the peptidyl transferase center on the large subunit[20].

The structure of the large subunit in complex with transition state analogues and substrate mimics shed light on the catalytic step of the elongation cycle, the peptidyl transfer reaction[7,21-23]. After accommodation, the 3'CCA end of the aminoacyl-tRNA is bound to the A loop in the PTC. A second P loop is base paired to the 3' CCA end of the peptidyl-tRNA (Fig. 4). The active site loops position the two acylated tRNAs precisely such that the α-amino group of the acceptor (A) site aminoacyl-tRNA is close enough to the carbonyl carbon of the peptidyl-tRNA (P) site for the nucleophilic attack. Subsequently, a tetrahedral intermediate is formed (Fig. 5). The Yarus inhibitor, which was soaked in the 50S crystal, mimics this transition state[7]. The formation of the peptide bond results in a deacylated tRNA in the P site, and the A site aminoacyl-tRNA becomes the new peptidyl-tRNA. The peptidyl transfer occurs spontaneously, i.e. without additional energy input. The ribosome catalyses this reaction by accurate positioning of the two substrates. Whether the peptidyl transfer reaction is further enhanced by acid-base catalysis involving the universally conserved adenosine A2451 (*E. coli* numbering) is still investigated[24-27]. Evidence has been presented that peptide bond formation in the PTC involves an ionizable group, which however could not be unambiguously assigned to A2451[28]. The uncatalyzed reaction occurs approximately at a very slow rate, whereas the ribosome enhances the peptide bond formation by a factor of 10^5 to a rate of 50^{-s}. At pH lower than 7, the ionizable group is protonated in the PTC, and the reaction rate is reduced by two orders of magnitude. This demonstrates that the positioning effect accounts for more than 90% of the reaction rate and that the acid base catalysis may be responsible for an additional 100-fold rate increase[28,29]. The conservation of the residues in the PTC suggests that the reaction mechanism is the same in all species.

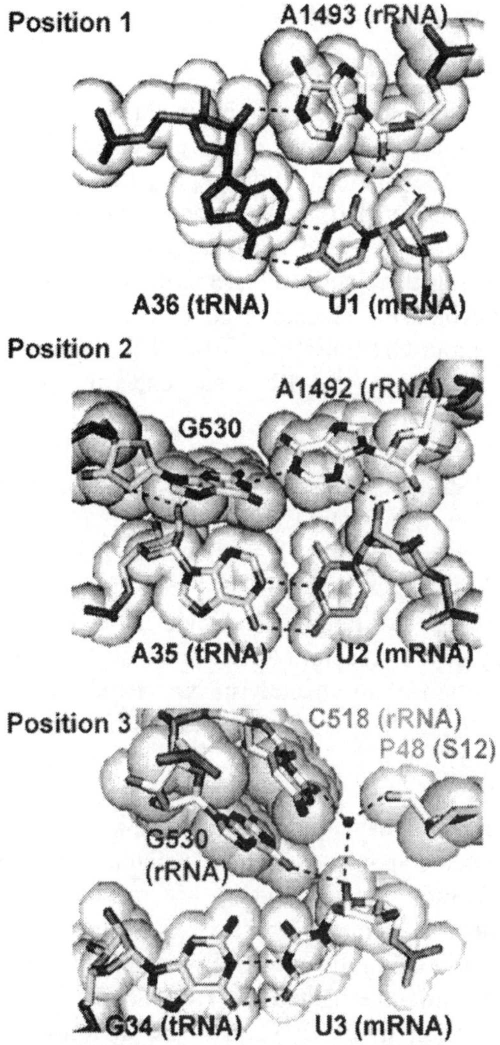

Figure 3. Codon recognition in the decoding center at the first, second and third base position. (1) Adenine1493 binds into the minor groove of the first A-U base pair in an A-minor motif type interaction. A1493 forms hydrogen bonds to the 2'OH group of the tRNA with its N1 atom and to the 2' OH group of the mRNA base with its 2'OH group. An additional hydrogen bond is formed between the 2'OH group of A1493 and the O2 group of uracil in the mRNA. (2) In the second position, A1492 and G530 both bind into the minor groove and form hydrogen bonds with their N1 atoms. A1492 interacts with the 2'OH group of the mRNA via its N3 and 2'OH group. G530 forms two hydrogen bonds with its 2'OH and N3 and the 2'OH group of the tRNA respectively. (3) The third base-pair is monitored less stringently (wobble position). O6 of G530 forms a hydrogen bond to the 2'OH group of the mRNA. A magnesium ion-mediated interaction occurs between the 2'OH group of the mRNA and O2 of C518, as well as the main chain carbonyl of proline48 of protein S12. The G-U sheared mismatch base-pair is not selected against at the third position.

The 3'CCA ends of A and P site tRNAs bound to the A and P-loops of 23S rRNA are related by approximately 180°. This relationship would result in a rotational movement of the peptide after peptidyl transfer and translocation, which has been suggested to aid in the transport of the nascent peptide through the tunnel[23].

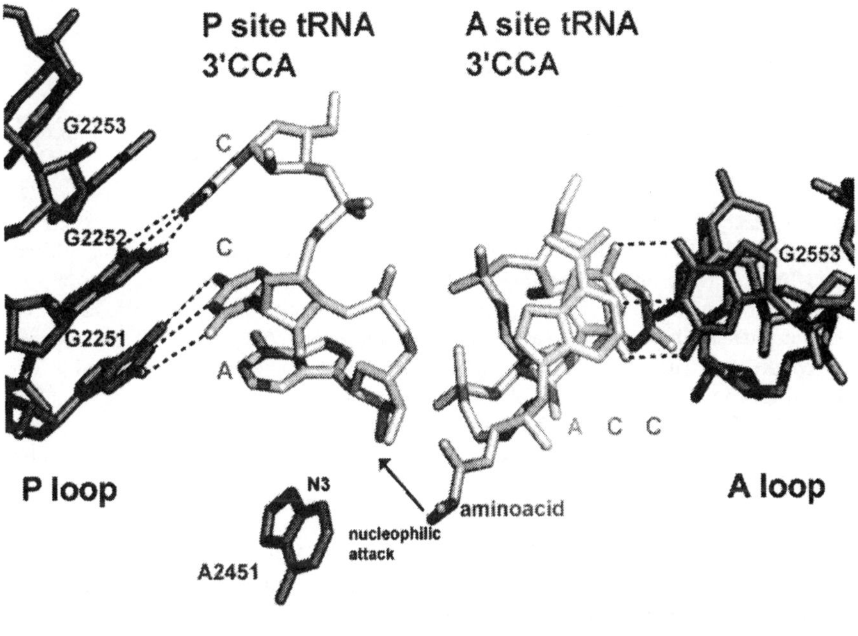

Figure 4. The peptidyl transferase center of the large subunit with P site and A site substrate analogues. In the P site, the cytosines of the peptidyl-tRNA 3'CCA end form a base pair with two guanines (G2251 and G2252) of the P-loop. In the A site, the second base of aminoacyl-tRNA 3'CCA end pairs with guanine 2553 of the A-loop. The structure demonstrates how the ribosome positions the two tRNAs relative to each other thereby facilitating the nucleophilic attack of the α-amino-group of the aminoacyl-tRNA. A2451 is part of the central loop in the peptidyl transferase center, and it may play a role in acid-base catalysis.

The exit tunnel for the growing nascent polypeptide chain is the second prominent feature of the large ribosomal subunit[1,7]. The tunnel protects the nascent chain from proteases. More recently, it was shown that some sequences interact with the tunnel and thereby influence translation elongation and termination[30]. These findings implicated that the tunnel is not a neutral environment for the peptide but is able to discriminate between sequences and may have a regulatory function in gating. The tunnel spans the entire body of 50S starting from the PTC to the bottom of the backside of

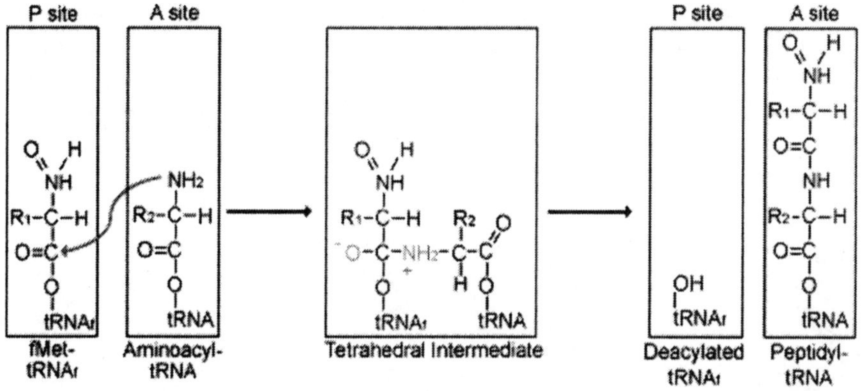

Figure 5. Mechanism of peptide bond formation catalyzed in the peptidyl transferase center of the large subunit. The substrates of the peptidyl-transferase reaction are aminoacyl-tRNA and peptidyl-tRNA (here: formyl-methionine-tRNA/ initiator-tRNA) binding to the ribosomal A site and P site respectively. The α-amino group of the aminoacyl-tRNA in the A site of the ribosome attacks the carbonyl carbon atom of the ester bond that links the nascent polypeptides to P site bound tRNAs. Thereby, a tetrahedral intermediate is formed at the carbonyl group with a chiral carbon atom and negatively charged oxygen. The 2' (or 3') OH group of the tRNA 3' CCA end is the leaving group of the tetrahedral intermediate, and a new peptide bond (amide bond) is generated, resulting in a peptidyl-tRNA in the A site and a deacylated tRNA in the P site.

the 50S subunit. It is approximately 100 Å in length and has a diameter of about 15 Å. The tunnel can accommodate 30-40 amino acid residues of a growing polypeptide chain. The polypeptide exit tunnel is largely formed by 23S rRNA, but has significant contributions from ribosomal proteins L4 and L22 (Fig. 6). L4 and L22 fence the most constricted part of the tunnel with a diameter of 10 Å[7]. Macrolide antibiotics have been soaked into the 50S crystals and shown to bind between the constriction and the PTC[14,15,31]. Thereby, these antibiotics obstruct the tunnel further and prevent passage of the growing nascent peptide chain. The tunnel exit is encircled by ribosomal proteins L19, L22, L23, L24, L29 and L31. Some of these proteins, L23 in particular, are implicated to play a role in cotranslational protein folding and can interact with molecular chaperones such as the trigger factor[32] and the signal recognition particle involved in targeting of proteins[33].

The last step in the elongation cycle is the translocation of the mRNA and the peptidyl-tRNA from the A site to the P site of both subunits. The deacylated tRNA in the P site moves to the empty (E) site. This concerted movement has to be very accurate in order to avoid frame-shifting or abortion of translation due to incorrect placement of the peptidyl-tRNA in the PTC. Translocation can occur spontaneously, but it is accelerated considerably by GTP hydrolysis of EF-G•GTP. Cryo-electron microscopy

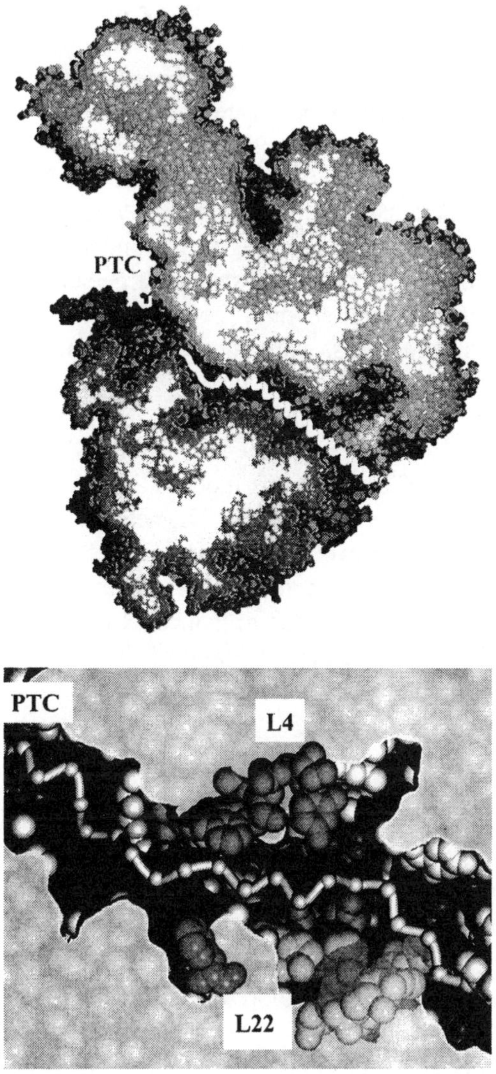

Figure 6. The polypeptide exit tunnel. In the upper panel, the 50S subunit is cut in half such that the tunnel is shown in its entire length. A nascent polypeptide chain is modeled into the tunnel. The polypeptide is still connected to the peptidyl transferase center and can adopt an α-helical conformation within the tunnel. In the bottom panel, a longitudinal section of the ribosomal tunnel is shown. The section includes the upper half of the tunnel, which is formed by 23S rRNA and proteins L4 and L22.

(cryo-EM) structures of 70S with EF-G stalled in the GTP conformation by the antibiotic fusidic acid show that EF-G is mimicking aminoacyl-tRNA in complex with EF-Tu by binding to the A site on the small subunit[34,35] (Fig. 2). On the large subunit, EF-G•GTP binds to the sarcin-ricin loop (SRL) and the L7/L12 stalk. Either the SRL or the ribosomal protein L11 function as a GTPase activator. The current hybrid state model of translocation[36] involves separate movement of the tRNAs in the two subunits, which requires conformational changes in the 70S ribosome (Fig. 2). Indeed, it has been observed by cryo-EM that the two subunits can rotate 6° relative to each other upon EF-G binding[37,38]. According to this model, tRNAs can occupy different tRNA binding sites on the small and large subunit, corresponding to a hybrid state. On the small subunit, the mRNA moves in complex with the tRNA anticodon stem-loops, while the acceptor stems of the tRNAs are fixed in their respective binding site of the large subunit. Thus, the A site on 30S is empty and EF-G can bind to this hybrid state ribosome where the A and P sites are occupied on 50S and the tRNAs moved to the P and E sites on 30S. Subsequently, GTP hydrolysis induces the movement of the tRNA acceptor stems on 50S to the P and E site. The deacylated tRNA stays in the E site because the L1 stalk prevents its release. Once the cognate aminoacyl-tRNA is bound and the 30S conformation changes, the E site tRNA dissociates. E site tRNA is implicated to induce a low affinity binding mode in the A site of 30S, which reduces binding of the ternary complex to the ribosome to codon-anticodon interactions. Thereby, only cognate and near-cognate aminoacyl-tRNAs can bind[29].

To date, no crystallographic study of the ribosome in complex with elongation or termination factors has been reported. Consequently, most structural information about the interaction of the ribosome with translation factors and the conformational changes of the ribosome during the elongation cycle has been obtained using cryo-EM. Current research on the ribosome and translation is directed towards understanding the interplay of the two subunits, interactions of the ribosome with translation factors and proteins involved in folding or export of the nascent polypeptide chain. Cryo-EM structures of many of these complexes have been solved, providing a good model for these interactions and evidence for the structural dynamics of the translating ribosome. Ultimately, however, all relevant intermediates, complexes and conformational states of the ribosome need to be characterized at high resolution to completely understand translation and associated processes such as cotranslational protein folding and translocation at the molecular level.

ACKNOWLEDGEMENT

C.S. is supported by a postdoctoral fellowship from the Ernst Schering Research Foundation. We gratefully acknowledge support by the Swiss National Science Foundation, NCCR Structural Biology of the SNSF and a Young Investigator Grant from the Human Frontier Science Program.

REFERENCES

1. Ban, N., Nissen, P., Hansen, J., Moore, P. B., and Steitz, T. A., 2000, The complete atomic structure of the large ribosomal subunit at 2.4 A resolution. *Science* **289:** 905-920.
2. Harms, J., Schluenzen, F., Zarivach, R., Bashan, A., Gat, S., Agmon, I., Bartels, H., Franceschi, F., and Yonath, A., 2001, High resolution structure of the large ribosomal subunit from a mesophilic eubacterium. *Cell* **107:** 679-688.
3. Wimberly, B. T., Brodersen, D. E., Clemons, W. M., Jr., Morgan-Warren, R. J., Carter, A. P., Vonrhein, C., Hartsch, T., and Ramakrishnan, V., 2000, Structure of the 30S ribosomal subunit. *Nature* **407:** 327-339.
4. Schluenzen, F., Tocilj, A., Zarivach, R., Harms, J., Gluehmann, M., Janell, D., Bashan, A., Bartels, H., Agmon, I., Franceschi, F., and Yonath, A., 2000, Structure of functionally activated small ribosomal subunit at 3.3 angstroms resolution. *Cell* **102:** 615-623.
5. Yusupov, M. M., Yusupova, G. Z., Baucom, A., Lieberman, K., Earnest, T. N., Cate, J. H., and Noller, H. F., 2001, Crystal structure of the ribosome at 5.5 A resolution. *Science* **292:** 883-896.
6. Vila-Sanjurjo, A., Ridgeway, W., Seymaner, V., Zhang, W., Santoso, S., Yu, K., and Cate, J. H., 2003, X-ray crystal structures of the WT and a hyper-accurate ribosome from Escherichia coli. *Proc. Natl. Acad. Sci. U. S. A.* **100:** 8682-8687.
7. Nissen, P., Hansen, J., Ban, N., Moore, P. B., and Steitz, T. A., 2000, The structural basis of ribosome activity in peptide bond synthesis. *Science* **289:** 920-930.
8. Ogle, J. M., Brodersen, D. E., Clemons, W. M., Jr., Tarry, M. J., Carter, A. P., and Ramakrishnan, V., 2001, Recognition of cognate transfer RNA by the 30S ribosomal subunit. *Science* **292:** 897-902.
9. Ogle, J. M., Murphy, F. V., Tarry, M. J., and Ramakrishnan, V., 2002, Selection of tRNA by the Ribosome Requires a Transition from an Open to a Closed Form. *Cell* **111:** 721-732.
10. Schmeing, T. M., Moore, P. B., and Steitz, T. A., 2003, Structures of deacylated tRNA mimics bound to the E site of the large ribosomal subunit. *RNA* **9:** 1345-1352.
11. Carter, A. P., Clemons Jr, W. M., Brodersen, D. E., Morgan-Warren, R. J., Hartsch, T., Wimberly, B. T., and Ramakrishnan, V. V., 2001, Crystal Structure of an Initiation Factor Bound to the 30S Ribosomal Subunit. *Science* **292:** 498-501.
12. Brodersen, D. E., Clemons, W. M., Jr., Carter, A. P., Morgan-Warren, R. J., Wimberly, B. T., and Ramakrishnan, V., 2000, The structural basis for the action of the antibiotics tetracycline, pactamycin, and hygromycin B on the 30S ribosomal subunit. *Cell* **103:** 1143-1154.
13. Carter, A. P., Clemons, W. M., Brodersen, D. E., Morgan-Warren, R. J., Wimberly, B. T., and Ramakrishnan, V., 2000, Functional insights from the structure of the 30S ribosomal subunit and its interactions with antibiotics. *Nature* **407:** 340-348.

14. Schlunzen, F., Zarivach, R., Harms, J., Bashan, A., Tocilj, A., Albrecht, R., Yonath, A., and Franceschi, F., 2001, Structural basis for the interaction of antibiotics with the peptidyl transferase centre in eubacteria. *Nature* **413:** 814-821.
15. Hansen, J. L., Ippolito, J. A., Ban, N., Nissen, P., Moore, P. B., and Steitz, T. A., 2002, The structures of four macrolide antibiotics bound to the large ribosomal subunit. *Mol. Cell* **10:** 117-128.
16. Nissen, P., Ippolito, J. A., Ban, N., Moore, P. B., and Steitz, T. A., 2001, RNA tertiary interactions in the large ribosomal subunit: the A-minor motif. *Proc. Natl. Acad. Sci. U. S. A.* **98:** 4899-4903.
17. Stark, H., Rodnina, M. V., Wieden, H. J., Zemlin, F., Wintermeyer, W., and van Heel, M., 2002, Ribosome interactions of aminoacyl-tRNA and elongation factor Tu in the codon-recognition complex. *Nat. Struct. Biol.* **9:** 849-854.
18. Valle, M., Sengupta, J., Swami, N. K., Grassucci, R. A., Burkhardt, N., Nierhaus, K. H., Agrawal, R. K., and Frank, J., 2002, Cryo-EM reveals an active role for aminoacyl-tRNA in the accommodation process. *Embo J.* **21:** 3557-3567.
19. Valle, M., Zavialov, A., Li, W., Stagg, S. M., Sengupta, J., Nielsen, R. C., Nissen, P., Harvey, S. C., Ehrenberg, M., and Frank, J., 2003, Incorporation of aminoacyl-tRNA into the ribosome as seen by cryo-electron microscopy. *Nat. Struct. Biol.* **10:** 899-906.
20. Ramakrishnan, V., 2002, Ribosome Structure and the Mechanism of Translation. *Cell* **108:** 557-572.
21. Schmeing, T. M., Seila, A. C., Hansen, J. L., Freeborn, B., Soukup, J. K., Scaringe, S. A., Strobel, S. A., Moore, P. B., and Steitz, T. A., 2002, A pre-translocational intermediate in protein synthesis observed in crystals of enzymatically active 50S subunits. *Nat. Struct. Biol.* **9:** 225-230.
22. Hansen, J. L., Schmeing, T. M., Moore, P. B., and Steitz, T. A., 2002, Structural insights into peptide bond formation. *Proc. Natl. Acad. Sci. U. S. A.* **99:** 11670-11675.
23. Bashan, A., Agmon, I., Zarivach, R., Schluenzen, F., Harms, J., Berisio, R., Bartels, H., Franceschi, F., Auerbach, T., Hansen, H. A. S., Kossoy, E., Kessler, M., and Yonath, A., 2003, Structural Basis of the Ribosomal Machinery for Peptide Bond Formation, Translocation, and Nascent Chain Progression. *Mol. Cell* **11:** 91-102.
24. Bayfield, M. A., Dahlberg, A. E., Schulmeister, U., Dorner, S., and Barta, A., 2001, A conformational change in the ribosomal peptidyl transferase center upon active/inactive transition. *Proc. Natl. Acad. Sci. U. S. A.* **98:** 10096-10101.
25. Polacek, N., Gaynor, M., Yassin, A., and Mankin, A. S., 2001, Ribosomal peptidyl transferase can withstand mutations at the putative catalytic nucleotide. *Nature* **411:** 498-501.
26. Thompson, J., Kim, D. F., O'Connor, M., Lieberman, K. R., Bayfield, M. A., Gregory, S. T., Green, R., Noller, H. F., and Dahlberg, A. E., 2001, Analysis of mutations at residues A2451 and G2447 of 23S rRNA in the peptidyltransferase active site of the 50S ribosomal subunit. *Proc. Natl. Acad. Sci. U. S. A.* **98:** 9002-9007.
27. Parnell, K. M., Seila, A. C., and Strobel, S. A., 2002, Evidence against stabilization of the transition state oxyanion by a pKa-perturbed RNA base in the peptidyl transferase center. *Proc. Natl. Acad. Sci. U. S. A.* **99:** 11658-11663.
28. Katunin, V., Muth, G., Strobel, S., Wintermeyer, W., and Rodnina, M., 2002, Important Contribution to Catalysis of Peptide Bond Formation by a Single Ionizing Group within the Ribosome. *Mol. Cell* **10:** 339-346.
29. Wilson, D. N., and Nierhaus, K. H., 2003, The ribosome through the looking glass. *Angew. Chem. Int. Ed.* **42:** 3464-3486.
30. Tenson, T., and Ehrenberg, M., 2002, Regulatory nascent peptides in the ribosomal tunnel. *Cell* **108:** 591-594.

31. Berisio, R., Schluenzen, F., Harms, J., Bashan, A., Auerbach, T., Baram, D., and Yonath, A., 2003, Structural insight into the role of the ribosomal tunnel in cellular regulation. *Nat. Struct. Biol.*, **10:** 366-370.
32. Kramer, G., Rauch, T., Rist, W., Vorderwulbecke, S., Patzelt, H., Schulze-Specking, A., Ban, N., Deuerling, E., and Bukau, B., 2002, L23 protein functions as a chaperone docking site on the ribosome. *Nature* **419:** 171-174.
33. Pool, M. R., Stumm, J., Fulga, T. A., Sinning, I., and Dobberstein, B., 2002, Distinct modes of signal recognition particle interaction with the ribosome. *Science* **297:** 1345-1348.
34. Agrawal, R. K., Heagle, A. B., Penczek, P., Grassucci, R. A., and Frank, J., 1999, EF-G-dependent GTP hydrolysis induces translocation accompanied by large conformational changes in the 70S ribosome. *Nat. Struct. Biol.* **6:** 643-647.
35. Stark, H., Rodnina, M. V., Wieden, H. J., van Heel, M., and Wintermeyer, W., 2000, Large-scale movement of elongation factor G and extensive conformational change of the ribosome during translocation. *Cell* **100:** 301-309.
36. Noller, H. F., Yusupov, M. M., Yusupova, G. Z., Baucom, A., and Cate, J. H., 2002, Translocation of tRNA during protein synthesis. *FEBS Lett.* **514:** 11-16.
37. Frank, J., and Agrawal, R. K., 2000, A ratchet-like inter-subunit reorganization of the ribosome during translocation. *Nature* **406:** 318-322.
38. Gao, H., Sengupta, J., Valle, M., Korostelev, A., Eswar, N., Stagg, S. M., Van Roey, P., Agrawal, R. K., Harvey, S. C., Sali, A., Chapman, M. S., and Frank, J., 2003, Study of the structural dynamics of the E coli 70S ribosome using real-space refinement. *Cell* **113:** 789-801.

The Relationship Between Protein Sequence, Structure and Function:

Protein Function Prediction

ANNA TRAMONTANO AND DOMENICO COZZETTO

Department of Biochemical Sciences, University of Rome "La Sapienza". P.le Aldo Moro, 5. 00185 Rome, Italy

1. INTRODUCTION

The organization of the information in genomes can be very complex, especially in eukaryotes[1,2]. For example, only 3% of the human genome seems to be involved in directing th synthesis of biological molecules which, in turn, interact with each other to perform those functions that we associate with the concept of life. Finding the coding regions in a higher organism genome can be much more difficult than finding a needle in a haystack. While the difference between a needle and a hay straw is obvious, the difference between coding and non coding regions is not equally evident. At first sight, both of them appear as a sequence of nucleotides without apparent regularities. Unless an evolutionary related sequence has been already experimentally characterized, we need sophisticated statistical analysis to distinguish between them.

Once the coding regions of a genome have been identified, they can be "translated" into the sequence of their products, for example proteins and RNA molecules so that we can address the question of which is their function.

Defining function in biology is also non trivial. Each gene product has a molecular, biological and cellular function. The function of thrombin, for example, can be defined as an enzyme (molecular level), part of the blood coagulation cascade (biological level) and as an extra-cellular protein (cellular level)[3]. Furthermore, molecular function can be defined at different

Supramolecular Structure and Function 8, Edited by Pifat-Mrzljak
Kluwer Academic/Plenum Publishers, New York 2004

levels of detail. The enzymatic molecular function of thrombin can be characterized (with increasing resolution) as a hydrolase, a peptidase, an endopeptidase, a protease, a serine protease, and, finally, as thrombin.

If we want to interfere with the activity of a gene product, for medical, biotechnological or investigative purposes, we need to know its function at the molecular, biological and cellular level. Only in this case we can try to interfere with the biological process by inhibiting the molecular function with a compound targeted to the appropriate cellular compartment.

By and large, there are three routes we can follow to elucidate the function of a gene product: we can perform experiments, look for evolutionary related gene products whose function has already been characterized, or try to compute its three-dimensional shape and use it to infer function.

We will not discuss here the experimental approaches and the methods to analyze and interpret the data they produce, although it is important to mention that new large-scale experiments have been made possible by the genomic knowledge and are fostering the continuous development of innovative specialized computational tools.

We will instead focus on the exploitation of our understanding of molecular evolution for assigning function to a gene product and for trying to "predict" its three-dimensional structure.

2. EVOLUTIONARY APPROACHES TO FUNCTION ASSIGNMENT

Life is the product of evolution: changes in the genetic material might produce changes in the encoding products and therefore in their function. Selective pressure acts at the functional level; therefore only changes that do not destroy essential functions and do not impair the reproductive capability of the individual carrying them are accepted in a population.

When a gene evolves, its function is generally preserved and therefore only mutational events compatible with the function are accepted. This implies that the knowledge of the function of a gene product in an organism can be transferred to the evolutionary correspondent gene product in another organism and the problem of assigning function is reduced to the detection of the evolutionary relationship.

However, novel biological functions are also developed as a consequence of evolution. Usually this is brought about by the duplication of a gene followed by its evolution that might result in a new advantageous function and therefore genes with different functions can also share an evolutionary relationship.

This represents a major problem in evolution-based assignment of function since we need to understand whether two genes diverged after a speciation (orthologous genes) or a duplication event (paralogous genes) since only in the former case we can be reasonably sure that they have a common function. We will come back to this problem, but we first need to discuss how an evolutionary relationship can be detected.

The problem can be stated as follows: given two protein sequences how likely it is that they have been generated from the same ancestral protein sequence via a set of accepted mutations?

We usually approach the problem using a maximum likelihood approach and formulate the question by counting the minimum number of evolutionary events needed to generate the two sequences from the same ancestral sequence or, almost equivalently, the maximum number of non-mutated amino acids that can·be detected between the two sequences. We then compute the expected probability that the observed similarity is not due to chance and is therefore significant.

The procedure used is:

- Find the correspondence between each amino acid of the first sequence and each amino acid of the second that maximizes the number of identical amino acids (optimal sequence alignment)
- Count the number of identical amino acids
- Assign the probability that they are evolutionary related by comparing the number of observed identical amino acids with that expected by chance alone.

The above assumes that each change is equally likely, however in biology this is not the case. Amino acids have chemical-physical properties that make some pair-wise substitutions more likely than others, the substitution of one positively charged amino acid with another positively charged one is more likely to be accepted than its substitution with a negatively charged or hydrophobic one.

How do we calculate the probability of each of the 380 possible replacements that is of the exchange of each of the twenty amino acids with any of the other nineteen?

We resort to empirical data, by counting how often each substitution has been observed between proteins that are clearly evolutionary related (homologous). The latter statement implies that we should derive these numbers from unambiguous alignments between related proteins, i.e. we need to analyze very similar protein sequences.

If we have several pairs of very similar sequences, so that the optimal alignments can be unequivocally deduced, we can calculate the frequency f_{ij} with which we find the amino acid pair (i,j) in corresponding positions in the alignment. This will be the number of cases where i and j are aligned to each other divided by the number of total pairs of aligned amino acids.

The $f_{i,j}/f_i f_j$ ratio is a measure of how frequently we observe i and j in aligned positions with respect to a random distribution. A symmetric matrix reporting these values for each pair of amino acids is called a substitution matrix and usually contains the base-2 logarithm of the values because this gives computational advantages. There are two widely used families of similarity matrices, PAM[4] and BLOSUM[5], and they differ in the type of alignment used to derive them.

PAM (Point Accepted Mutation) is a measure of evolutionary distance between protein sequences. An accepted point mutation is the substitution of one amino acid with another that is still compatible with the protein function and has therefore been fixed in the population. Two protein sequences are at 1 PAM distance if there exist a set of point accepted mutation that can convert one in the other with an average of one accepted point mutation each 100 amino acids.

For each pair of amino acids (a_i, a_j) the value of the cell PAM*N*(i,j) is the expected frequency with which a_i is substituted by a_j in protein sequences that are at a distance of *N* PAM. PAM*1* is derived using very similar sequences, values for higher PAM matrices are extrapolated from PAM*1*.

The BLOSUM (Blocks Substitution Matrix) matrices are derived using only local alignments of highly conserved regions in homologous protein families. Several sets of sequences are considered to derive a set of matrices; a BLOSUM-*x* matrix is derived from a set of sequences not sharing a percent identity higher than *x*.

The substitution matrices specify only one type of evolutionary events, the amino acids substitution, but do not take into account the possibility of insertions and deletions of amino acids (gaps), events which are rare in closely related proteins, but that are observed in distantly related ones.

Not only a gap in an alignment has to be penalized with respect to a substitution, since it is a rarer event, but we also have to consider that the probability of having a contiguous insertion or deletion of N amino acids is higher than the probability of having N insertions or deletions in different sites of the sequence since a gap cannot be accommodated in every structural position.

The most widely used methods for gap treatment assign to each gap of length N a penalty higher for the first event and a lower one for each of the N-1 subsequent insertions or deletions. Some methods assign to each element of a gap a penalty lower than the penalty assigned to the previous one.

Given a substitution matrix and a gap penalty scheme we can now calculate the alignment between two protein sequences that maximizes their similarity.

In principle we could generate all possible sequence alignments between the two strings, calculate the score (that is the sum of the substitution values for each pair of aligned amino acids diminished by the penalty values of the gaps) and select the "best", that is the alignment with maximum score that most likely reflects the evolutionary relationship between the two sequences. This would be highly inefficient and therefore we use algorithms, usually dynamic programming ones, to perform the task [6,7].

A score S for the alignment can be calculated by adding the similarity values of all the pairs of aligned amino acids diminished by the gap penalties. This has to be compared with the score expected by chance alone. The random distribution of scores is obtained by comparing a very large number of pairs of random sequences generated using the same amino acid composition of our two original sequences. The probability that the two sequences are evolutionary related can be calculated by computing the value:

$$Z = (S\text{-m})/\sigma$$

Where S is the score obtained for the two target sequences, and m and σ are the average and standard deviation of the random distribution, respectively. In the hypothesis that the random scores are normally distributed, the Z-score can be directly related to the probability that the two sequences are evolutionary related.

The method described above will detect many evolutionary relationships, but will not be sufficiently sensitive for very distantly related sequences. However evolutionary relationships are transitive, that is two sequences evolutionary related to a third one are evolutionary related and this observation has extremely important consequences in computational biology. It implies that the alignment of more than two sequences can highlight evolutionary relationships otherwise undetectable.

A simplified example of the concept is shown in Figure 1. In part a) we show a segment of a multiple sequence alignment of the flavodoxin family. The percent of identical amino acids between each pair of sequences is calculated and shown in part b) of the figure. Let's make the simplifying assumption that an evolutionary relationship between two proteins can be inferred if they are more than 40% identical. In this hypothesis, the pair-wise alignment of protein 1 with all the other sequences would not detect any relationship with sequences 4,5,6,8,9,12 and 13. This is shown graphically in part c) of the figure where each sequence is depicted as a diamond and the distance between them is roughly inversely proportional to their similarity.

The radius of the circle represents a similarity equal to 40% and therefore all sequences outside the circle are not sufficiently similar to be identified as homologous. Sequence 2 is evolutionary related to sequence 1, so we can safely assume that sequences related to sequence 2 (that is sharing at least 40% sequence identity) are also homologous to sequence 1. This, as shown graphically in part d) of the figure, allows sequences 6,8,12 and 13 to be included in the protein family.

As mentioned before, evolutionary relationships can be detected between orthologous and paralogous sequences and there are several empirical studies that have tried to evaluate the extent of sequence similarity that guarantees orthology, that is function conservation, between protein sequences.

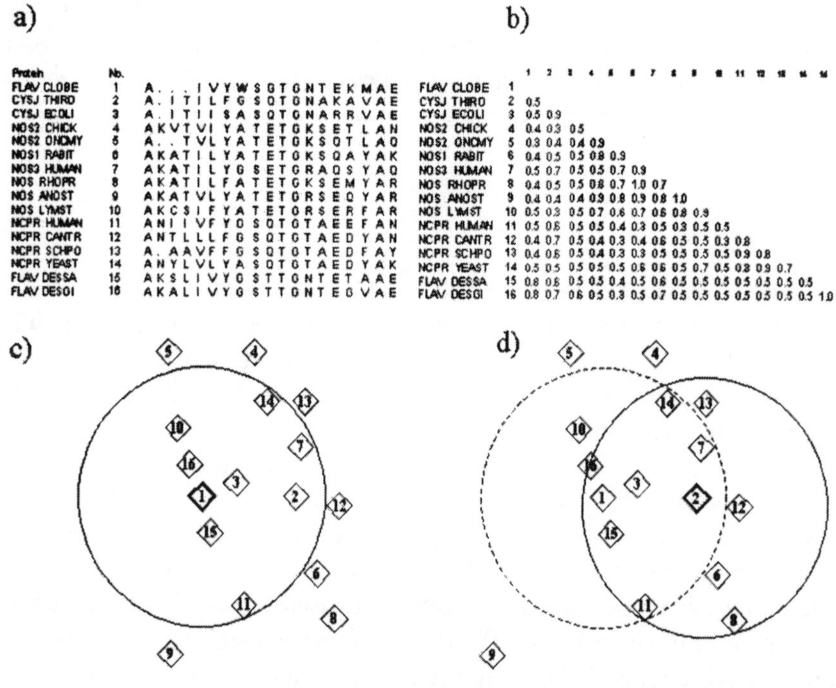

Figure 1. Example of the use of multiple sequence alignment in detecting evolutionary relationships.

It is generally accepted that a sequence identity above 50% guarantees conservation of function, but a recent report challenges this view and asserts that less than 30% of the pair fragments above 50% sequence identity have identical molecular function[8]. However, it should be mentioned that most misclassifications originate from similarities in relatively short regions and/or from transferring annotations for different domains. Although both problems can be easily detected in a manual assignment, avoiding them in genome-wide automatic function assignments is much harder. Work in this area is flourishing and we expect to see relevant progress in the near future.

3. STRUCTURE BASED FUNCTION ASSIGNMENT

3.1 Structure prediction

The molecular function of a protein depends on its three-dimensional structure, since it is ultimately brought about by the interaction of specific groups precisely located in space, be them the reactive atoms of an enzymatic active site or the atoms responsible for specific intra-molecular interactions. It is therefore not surprising that a large part of the computational biology community tries to develop methods able to deduce or compute the three-dimensional structure of a protein from its amino acid sequence.

Each protein folds to a unique three-dimensional conformation, its native structure, and this implies that nature has an algorithm that determines such structure only on the basis of the amino acid sequences. Methods aimed at reproducing this natural process, known as *ab initio* methods, and based only on physical principles have been extensively tried but with very limited success.

These methods try to compute the interactions between protein atoms and to define a computable function able to associate an energy value to each possible protein conformation. In other words they ultimately try to find the conformation of minimum free energy that the protein can kinetically reach.

From a computational point of view, the complexity of the problem is enormous since the conformational space available to a protein, even a very small one, is enormous.

Exploring the whole conformation space available to a protein with computational methods is clearly unfeasible. What is most important is that a simple "back of the envelope" calculation can easily convince the reader that this is not the route followed by nature either, as stated by Cyrus Levinthal in his well known paradox[9]. The reasoning goes as follows:

Given a protein of 100 amino acids and the assumption that each of them can only assume 4 different conformations (a very low estimate of the actual number of states an amino acid can assume in a protein chain), then the number of conformations available to the protein would be 4^{100}. Even if each conformation were explored in 1 femtosecond (that is the time of a molecular vibration), then the time needed to fold the protein would be:

$$10^{-15} \times 4^{100}\ \text{s} > 10^{79}\ \text{year}$$

that is longer than the life of the universe!

Recent theories have at least partially solved Levinthal paradox. Although a detailed treatment of these theories is beyond the scope of this paper, we will just mention that they are based on the idea that the unfolded chain will proceed to its folding state via several paths and, along them, some interactions between amino acids distant in sequence will form, most of them reflecting the final native interactions, and be sufficiently energetically favorable to be maintained during the folding process thus reducing the conformational space available to the rest of the polypeptide chain[10-15] in the subsequent steps and "funneling" the chain towards the native structure.

Even if computational simulations have demonstrated that it is possible in principle to simulate this "funnel" process, the time when we will be able to correctly identify the native structure of a protein by *ab initio* methods seems still quite far in the future. The real hurdle lies in the calculation of the energetic terms that govern folding. A protein folds if the difference between the large entropic loss associated with its transition from the unfolded to the folded state and the energy gain due to internal interactions in the folding state and to the entropic contribution of the hydrophobic effect, that is of burying hydrophobic amino acid side chains from the polar solvent, is favorable. However, proteins are only marginally stable and this energy difference is a very small number. Furthermore, the contributions of each individual interaction to the final stability are of the same order of magnitude and quite small. Therefore none of them can be neglected in the calculation and they need to be computed with a very high precision that we cannot achieve today[16].

Computational biology needs to exploit alternative methods based on the analysis of experimentally determined protein structures.

Because protein stability is brought about by a large number of weak interactions, the replacement of one amino acid with another can have one of two effects: destabilize the native structure to an extent that does not permit folding or be accommodated in the structure producing only local effects. In other words the probability that a single amino acid change destabilizes the native protein structure and stabilizes a completely different one is neglectable[17]. This observation is at the basis of a method known as "comparative modeling" that aims at predicting the effect of changes in the amino acid sequence between evolutionary related proteins so that the knowledge of the structure of one of the proteins of the family (template), allows the structure of the others (targets) to be predicted by modeling the effect of the evolutionary accepted sequence changes.

Clearly the better we are at recognizing evolutionary related proteins, the more models we will be able to produce. As we mentioned before, multiple alignments of members of a homologous protein family can help detecting more distant relationships [18-20].

Even when evolutionary relationships are undetectable from the comparison of the sequences, we can still try to build several putative models for the target protein sequence using the structure of each protein of known structure as template and then evaluate the sequence-structure fitness in each case[21-23]. The rationale for this method, known as fold recognition, is that structures are more conserved than sequences during evolution and therefore an evolutionary relationship can be more easily detected from structures than from sequences. This method has the added bonus that should, in principle, be able to detect sequence - structure fitness also between non-evolutionary related proteins, i.e. between proteins that have converged to the same architecture rather than diverged from a common ancestor.

There are several algorithms used by fold recognition methods and new ones are continuously developed, however they can be roughly divided into "profile" and "threading" based methods.

The former first transforms each known protein structure in a string where each amino acid position is encoded by a symbol that describes its structural properties (for example secondary structure, solvent exposure, etc.) thus recasting the database of known protein structures into a database of mono-dimensional strings. Each amino acid of the target sequence is then replaced by a symbol describing its propensities (for example preference for secondary structure elements, solvent exposure, etc.). The comparison between the latter string of symbols and the linearly transformed structure data base can be performed using any of the sequence alignment tools developed for the detection of sequence similarities. Often a multiple sequence alignment of the target sequence is used rather than a single sequence.

Threading methods explicitly build a three-dimensional model of the target sequence using each known structure as a putative template and calculate an approximate energy value for each sequence - structure pair.

The energy function is usually an empirically derived pair-wise potential[24]: considering a large number of protein structures we can define N^s_{ij} as the number of amino acid pairs at distance s in sequence and $N^s_{ij}(r)$ the number of pairs (i,j) at a distance r in space. According to Boltzmann equation we have:

$$N^s_{ij}(r) = AN^s_{ij}M^s(r)\exp(\Delta E^s_{ij}(r)/RT_c)$$

Where:

A = normalization constant
$M^s(r)$ = probability of finding residues at a distance s in sequence and r in space in the considered data set
T_c = temperature of freezing of native folds (~300 K)
R = gas constant
ΔE^s_{ij} (r) = direct interaction energy between the amino acids i and j at a distance r.

The residue - residue interaction energy can therefore be estimated as:

$$\Delta E^s_{ij}(r) = -RT_c \ln [N^s_{ij}(r)/N^s_{ij}M^s(r)] + RT_c \ln A - E^s_{ij}(r)$$

The energy associated with a given sequence - structure combination is evaluated by adding the contribution of each residue pair plus a term taking into account the solvation propensity of surface amino acids.

Both comparative modeling or fold recognition methods rely on the existence of a known structure similar to that of the target. Recently, empirical methods for the prediction of proteins the structure of which is dissimilar from any known protein structure are also being developed. They come in different flavors, but they all have in common the idea of constructing several plausible folds for a given protein sequence and then scoring each sequence structure combination on the basis of an approximate energetic calculation similar to that employed by fold recognition methods. The most successful ones [25] construct the plausible folds by dividing the sequence of the target protein in short segments and then searching the data base of known protein structures for fragments with the same or similar sequence. Several known structure fragments will be selected for each target sequence segment and they will later be combined using some stochastical approach such as Monte Carlo, Genetic Algorithms or Simulated Annealing.

3.2 Reliability of methods for structure prediction

How well do these methods work and, more importantly, how useful are the information they provide for structure based function assignment?

Evaluating structure prediction methods is far from trivial: until a few years ago information about the reliability of each method was generally based on results obtained on predicting already known protein structures and provided by the method developers themselves. This is not ideal since each method is tested on a different set of cases. Furthermore most methods are heuristic, i.e. based on parameters derived from known protein structures, and therefore it could not be excluded that information on the known target structure is embedded in the method.

Nowadays it is generally accepted that a reliable assessment system of the state of protein structure prediction methods is the one provided by the CASP (Critical Assessment of Techniques for Protein Structure Prediction) experiments[26-30].

In these experiments, structural biologists publicly announce the amino acid sequence of the proteins whose structure they are about to solve experimentally. Computational biologists submit models for these proteins before their structure is known and, later, models and structures are compared. Three independent experts analyze the comparison data for comparative modeling, fold recognition and new fold targets and evaluate the state of the art in these three areas. CASP has been repeated every two years since 1994.

Drawing general conclusions from the wealth of data generated by each CASP experiment is not trivial, and several reports describe the results at different level of technical details. Here we will shortly mention only the aspects that are related to function prediction.

It should first be mentioned that even the knowledge of an experimental protein structure might not be sufficient to infer function, because of the difficulties in detecting the active site and in associating it with a molecular function. It is however obvious that if a model does not predict correctly the general features of the structure and the details of the active site, there is no possibility to proceed through the process of function identification. Therefore the question that we will address here is: for a given evolutionary distance between a target and a template, how well can the active site of a protein be predicted?

We will limit our analysis to comparative modeling targets, since other methods are known to provide less detailed models.

Fig. 2 shows a plot of the root mean square deviation between the atoms of the active sites of models and experimental structures in the CASP4 experiment as a function of the evolutionary distance between the target and the template used to build the model[32]. The evolutionary distance d is defined here as the lowest sequence identity between members of the protein family that includes target and template. It is clear from the plot that for $d>20\%$ the position of the active site atoms can be predicted with an error lower than 1 Å, a precision sufficiently high to attempt function prediction in many cases.

An equivalently detailed analysis for the prediction of active sites is not available for other CASP experiments, however visual inspection allowed the assessor of the fifth edition of CASP to conclude that the relationship highlighted in Figure 1 holds in most cases[31]. It should be mentioned that, since comparative modeling is based on structural conservation between related proteins, the stronger evolutionary constraints imposed on functional sites guarantee that these will be the regions more accurately predicted.

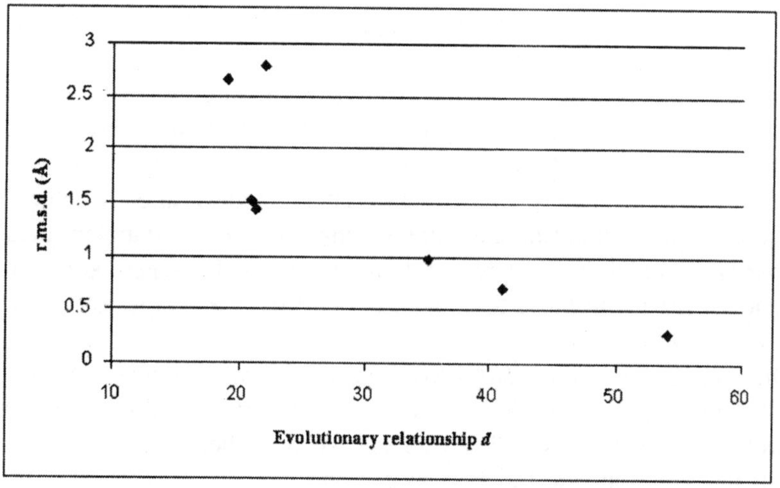

Figure 2. Quality of the prediction of active sites in the CASP4 experiment. See text for an explanation of the meaning of *d*.

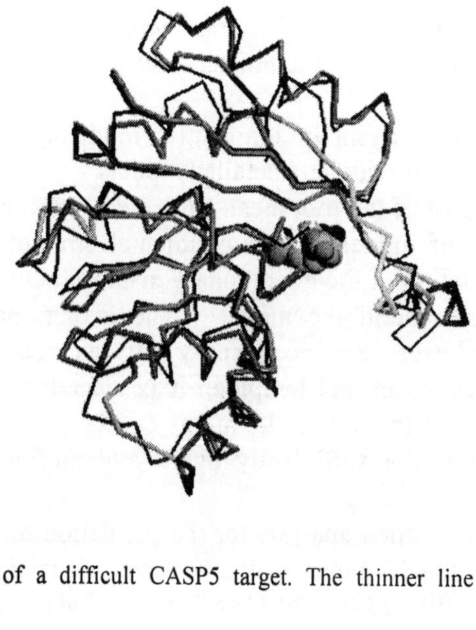

Figure 3. Prediction of a difficult CASP5 target. The thinner line is the experimental structure.

An example is given in Fig 3, where a model for a very difficult target (sequence identity with the evolutionary related template of about 15%) is shown. Even in this case, the region of active site (that is the region binding the ligand) is predicted with about 1 Å accuracy[31].

4. CONCLUSIONS

The holy grail of computational biology is the elucidation, at a molecular, cellular and biological level, of function.

As we tried to illustrate here, several routes are being followed to address the problem, each having advantages and pitfalls. The present challenge is to use our understanding of protein evolution to develop accurate methods for large-scale automatic assignment of function. At present, the level of accuracy with which we can predict function on the basis of genomic sequences is still dependent on the specific case considered, namely on the size of the family of the target protein and on its evolutionary distance from members of known function. The continuous growth of the number of available protein sequences and the coordinated effort of many research groups hold the promise that experimental biologists will soon be able to add reliable functional assignment tools to the suite of the many computational methods that are already part of their laboratory toolbox.

ACKNOWLEDGMENTS

This work was partially supported by "Progetto Strategico Genetica Molecolare L. 449/97". AT is grateful to Nelia Lopez, Rolando Rodriguez and Gabriel Padron of the Centro de Ingenería Genética y Biotecnología in La Habana (CU) for their kind hospitality during the preparation of this manuscript.

REFERENCES

1. Lander, E.S. et al. Initial sequencing and analysis of the human genome. Nature 409, 860-921 (2001).
2. Venter, J.C. et al. The sequence of the human genome. Science 291, 1304-1351 (2001).
3. The Gene Ontology Consortium. Gene Ontology: tool for the unification of biology. Nat Genet 25, 25-29 (2000).
4. Dayhoff, M.O., Schwartz, R.M. & Orcutt, B.C. A Model for Evolutionary Change. In Atlas of Protein Sequence and Structure. in Atlas of Protein Sequence and Structure, Vol.

5 (ed. Dayhoff, M.O.) 345-358 (National Biomedical Research Foundation, Washington, 1978).
5. Henikoff, S. & Henikoff, J.G. Amino Acid Substitution Matrices from Protein Blocks. Proc. Natl. Acad. Sci. USA 89, 10915-10919. (1992).
6. Needleman, S.B. & Wunsch, C.D. A General Method Applicable to the Search for Similarities in the Amino Acid Sequence of Two Proteins. J. Mol. Biol. 48, 442-453. (1970).
7. Smith, T. & Waterman, M. Identification of common molecular subsequences. J. Mol. Biol. 147, 195-197 (1981).
8. Rost, B. Enzyme function less conserved than anticipated. J. Mol. Biol. 318, 595-608 (2002).
9. Levinthal, C. Molecular model-building by computer. Scientific American 214, 42-52 (1966).
10. Oliveberg, M., Tan, Y.J., Silow, M. & Fersht, A.R. The changing nature of the protein folding transition state: implications for the shape of the free-energy profile for folding. Journal of Molecular Biology 277, 933-943 (1998).
11. Doyle, R., Simons, K., Qian, H. & Baker, D. Local interactions and the optimization of protein folding. Proteins 29, 282-291 (1997).
12. Onuchic, J.N., Luthey Schulten, Z. & Wolynes, P.G. Theory of protein folding: the energy landscape perspective. Annual Review of Physical Chemistry 48, 545-600 (1997).
13. Onuchic, J.N., Socci, N.D., Luthey Schulten, Z. & Wolynes, P.G. Protein folding funnels: the nature of the transition state ensemble. Folding & Design 1, 441-450 (1996).
14. Wolynes, P.G. Folding funnels and energy landscapes of larger proteins within the capillarity approximation. Proceedings of the National Academy of Sciences of the United States of America 94, 6170-6175 (1997).
15. Onuchic, J.N., Wolynes, P.G., Luthey Schulten, Z. & Socci, N.D. Toward an outline of the topography of a realistic protein-folding funnel. Proceedings of the National Academy of Sciences of the United States of America 92, 3626-3630 (1995).
16. Finkelstein, A. Protein structure: what is it possible to predict now? Current Opinion in Structural Biology 7, 60-71 (1997).
17. Chothia, C. & Lesk, A. The evolution of protein structures. Cold Spring Harb Symp Quant Biol 52, 399-405 (1987).
18. Altschul, S.F., Gish, W., Miller, W., Myers, E.W. & Lipman, D.J. Basic local alignment search tool. Journal of Molecular Biology 215, 403-410 (1990).
19. Barton, G.J. Protein multiple sequence alignment and flexible pattern matching. Methods Enzymol. 183, 403-428. (1990).
20. Teichmann, S.A., Chothia, C., Church, G.M. & Park, J. Fast assignment of protein structures to sequences using the intermediate sequence library PDB-ISL. 16, 117-124 (2000).
21. Bowie, J.U., Luthy, R. & Eisenberg, D. A method to identify protein sequences that fold into a known three-dimensional structure. Science 253, 164-170 (1991).
22. Jones, D.T., Taylor, W.R. & Thornton, J.M. A new approach to protein fold recognition. Nature 358, 86-89 (1992).
23. Sippl, M.J. & Weitckus, S. Detection of native-like models for amino acid sequences of unknown three-dimensional structure in a data base of known protein conformations. Proteins 13, 258-271 (1992).
24. Sippl, M.J. Knowledge-based potentials for proteins. Current Opinion in Structural Biology 5, 229-235 (1995).
25. Bonneau, R. et al. Rosetta in CASP4: Progress in ab initio protein structure prediction. Proteins Suppl. 5, 119-126 (2001).

26. Moult, J., A., Z., Fidelis, K. & Hubbard, T. Critical assessment of methods of protein structure prediction (CASP)-round V. Proteins Suppl. 6, 334-339 (2003).
27. Moult, J., Fidelis, K., Zemla, A. & Hubbard, T. Critical assessment of methods of protein structure prediction (CASP): round IV. Proteins Suppl 5, 2-6 (2001).
28. Moult, J., Hubbard, T., Fidelis, K. & Pedersen, J. Critical assessment of methods of protein structure prediction (CASP): round III. Proteins Suppl 3, 2-6 (1999).
29. Moult, J., Hubbard, T., Bryant, S., Fidelis, K. & Pedersen, J. Critical assessment of methods of protein structure prediction (CASP): round II. Proteins Suppl 1, 2-6 (1997).
30. Moult, J., Pedersen, J., Judson, R. & Fidelis, K. A large-scale experiment to assess protein structure prediction methods. Proteins 23, ii-v (1995).
31. Tramontano, A. & Morea, V. Assessment of homology-based predictions in CASP5. Proteins 53 Suppl 6, 352-368 (2003).
32. Tramontano, A., Leplae, R. & Morea, V. Analysis and assessment of comparative modeling predictions in CASP4. Proteins 45 Suppl 5, 22-38 (2001).

Differences in Binding of Stereoisomers to Protein Active Sites

GERHARD KLEBE
Institute of Pharmaceutical Chemistry, Philipps-University of Marburg, Marbacher Weg 6, D-35032 Marburg, Germany

1. INTRODUCTION

The spatial structure of a drug molecule determines and in consequence, evolves its biological properties at a particular target receptor of therapeutic relevance. Its geometry is determined by the connectivity of the atoms composing the molecule under consideration. If this spatial configuration encodes for asymmetry, the molecule gives rise to optical activity and pairs of enantiomers exist. This asymmetry can either result due to the presence of stereogenic centers (e.g. tetrahedral atoms bound to four chemically distinct substituents) or the entire molecular skeleton gives rise to an asymmetric overall structure (e.g. twistane 1 or substituted allenes 2, Fig. 1). Without breaking bonds such pairs of chiral molecules cannot be transformed from one to the other form. At this point we want to neglect special situations showing slowly converting atropisomers, such as ortho-substituted biphenyl systems (3) or tricyclic skeletons adopting a butterfly-type overall shape (4, Fig. 1). As long as such enantiomers are exposed to an achiral environment they possess identical properties. However, once presented to a handed surrounding, image and mirror-image will be recognized differently and will produce along with the local environment distinct diastereomeric effects. Receptor proteins, for example enzymes, often being target of drug molecules, create a chiral environment in their binding pockets. They are constructed themselves from chiral building blocks, such as L-amino acids. Other targets, e.g. DNA, RNA or ribozymes, are composed by D-ribose or D-desoxyribose as chiral building blocks. Furthermore, as a whole, these

Supramolecular Structure and Function 8, Edited by Pifat-Mrzljak
Kluwer Academic/Plenum Publishers, New York 2004

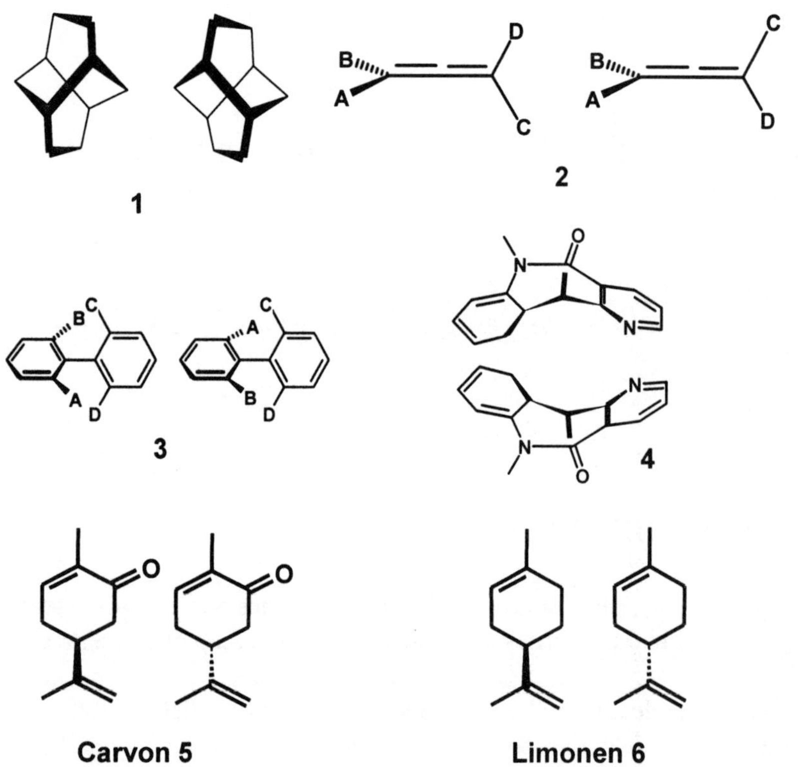

Figure 1. Twistane (1) and substituted allenes (2) possess molecular skeletons that give rise to overall molecular asymmetry. In case of orthosubstituted biphenyls (3) racemication can be achieved via rotation about the central single bond. The barrier to rotation and accordingly the atropisomerization depends on the steric bulk of the ortho substituents. The molecular skeleton of an annulated tricyclic system (4) with a central seven-membered ring adopts a butterfly-type geometry. Appropriately substituted, such a system gives rise to molecular asymmetry. Dependent on the barrier of ring inversion such systems can interconvert and succeed in racemization. The enantiomers of carvon (5) and limonen (6) create quite different smelling characteristics via binding to their receptor. This difference results from distinct chiral recognition of the enantiomers at a G-protein coupled receptor constructed from seven transmembrane helices.

molecules adopt conformations and spatial arrangement (e.g. helices) that correspond to handed objects in space. Accordingly, different biological response can be expected once a pair of mirror-symmetrical molecules binds to such chiral receptors. For example both enantiomeric forms of carvone and limonen (5 and 6 in Fig.1) create quite different smelling characteristics once exposed to the corresponding receptors on the olfactory hairs in our nose[1]. They belong to the class of G-protein coupled receptors, large protein assemblies spanning the cell membrane by seven transmembrane helices. Many examples are known that demonstrate the deviating biological properties of chiral molecules, producing distinct profiles either in the intensity or quality of their bio logical response. Many of the presently available major-selling drug molecules (Fig. 2) possess one or more stereogenic centers giving rise to several enantiomeric pairs of diastereomers. For example, the R-form of N-methyl-phenyl-propyl barbituric acid 7 (Fig. 3) shows narcotic action whereas the S-enantiomer possesses convulsive properties[2]. Propoxyphen 8 (Fig. 3) exhibits two adjacent stereogenic centers. The S, R-isomer shows potential as pain relief whereas the mirror image is known for antitussive properties. For the dihydropyridine Bay K8644 9 the R-enantiomer is described to block the calcium channel. In contrast the S-enantiomer opens the channel. In the late fifties, the sleeping pill Contergan® containing thalidomide 10 as active ingredient was introduced to market. The application of this drug is perhaps the most devastating example of serious drug side effects since obviously one of the two enantiomers shows strong teratogenic effects, causing fatal abnormalities such as severely underdeveloped limbs. Whether the application of the correct enantiomers would have been sufficient to avoid this catastrophy is in question since both enantiomers racemize under in-vitro and in-vivo conditions. Contemporary drug therapy is meanwhile sensitized towards the release of racemic versus stereochemically "pure" drugs. For example, the world-wide sales of chiral drugs in single-enatiomer dosage form has grown to more than 13% and the market section amounted in 2000 to 133 billion USD[3]. However, it has to be critically assessed whether the release of single-enantiomeric forms is relevant. For example, the proton pump blocker omeprazole, recently introduced to market in single enantiomer dosage form, possesses a stereogenic center that is lost upon degradation of the applied prodrug form[4]. Both enantionomeric forms of the inflammatory drug ibuprofen show clearly deviating in vitro activity, however the in-vivo inactive R-form is metabolized through inversion of its stereocenter into the active S-enantiomer (Fig. 3).

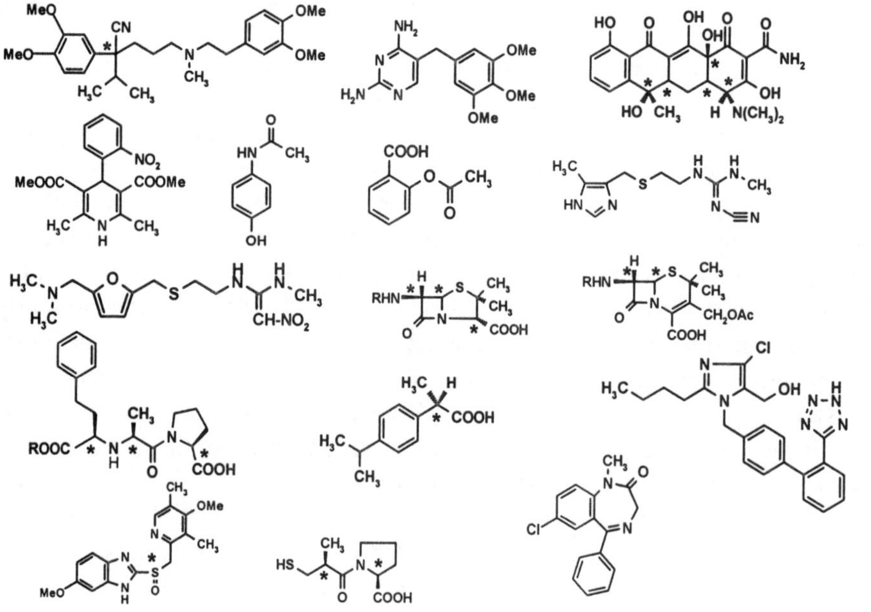

Figure 2 Chemical formulae of some of the best selling drugs. Many of them possess stereogenic centers giving rise to molecular asymmetry (*).

2. CHIRAL MOLECULES IN A CRYSTAL ENVIRONMENT

Crystallography is the most powerful method to elucidate the three-dimensional structure of molecules. Exploiting the effects of anomalous scattering of particularly the heavier atoms, this technique allows to determine the absolute configuration of molecules.

The crystallization of a racemate can either result in the formation of racemic crystals showing both enantiomers side-by-side in the same unit cell, or spontaneous resolution is observed leading to an agglomerate of morphologically enantiomorphic crystals. The latter crystals are composed either by the image or the mirror-image. They do not possess any symmetry operations corresponding to a mirror or a glide plane or involve an inversion center. Due to these restrictions out of the 230 possible space groups, only 65 can accommodate chiral molecules. In a crystal, molecules have to pack densely against each other. The obtained self-recognition of a racemate or an enantiomeric pure compound must in consequence produce quite different packing patterns. They can induce different conformations to be adopted in the solid state. For example the amino acid histidine crystallizes in its racemic form in space group P21/c whereas the enantiomeric-pure crystal corresponds to P21[5-7]. Distinct packings with deviating comformations of the amino acid are observed. Proteins being chiral objects and found in nature as single enantiomers can only crystallize in one of the 65 enantiomorphic space groups. As a curiosity the polypeptide chain of Rubredoxin has been synthesized from D-amino acids only[8]. A 1:1 mixture with the natural L-protein resulted in a racemate forming racemic crystals in a centrosymmetrical space group.

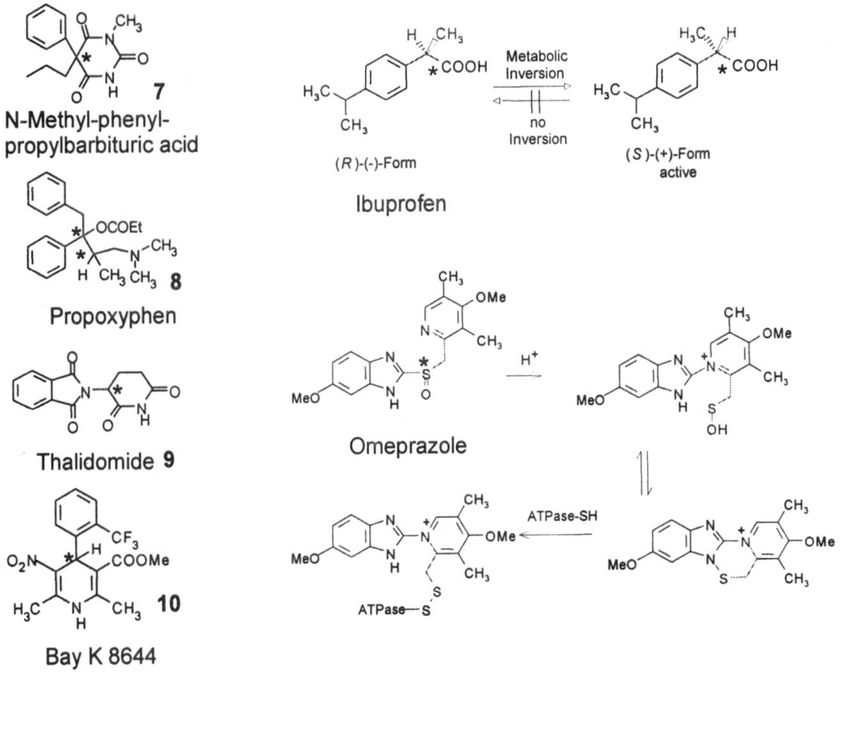

Figure 3: The R-enantiomer of N-methyl-phenyl-propyl-barbituric acid (7) possesses narcotic properties, whereas the S-enantiomer exhibits convulsive action. S,R-isomer of propoxyphen (8) has pain reliefing properties whereas the R,S derivative shows anti-tussive effects. The dihydropyridine Bay K8644 (10) is in its R-form a calcium channel blocker, whereas the mirror image opens this channel. Only one of the enantiomeric forms of thalidomide (9) has been reported to have teratogenic side effects. The inactive R-enantiomer of ibuprofen is metabolically converted into the active S-enantiomer. The stereocenter at sulfur in omeprazole is lost upon activation of the drug from its administered prodrug precursor form.

2.1 Recognition of Chiral Ligands at Protein Binding Sites

Due to diastereomeric interaction properties at a protein binding site, enantiomers are recognized as distinct species. Accordingly, deviating affinities are the rule, similar affinities the exception. HIV protease is a key enzyme in the replication of the human immunodeficiency virus[9]. It cleaves the produced polypeptide chain into functional proteins. This protease is a homodimer of two 99 amino-acid peptide chains. Also this enzyme has been synthesized purely from D-amino acids[10,11]. Interestingly enough, the L-protein can only cleave the natural L-substrate,

whereas the D-protein operates only on the all D-substrate. The enantiomers of chiral inhibitors discriminate between both handed forms of the enzyme whereas an achiral inhibitor reveals the same inhibitory potency towards both forms.

Binding modes of ligands can be studied by determining their crystal structures when complexed with a protein. Usually the affinities of two enantiomers towards a given protein differ by some orders of magnitude. Thus, only the binding geometry with one of the mirror images can be studied successfully, since only in this situation the bound inhibitor is sufficiently populated at the binding site to allow for conclusive interpretations of the electron density maps. Therefore, of particular interest are cases where both species reveal comparable affinities. In such cases, crystallography can provide some insight of how both forms can be accommodated at the binding site. The serine protease inhibitor Daiichi DX 9065 (11, Fig. 4) has been cocrystallized with trypsin[12]. Affinity measurements of both enationomers in solution revealed comparable values. Two distinct crystal forms have been discovered showing deviating crystal packings in the solid state (Fig. 4, a,b). In the first form, the binding pocket opens towards a solvent channel in the crystal. In the second form, a neighboring symmetry-related molecule packs towards the binding site of the first imposing some geometry constraints in this region. Crystallization of the complex has been performed with the racemate of the inhibitor, which possesses a stereogenic center next to its carboxylate group. Interestingly enough, in crystals of the first form exhibiting access to the solvent, clear evidence for binding of the racemate is indicated by the difference electron density. In the second form with the geometrically constrained binding site, only one enantiomer can be detected. Obviously, the contacts to the neighboring molecule in the crystal packing generate a binding site that favors binding of only one of the two enantiomers. Apparently, the conditions experienced in the first form, equivalently accommodating both enantiomers, resembles better the situation in solution and, most likely, under in-vivo conditions. However, this example demonstrates that chiral recognition requires pronounced and well-discriminating molecular interactions with the close environment of the stereo center thus creating local molecular asymmetry for the considered ligands.

Another example for inhibitor binding of distinct enantiomers has been studied with human carbonic anhydrase II (Fig. 5)[13]. Both enantiomeric forms of a potent sulfonamide (12) show quite similar binding modes, although their affinity differs by approximately two orders of magnitude. Since solvation/desolvation properties are identical for both enantiomers, the affinity difference must result from distinct interactions at the binding site. Both molecules bind with their sulfonamide group to the catalytic zinc and the neighboring Thr199. Furthermore, the SO2 group in the six-membered rings interact identically with Gln92. The iso-butylamino group at the stereogenic center orients into the same hydrophobic pocket, but the conformation required to adopt this spatial arrangement has been computed as less favorable for the enantiomer showing lower affinity. This difference provides an explanation for the observed affinity discrepancy.

Another structure determination has been reported on the enantiomeric discrimination of two agonists (13, Fig. 6) binding to the human nuclear retinoic acid receptor[14]. Both enantiomers bind to the conformationally unchanged binding pocket of the receptor. With their benzoic acid moiety, both form a network of hydrogen bonds to Arg278, Ser289, a water and Leu233. The fluoro-substituted phenyl ring adopts in both enantiomers a mutually flipped orientation placing the fluorine atom into different local environments. The orientation of the hydroxy group in the central bridge gives rise to both enantiomeric forms. In both, this OH group forms a H-bond with good geometry to the sulfur of Met272. Interestingly enough, the tetrahydronaphthalene moiety adopts in both situations a different conformation with respect to the bond next to the hydroxymethylene bridge. The rotated orientation of the tetrahydronaphthalene moiety together with the rearrangement of the amide group occurs to maintain the H-bond to Met272, simultaneously allowing favorable accommodation of the naphthalene moiety. A comparison of the bound conformations of both enantiomers with the energetically most favorable arrangement in the free state suggests that the largely inactive S-enantiomer binds with an energetically less favorable conformation compared to the active R-enantiomer. It is reported that the affinity of both compounds towards the receptor differs by three orders of magnitude. Accordingly, and in analogy with the example of carbonic anhydrase binding, the affinity difference and thus chiral discrimination result from an adaptation of the less active enantiomer in a rather strained unfavorable conformation to fit into the structurally unchanged binding pocket.

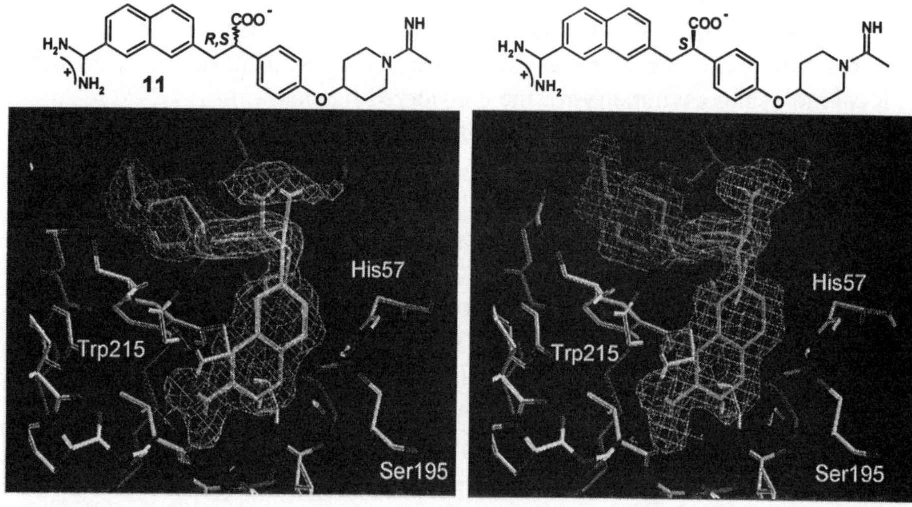

Figure 4: Binding geometry of the serine protease inhibitor DX9065a (11) as bound to two different crystal forms of trypsin. On the left, the structure of a crystal form with the binding pocket opening to a solvent channel is shown. It accommodates both enantiomers by giving a different orientation to the carboxylate group. A second crystal form, displayed on the right, closes up the binding site due to contacts with a neighboring molecule in the densely packed crystal. The residual electron density indicates that this crystal form is only occupied by one of the enantiomers.

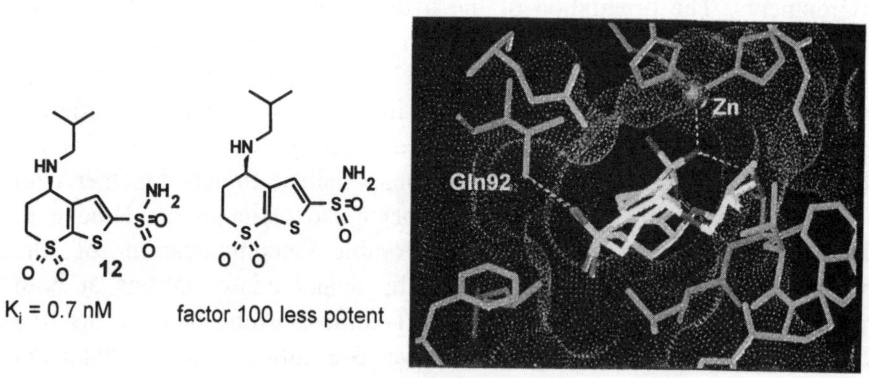

Figure 5: The two enantiomers of the sulfonamide (12) bind with comparable orientation to carbonic anhydrase. They coordinate to zinc through their sulfonamide groups. The hydrophobic isobutylamino groups, attached to the stereogenic center, orient similarly into a hydrophobic pocket. Energetically distinct ligand conformations are adopted to achieve this comparable binding mode, however they parallel a difference in affinity by two orders of magnitude.

2.2 Recognition of Chiral Building Blocks in Stereoisomers at the Protein Binding Site

The above-mentioned HIV protease belongs to the class of aspartyl proteinases. Its catalytic center is composed by two aspartates, each residing in one of the two C2-symmetrical homodimers. A set of four inhibitors (14, Fig. 7) has been studied with this enzyme[15,16]. The genuine skeleton of these inhibitors has been designed to resemble the C2-symmetry of the protein (Fig. 7). To mimic possible transition state analogs, OH groups with varying stereochemistry were attached, resulting in four different compounds. The inhibitors 14a with X = S-CH(OH)CH_2 and 14b with X = S-CH(OH)-R-CH(OH) were found to adopt nearly identical binding modes and to have very similar affinities. In both inhibitors the S-CH(OH) groups are recognized by Asp25' and Asp25. The additional OH group, present in

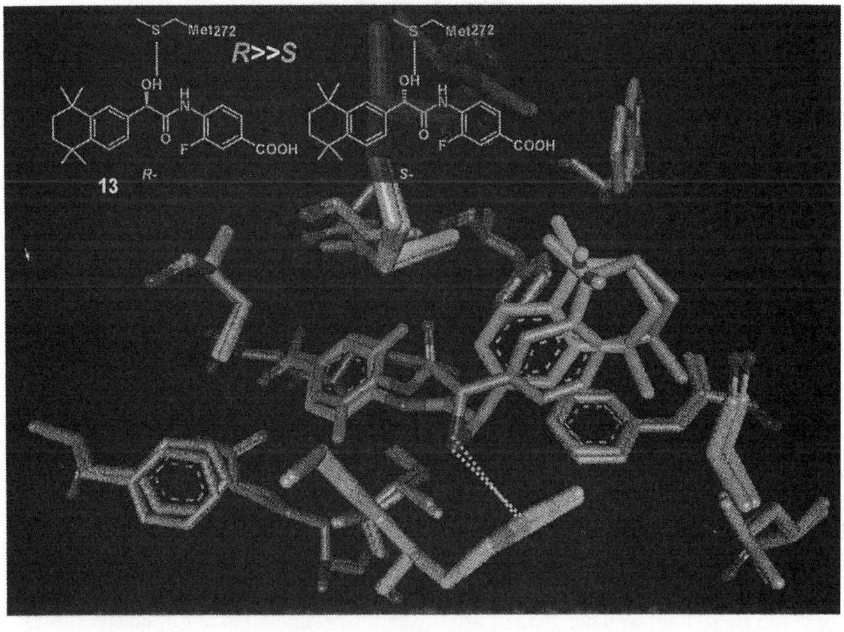

Figure 6: Binding modes of two enationmers (13) to the human nuclear retenoid acid receptor. Both forms of the inhibitor are hydrogen-bonded via their OH groups to the sulfur atom of Met272. The benzoic acid group is similarly recognized whereas the fluoro substituted phenyl moiety adopts a flipped orientation in the two cases. Due to conformational differences along the central amide bond, the terminal tetrahydronaphthyl group is differently accommodated for both enantiomers.

the second inhibitor 14b, only interacts with Asp25. Obviously, this additional hydrogen bond does not contribute to binding affinity and is even compensated by partially unfavorable solvation effects resulting in a slightly reduced binding affinity (Fig. 7).

Two other stereoisomers 14c and 14d, being mirror-symmetrical to each other in their central diol moiety, were studied. The S,S and R,R derivatives differ in affinity by a factor of twelve. The reversed orientation of the diol portion is believed to allow for a more efficient hydrogen-bonding network in the case of the S,S-derivative 14c compared to the R,R-analog 14d (Fig. 7).

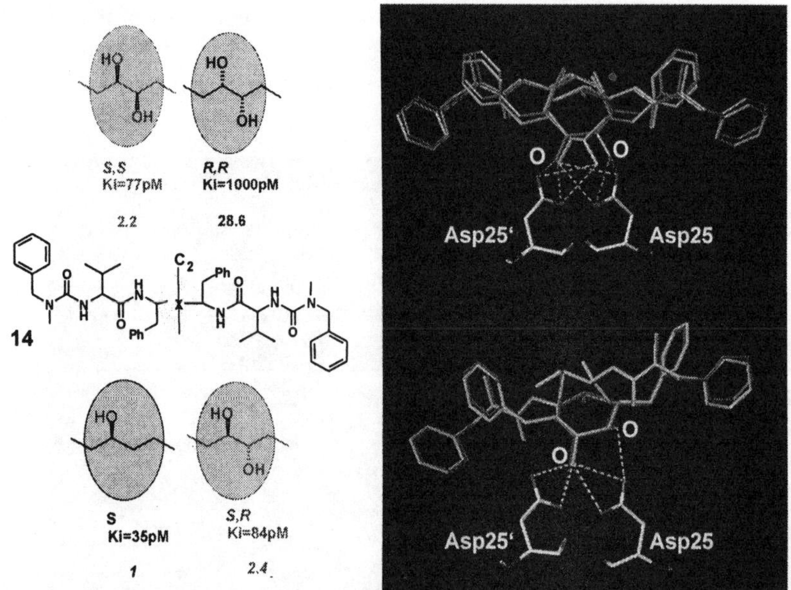

Figure 7: HIV protease is a homodimer with C2 symmetry. Inhibitors such as 14 have been synthesized and studied crystallographically, reflecting this C2 symmetry. The two enantiomers S,S (14c left) and R,R (14d right) (upper row) differ by a factor of 13 in affinity. This difference can be explained by a less efficient hydrogen-bonding pattern formed by the central R,R-diol unit compared to the S,S-unit. The pseudo-symmetrical S- (14a left) and S,R-inhibitors (14b right) bind with comparable affinity and very similar binding geometry (lower row). Obviously, the additional H-bond forming OH group in the S,R-diol does not improve binding affinity. The gain of one hydrogen-bond is compensated by less favorable solvation/desolvation properties of this group compared to the unbound situation in a water environment

Inversion of configuration at C1 in N-acetylglucosamide (15, Fig. 8) results in different diastereomers. This local change strongly modifies the binding mode of the ligand[17]. Although still recognized by Asp59 and Ala107 through their amide groups, the glucose moieties of both isomers orient in opposite directions of the binding pocket, thus interacting with different binding-site residues. Changes in the molecular skeleton of a ligand can also require changes in the local stereochemistry of a ligand. The two thrombin inhibitors Napap (16) and Argatroban (17) (Fig. 8) bind with their amidino or guanidino group, respectively into the specificity pocket of this serine protease[18]. The altered topology of the basic side-chain along with the modified length of the sulfonamide linker requires inverted stereochemistry at the central Cα atom for potent enzyme inhibition.

Without knowing the binding mode of a particular substrate, the stereochemical prerequisites for a competitive inhibitor are difficult to estimated. The natural substrate of thrombin, fibrinogen, has a L-phenylalanine at its N-terminal part, just next to the cleaved peptide bond (Fig. 8). The inhibitor Gyki N-methyl-D-phenyl-prolyl-arginine (18), exhibits a phenylalanine with inverted stereochemistry. The role of this inversion becomes obvious when examining the crystal structures of the enzyme with the substrate and 18, indeed the benzyl moiety of D-phenylalanine mimics a hydrophobic surface of the substrate formed by a L-phenylalanine and L-leucine residue[19,20].

3. CHEMICAL REACTIONS IN PROTEINS USING STEREOISOMERIC SUBSTRATES

Nature has provided several classes of enzymes to cleave polypeptide chains. Usually these catalysts operate efficiently only on peptides composed by L-amino acids. Nevertheless, some naturally occurring peptide antibiotics contain D-amino acids, resulting in metabolically more stable compounds. For the same reasons, D-amino acids have been introduced in some synthetic drugs of peptidic origin. These often exhibit higher activity and better metabolic stability. A special case are synthetic peptides with retro-inverse configuration, where the sequence of amino and carboxy groups is interchanged (Fig. 9). In order to retain the same relative configuration, L-amino acids are replaced by their D-analogs. Apparently, this exchange can mislead some enzymes or receptors and recognition similar to the natural peptide is observed. As an advantage the retro-inverse peptides are usually more resistant to metabolic degradation. The experimentally observed binding modes of thiorphen (19) and retro-thiorphen (20) at the binding site of thermolysin[21] possibly provide a first glimpse of how the similar recognition of peptides with reversed sequence could be achieved (Fig. 9). Both inhibitors are equally potent. At the binding site they adopt a geometry that places their peptide bonds in similar orientation even so being reversed in the two cases. The backbone amide

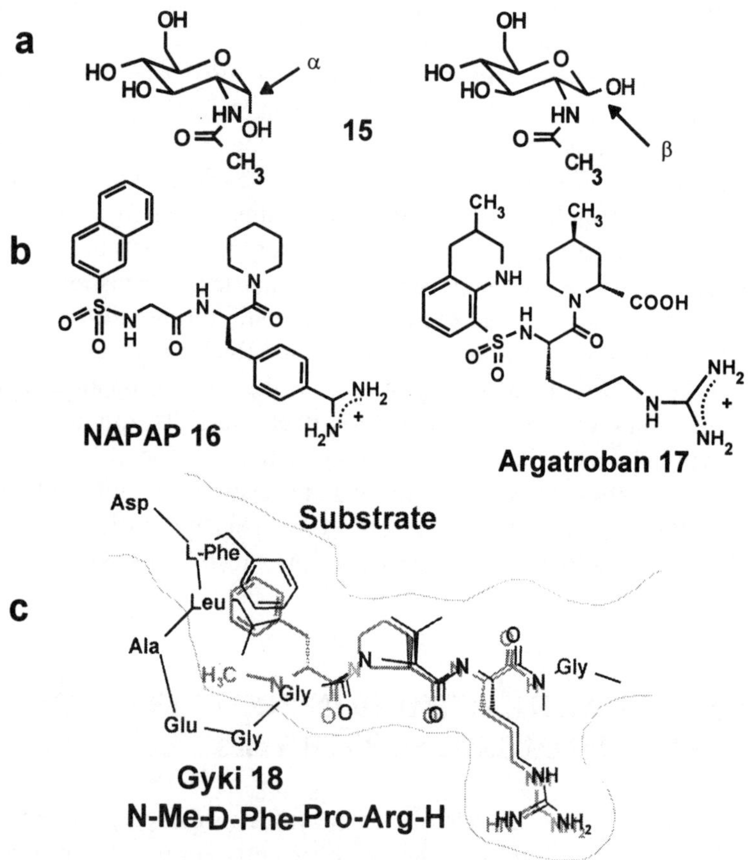

Figure 8: a: The orientation of the OH group at C1 of N-acetylglucosamide (15) gives rise to different diastereomers. Due to this difference the recognition properties are substantially modified and binding modes with reversed orientation are observed. Only the amide groups are equivalently recognized by Asn59 and Ala107. The remaining parts of the molecules, including the distinct C1 OH group, are placed to opposite sides of the binding pocket of lysozyme.
b: Napap 16 and Argatroban 17 are recognized with their basic amidino- or guanidino group by Asp189 in the specificity pocket of thrombin, the piperidine and naphthyl or tetrahydroquinoline moiety packed together and occuping similarly the proximal and distal hydrophobic S2/S3 pockets in thrombin. Due to the recognition differences of the basic groups, the rather similar placements of the hydrophobic molecular portions are only achieved with inverted stereochemistry at Cα carbon.
c: Compared to the natural substrate fibrinogen of thrombin which places a L-phenylalanine into the S2/S3 pocket of thrombin, the small peptidic inhibitor Gyki, N-methyl-D-phenyl-prolyl-arginine (18, grey) requires reversed orientation at the Cα atom of its phenyl alanine residue to achieve proper placement into this part of the thrombin binding pocket. The stereochemical differences can be explained by the actually observed binding mode of the natural substrate that leaves the binding pocket with a stretch of four amino acids and return with its L-Phe into the addressed specificity pocket of thrombin.

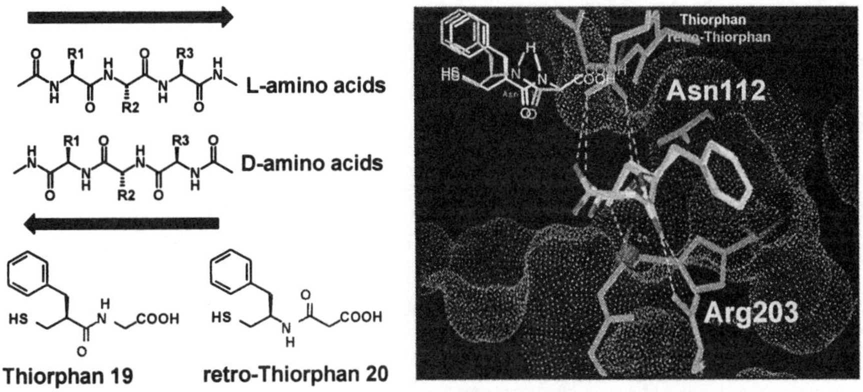

Figure 9: Thiorphen (19) and retro-thiorphen (20) differ by a reverse sequence of the amino- and carbonyl group in the central peptide bond. However, the side-chains and the reversed peptide bond are similarly recognized by the residues exposed to the binding site of thermolysin. This example gives an idea how retro-inverse peptides, usually more resistant to metabolic degradation, could be recognized equivalently at protein binding sites.

and carbonyl groups of both ligands can form equally stable H-bonds with the adjacent residues Asn112 and Arg203 (Fig. 9).

Catalytic pathways in enzymes are stereospecific. This property qualifies them as valuable stereoselective catalysts for organic synthesis. Organic reactions are often performed in non-aqueous solvents, however most biomolecules degrade under such conditions. Due to their biological function to operate next to hydrophobic interfaces, lipases usually remain intact in such a milieu[22]. Accordingly, they have frequently been used as stereospecific catalysts in organic synthesis. They belong to the class of serine hydrolases and their prominent biological function is the cleavage of ester bonds in triglycerides. The catalytic center in these lipases is composed by a triad formed by a serine, histidine and aspartate or glutamate residue. The cleavage reaction proceeds via a nucleophilic attack of the serine OH group on the ester carbonyl carbon, resulting in a tetrahedral transition state (Fig. 10). Subsequently, the weakened C-O ester bond breaks and a covalent acyl intermediate is formed. Usually in the following step a water molecule attacks this acyl intermediate as nucleophile and displaces the remaining part of the cleaved substrate from the catalyst. Under non-aqueous conditions, other nucleophiles can also be used to react with the acyl intermediate, e.g. alcohols or amines. These reagents will result in the formation of a new ester or amide bond. It has been shown that different nucleophiles react with deviating efficacy. Accordingly, many examples exist of lipases discriminating specifically between different nucleophiles.

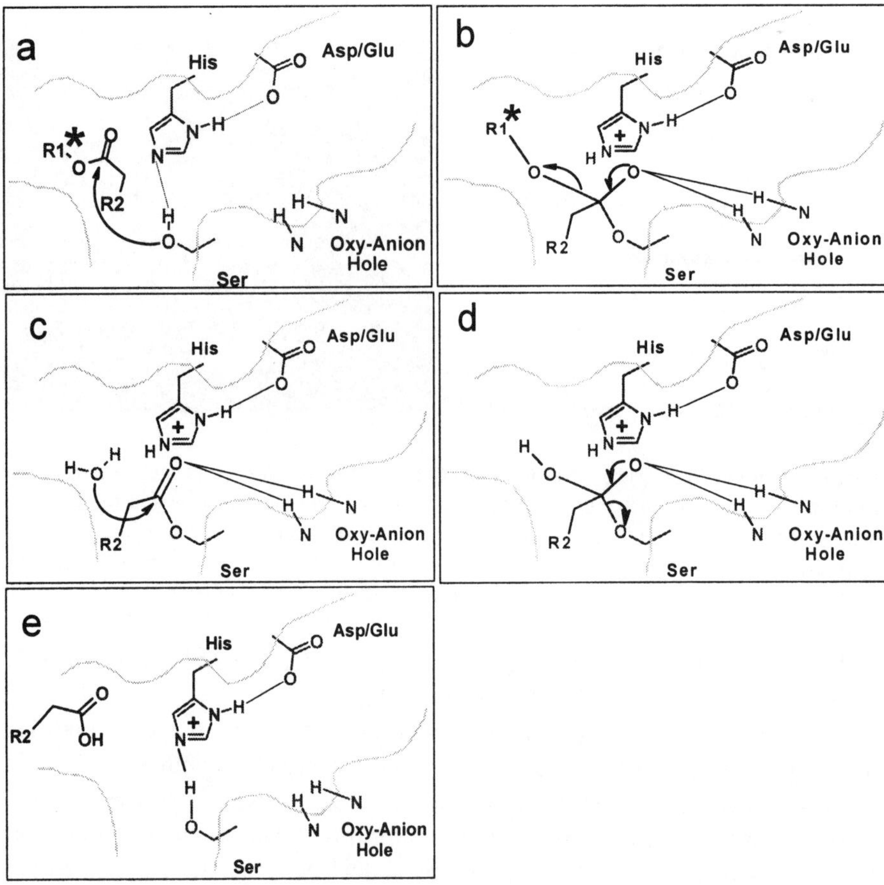

Figure 10: Schematic sketch of the ester cleavage reaction performed in a lipase. The reaction proceeds via a nucleophilic attack of the serine OH (a) on the ester carbonyl goup supported by the neighboring His and Asp. The intermediate tetrahedral transition state (b) decomposes forming a covalent acyl enzyme intermediate (c). In a subsequent nucleophilic attack of a water molecule, acyl form is cleaved via a second tetrahedral transition state (d). The cleaved product is released from the binding site (e).

Since enantiomers are recognized as distinct species by a protein catalyst, lipases can discriminate between the chiral forms of a nucleophile attacking the intermediately formed acyl complex [23,24].

3.1 Structural Basis for Chiral Resolution in Lipases

Various molecular properties are responsible for chiral discrimination in the decomposition step of the acyl intermediate. The reaction is performed on a particular time scale and differences in the reaction velocity will result in chiral discrimination. Since the reaction at a binding site of a protein involves several consecutive steps, it is difficult to estimate which of the various steps is the rate-determining one and whether or how this provokes chiral discrimination. Not necessarily the same step represents the rate-determining one for different reagents. Kinetic, energetic, and structural considerations are involved. To obtain some structural evidence, the binding of tetrahedral transition state has to be investigated. Since transition states are reactive short-living species more stable analogs have to be sythesized. Usually a phosphorous atom is selected to mimic the intermediately formed tetravalent carbon.

As an example, the enantio-preference of *Candida rugosa* lipase to hydrolyze both forms of menthyl pentanoate (21, 22 in Fig. 11) has been studied by Cygler et al.[25]. Kinetic evidence has been collected that the R-enantiomer (23) is favored. To mimic the tetrahedral transition state of the hydrolysis step, the phosphonate analog of menthyl heptanoate (23, 24) has been synthesized and covalently linked to Ser209 in the active site. The corresponding 1R-isomer mimics the transition state for the fast-reacting enantiomer while the 1S-isomer resembles the transition state for the slow-reacting enantiomer. Crystals were obtained for the inactivated enzyme containing either inhibitor 23 or 24 (Fig. 11). Subsequently, the structures with both isomers have been determined. Both phosphonate groups bind similarly with a covalent link to the terminal oxygen of Ser209, and show S-configuration at their phosphorous. The terminal phosphonyl oxygens of both inhibitors occupy the oxyanion hole with hydrogen bonds to the amide NH groups of Ala210 and Gly124. The hexyl chain orients along a hydrophobic channel that extends towards the center of the enzyme. For both inhibitors, the menthyl portion is positioned towards the entrance of the binding pocket where it opens to the solvent. In both isomers the iso-propyl substituents at the six-membered ring are placed in a similar spatial area next to the catalytic histidine. However, in the 1S-derivative this group is closely directed towards Phe344 and displaces the rings of Phe345 and His449. Compared to the apoenzyme and the structure of the 1R-isomer this rotation amounts to 60°. The reorientation of the His449 imidazole ring has been linked to the differences in reactivity for the two esters towards hydrolysis.

In the complex with the 1R-isomer, which mimics the transition state of the fast-reacting enantiomer, the imidazole ring of His449 adopts the same orientation as in the uncomplexed state and Nε2 forms a bifurcated hydrogen bond to Og of the catalytic Ser209 and to O1 of the mentyl moiety. This arrangement is in agreement with the usual assumptions about the catalytic mechanism in serine-type hydrolases. For the 1S-derivative, the transition-state-analog pushes His449 out of place and thus perturbs the hydrogen bond between Nε2 and O1 of the menthyl moiety. Simultaneously,

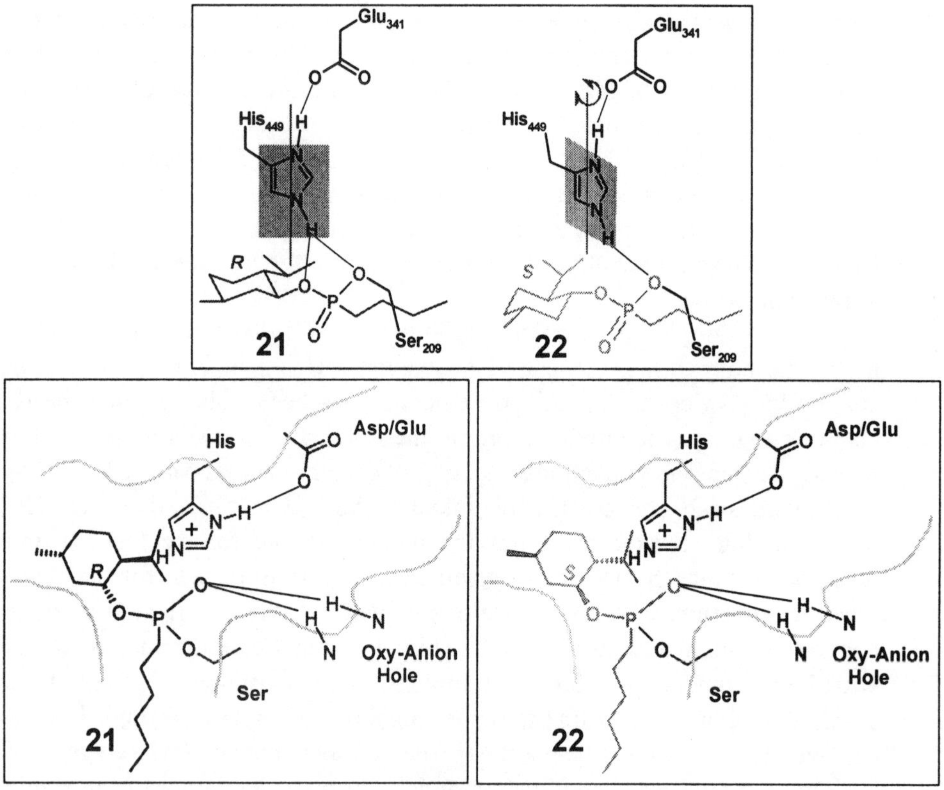

Figure 11: The two enantiomers of menthyl phosphonate transition state analogs (21, 22) have been cocrystallized with *Candida rugosa* lipase. Both inhibitors adopt slightly different orientations at the catalytic center. The isopropyl groups interfere with the neighboring residues. In the 1S-derivative in particular this group pushes the catalytically important His449 out of place. This reorientation has been linked to the lower reactivity of the ester corresponding to the related translation state analog.

conformational rearrangement modifies the torsion angles around phosphorous and increases the latter distance by 0.3 Å compared to the 1R-derivative. The interaction geometries formed by the tetrahedral intermediates along the reaction path are less favorable for the 1S-isomer. These differences are assumed to be responsible for the experimentally observed differences in reactivity along the two stereoisomeric enzyme reactions.

3.2 Energetic and Structural Determinants for Enantiopreference

Recently we could study the energetic and structural determinants of kinetic resolution of (Rc)- and (Sc)-1-phenylethylamine using *Candida antarctica* lipase B (CaL B)[26]. The lipase has been preacylated with an ethoxyacetyl group. Addition of the amine results in a nucleophilic attack on the acyl enzyme under non-aqueous conditions to produce the corresponding amide (Fig. 12). By stopped-flow kinetics we could show that neither substrate binding nor deacylation of the acyl enzyme is rate limiting, accordingly formation of the tetrahedral transition state is most likely rate limiting. In this crucial step, both enantiomeric forms of the (Rc)– and (Sc)-

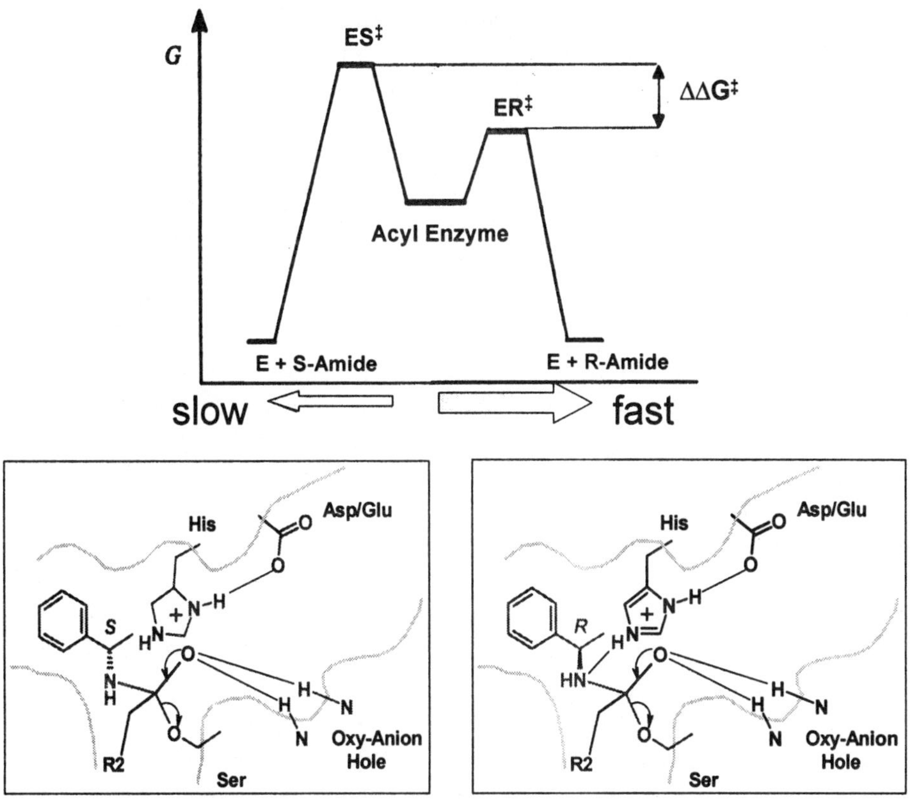

Figure 12: The acyl enzyme intermediate reacts with the R- or S-phenylethylamine by forming a tetrahedral transition state. Kinetics and thermodynamics show that the faster reacting R-enantiomer is favored by an enthalpic contribution of 33.1 kJ/mol whereas the S-enantiomer benefits from an entropic advantage of –13.7 kJ/mol.

phenylethylamine are exposed to the enzyme and the reaction is proceeded with remarkable stereodiscrimination in favor of the (Rc)-form (>99.9% ee). Studying the temperature dependence of the enantiomeric discrimination step reveals a differential activation free energy of 19.4 ± 6kJ/mol in favor of the fast-reacting Rc-amine. Factorizing $\Delta\Delta G^{\neq}$ into enthalpic and entropic contributions demonstrates the fast-reacting substrate to be enthalpically favored by 33.1±3 kJ/mol. Although the $\Delta\Delta H^{\neq}$ contribution dominates, the entropic portion points in opposite direction: it favors the slow-reacting enantiomer by $T\Delta\Delta S^{\neq}$ of -13.7±3kJ/mol. This is consistent with an overall reduction of enantio-preference with increasing temperature.

To obtain detailed structural insights into the observed discrimination we synthesized enantiopure phospho-transition state analog (TSA) inhibitors and reacted them with the enzyme. Product formation was confirmed by mass spectrometry. Subsequently, the crystal structures of the Rc- and Sc-phospho-TSAs complexes have been determined. The electron density for the fast-reacting Rc-phospho-TSA elucidates the binding mode of the inhibitor covalently attached to the catalytic serine Oγ. The oxyanion hole is filled by one of the terminal P-oxygen atoms, while the amide NH forms a hydrogen bond to the catalytic His residue (Fig. 13). The methyl group at the stereogenic center occupies the stereospecificity pocket and the terminal phenyl ring accommodates according to the well-defined difference electron density an orientation perpendicular to the adjacent Trp104 moiety. In total, the fast-reacting Rc-P-TSA shows virtually perfect complementary to the binding pocket of CaL B. In contrast, for the slow-reaction Sc-P-TSA pronounced residual mobility for the phenyl moiety has to be assumed, indicated by the partically ill-defined difference electron density in this region. To support the hypothesis of distinct residual mobility of both TSAs, molecular dynamics simulations have been performed. They clearly indicate that the molecular arrangements sampled for the fast-reacting Rc-TSA all cluster closely around the geometry found in the crystal structure (Fig. 13), simultaneously conserving the hydrogen bond between the amide NH and the catalytic His. In case of the slow-reacting Sc-TSA a virtually unrestricted tumbling of the terminal phenyl can be registered. It also involves strong fluctuations of the amide NH group, resulting in a frequent rupture of the H-bond to the histidine. As this hydrogen bond is assumed to be essential for the enzyme reaction, we suggested that the slow-reacting substrate achieves less frequently a transition state geometry productive for the reaction than the firmly fixed fast-reacting substrate. In addition, the latter is immobilized and stabilized by numerous favorable van der Waals contacts. Such firm immobilization is synonymous to favorable enthalpic interactions however they parallel a significant loss in motional degrees of freedom. Accordingly, in agreement with kinetic data, the binding of the fast-reacting Rc-TSA is entropically disfavored. In contrast the slow-

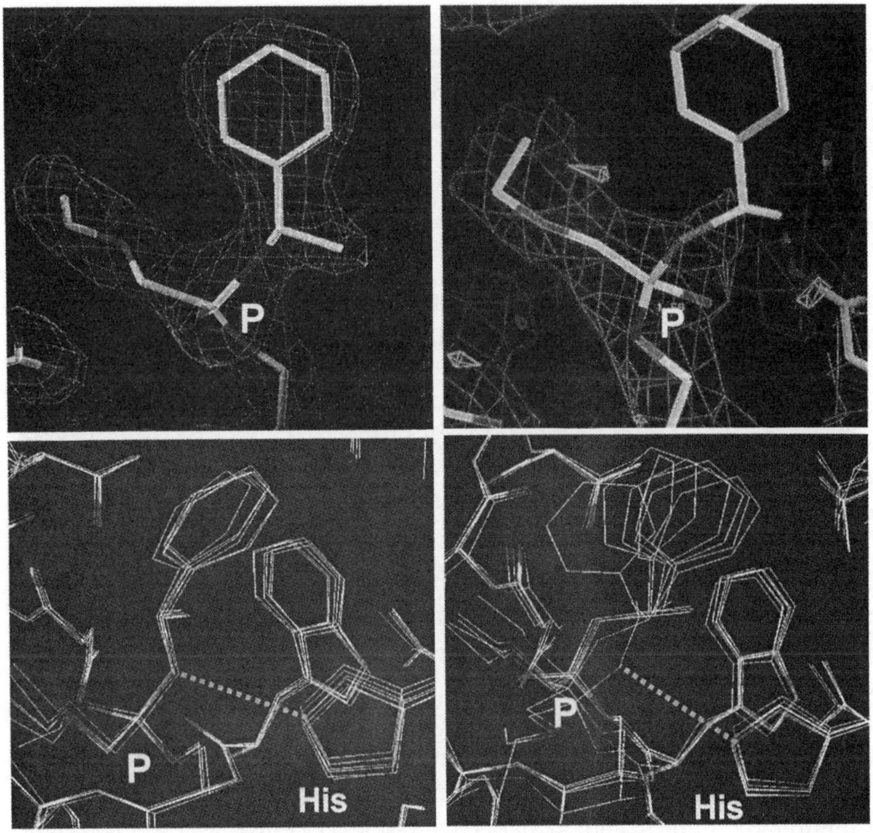

Figure 13: The crystal structure determined with Rc-phospho-TSA inhibitor (upper left) shows fully defined electron density indicating firm binding of this species. The slow-reacting Sc-phospho-TSA inhibitor (upper right) indicates partial disorder around the phenyl position in agreement with high residual mobility of the TSA at the catalytic center. MD simulations performed on both TSA (lower row) confirm the crystallographic evidence. The fast-reacting TSA resides mainly in the conformation indicated by crystal structure analysis. It thus adopts most of the time a geometry productive for the enzyme reaction. In contrast, the slow-reacting TSA indicates virtually unrestricted tumbling of the phenyl portion coupled with a frequent rupture of the NH to His 224N hydrogen bond. This residual mobility suggests that the Sc-TSA resides less frequently in an orientation productive for the enzyme reaction. Thus a reduced reaction rate can be expected.

reacting Sc-TSA shows significant residual mobility, thus it should loose less entropy upon binding. This matches with the entropically more favorable binding of the latter TSA. Enantiopreference is reduced with increasing temperature. Augmenting the temperature will result in higher overall mobility, thus reducing the advantage of the firmly fixed fast-reacting substrate and simultaneously diminishing the entropic differences between both substrates, in particular the beneficial entropic contribution for the slow-reaction Sc-TSA. This example of kinetic resolution of enantiomeric substrates using lipases shows that the actual discrimination step results from a complex picture of structural and dynamic aspects giving rise to opposing enthalpic and entropic contributions.

4. CONCLUSIONS

Biological macromolecules frequently being receptors of small molecule ligands such as drugs create a chiral molecular environment. They are themselves constructed from chiral building blocks and spatially arranged in handed conformations. Accordingly, enantiomers will be recognized as different species at the binding sites of such receptors. Some rare cases have been reported on the crystallographically determined binding mode of both enantiomers together with the protein.

Usually quite distinct binding affinities are observed for both forms. However, since solvation/desolvation properties are identical in these cases the energy difference can been attributed to conformational differences of local interaction geometries. A more strained and thus less favorable binding configuration is detected for the less active enantiomer. The distinct immobilization of stereoisomers in the handed environment of a binding pocket can also be exploited in enzyme kinetics, e.g. to achieve more favorable metabolic stability of retro/inverse peptides or to succeed in chiral resolution of amines and alcohols with lipases. Further crystallographic work with carefully selected and optimized crystallization conditions will hopefully reveal a more detailed insight into the spatial discrimination of stereoisomers at the binding site of proteins.

REFERENCES

1. Friedman, L. and Miller, J., 1971, Odor Incongruity and Chirality. *Science* **172**: 1044-1046
2. Böhm, H. J., Klebe, G. and Kubinyi, H.: Wirkstoffdesign, 1996, Heidelberg, *Spektrum*, Akademischer Verlag.
3. O'Brien, X. M., Parker, J., Lessard, P. and Sinskey, A., 2002, Engineering an indene bioconversion process for the production of cis-aminoindanol: a model system for the production of chiral synthons. *Appl. Microbiol. Biotechnol.* **59**::389-399
4. Dorey, E., 2000, Chiral Drugs viable, despite failure. *Nat. Biotechn.* **18**: 1239-1240
5. Edington, P. and Harding, M., 1974, The Crystal Structure of DL-Histidine. *Acta Crystallographica* Section **B 30**: 204-206
6. Madden, J., McGandy, E. and Seemann, N., 1972, The Crystal Structure of the Orthorhombic Form of L-(+)-Histidine. *Acta Crystallographica* Section **B28**: 2377-2383

7. Madden, J., McGandy, E., Seeman, N., Harding, M. and Hoy, A., 1972, The Crystal Structure of the Monoclinic form of L-Histidine. *Acta Crystallographica* Sect. **B 28**: 2382-2389
8. Zawadzke, L. and Berg, J., 1993, The Structure of a Centrosymmetric Protein Crystal. *Proteins* **16**: 301-305
9. West, M. and Fairlie, D., 1995, Targeting HIV-1 Protease: A Test for Drug Design Methodologies. *Trends Pharm. Sci.* **16**: 67-74
10. Milton, R., Milton, S. and Kent, S., 1992, Total Chemical synthesis of a D-Enzyme: the Anantionmers of HIV-1 Protease show Reciprocal Chiral Substrate Specificity. *Science* 256: 1445-1448
11. Jung, G., 1992, Proteine aus der D-chiralen Welt. Angew. Chem. Int. Ed. 104:1484-1486
12. Stubbs, M. T., Huber, R. and Bode, W., 1995, Crystal structures of factor Xa specific inhibitors in complex with trypsin: structural grounds for inhibition of factor Xa and selectivity against thrombin. *FEBS Lett* **375**: 103-107.
13. Greer, J., Erickson, J. W., Baldwin, J. J. and Varney, M. D., 1994, Application of the three-dimensional structures of protein target molecules in structure-based drug design. *J. Med. Chem.* **37**: 1035-1054
14. Klaholz, B., Mitschler, A., Belema, M., Zusi, C. and Moras, D., 2000, Enantiomer Discrimination Illustrated by high Resolution Structures of the Human Nuclear Receptor hRAR. *Proc. Nat. Acad Sci* **97**: 6322-6372
15. Kempf, D., Norbeck, D., Codacovi, L., Wang, X., Kohlbrenner, W., Wideburg, N. et al., 1990, Structure-based, C2 Symmetric Inhibitors of HIV Protease. *J. Med. Chem.* **33**: 2687-2689
16. Hosur, M. V., Bhat, T. N., Kempf, D. J., Baldwin, E. T., Liu, B., Gulnik, S. et al., 1994, Influence of Stereochemistry on Activity and Binding Modes for C_2 Symmetry-Based Diol Inhibitors of HIV-1 Protase. *J. Am. Chem. Soc.* **116**: 847-855
17. Imoto, T., Johnson, L., North, A., Phillips, J. and Rupley, J., 1972, Vertebrate Lysozyme. *The Enzymes* :665-868
18. Shafer, J. and Gould, R., 1993, Design of Antithrombotic Agents. *Persp. Drug, Discov. Design* **1**: 419-548
19. Bajusz, S., 1993, Chemistry and Biology of the Peptide Anticoagulant D-MePhe-Pro-Arg-H (GYKI-14766). *Adv. Exp. Med. Biol.* **340**: 91-108
20. Martin, P., Robertson, W., Turk, D., Huber, R., Bode, W. and Edwards, B., 1992, The Structure of Residues 7-16 of the Alpha chain of Human Fibrinogen bond to Bovine Thrombin at 2.3 Angstroms Resolution. *J. Biol. Chem.* **267**: 7911-7920
21. Roderick, S. L., Fournie-Zaluski, M. C., Roques, B. P. and Matthews, B. W., 1989, Thiorphan and retro-thiorphan display equivalent interactions when bound to crystalline thermolysin. *Biochemistry* **28**: 1493-1497.
22. Schmid, R. and Verger, R., 1998, Lipases: Interfacial Enzymes with Attractive Applications. *Angew. Chem. Int. Ed.* **37**: 1608-1633
23. Chen, C. and Sih, C., 1989, General Aspects and Optimization of Enantioselective Biocatalysis in Organic Solvents. The Use of Lipases. *Angew. Chem. Int. Ed. Engl.* **28**: 695-708
24. Gutman, A. and Shapira, M., 1995, Synthetic Applications of Enzymatic Reactions in Organic Solvents. ADV Biochem. *Eng. Biotechnol.*
25. Cygler, M., Grochulski, P., Kazlauskas, R., Schrag, J., Bouthillier, F., Rubin, B. et al., 1994, A Structural Basis for the Preferences of Lipases. *J. Am. Chem. Soc.* **116**: 3180-3186

26. Bocola, M., Stubbs, M. T., Sotriffer, C., Hauer, B., Friedrich, T., Dittrich, K. et al., 2003, Structural and energetic determinants for enantiopreferences in kinetic resolution of lipases. *Protein Eng.* **16**: 319-322.

Analytical Centrifugation: Looking at Aggregation in Free Solution

P. JONATHAN G. BUTLER
MRC Laboratory of Molecular Biology, Hills Road, Cambridge, CB2 2QH, UK

1. INTRODUCTION

The relative ease of DNA sequencing, leading to our knowledge of the complete genome of a number of organisms (including humans), results in the exact monomer mass either being already known, or else readily determined, for any protein of interest. However, this does not mean that one knows the more interesting mass – that of the protein in solution – unless one also knows the aggregation state under the relevant conditions. Frequently workers use a gel-based method, preferably gel permeation chromatography, but sometimes even non-denaturing gel electrophoresis, to try to determine the mass of the isolated protein, and hence its aggregation. The problem with this approach is that the parameter which actually determines the elution volume, or (together with charge) the electrophoretic mobility, is the Stokes radius, which is only indirectly related to the mass. To get round this problem, it is usual to calibrate the gel by also running a number of "marker proteins" and drawing a calibration curve, but this is only valid if the protein of interest and the marker proteins have similar overall molecular shapes. Typical marker proteins are globular, often close to spherical, while proteins of interest may well be more extended or have a non-ellipsoidal shape. A further serious problem arises for membrane proteins, which will usually have been solubilised by detergent extraction and are in solution only as a protein/detergent micelle – whose outer dimensions will not correspond to those of the isolated protein.

Supramolecular Structure and Function 8, Edited by Pifat-Mrzljak
Kluwer Academic/Plenum Publishers, New York 2004

A similar problem arises in the use of light scattering, even with auto-correlation analysis using a laser-based instrument. Again the measurement gives a radius-related parameter, either the mean radius directly or, in the case of auto-correlation analysis, the translational diffusion coefficient, and it is necessary to make assumptions about the molecular shape, and protein density, in order to convert this into a molecular mass. The problem of detergent micelles also remains unresolved.

These problems can be avoided by the use of either osmotic pressure measurements or, as I intend to discuss here, analytical ultracentrifugation. These techniques, while very different in their practice, are closely related during the analysis as both essentially count particle concentrations in the solution, and the molecular mass is then obtained by use of a concentration-related factor; the density increment in the case of the analytical ultracentrifuge. In practice biochemists have usually incorporated this into the equations using the approximation derived from the partial specific volume and solution density (often further approximated to solvent density), and I shall be discussing this later.

The analytical ultracentrifuge was originally developed by Thé Svedberg and his collaborators[1] and at first was driven by an air turbine, and was not convenient to use. However, an electrically driven machine was developed by Spinco, and this company was subsequently taken over by Beckman, who have now developed a very convenient machine which allows scanning of the samples in the rotor by either absorption or interference optics, recording the resulting scans in a computer readable form. In comparison to the early days, when all imaging was done photographically and the resulting plates had to be measured (and possibly densitometered) before any analysis could be carried out, this has enabled much more sophisticated analysis to be performed on a routine basis.

2. TYPES OF ANALYSIS AND THEIR INTERPRETATION

2.1 Sedimentation Velocity

The earliest work by Svedberg and his colleagues involved looking at the speed of sedimentation of macromolecules, by following the movement of the boundary generated as the molecule sediments away from the meniscus, and measuring the sedimentation coefficient (now denominated in "Svedberg units" (S), with dimensions of $10^{-13}.sec^{-1}$) for the molecule from the equation:

$$s \equiv \frac{dr / dt}{\omega^2 r} \tag{1}$$

where r is the radius of the molecule at time t, and ω is the angular velocity (in radians.sec^{-1}). This can be rearranged into the more amenable form:

$$s = \frac{1}{\omega^2} \frac{d\ln(\bar{r})}{dt} \tag{2}$$

where $\bar{r}$ is the average radius of the boundary of macromolecules.

The sedimentation coefficient is a useful parameter to characterise a macromolecule, depending as it does upon the molecular mass, size and shape, with the latter two coming into consideration in terms of the frictional drag as the molecule sediments through the solution. For this reason, the solvent viscosity is also important, as is the solvent density which influences the buoyancy. Since these will vary with both solvent composition and, particularly the viscosity, temperature, it is general practice to normalise the sedimentation coefficient to that which would (theoretically) be found in water at 20°C, using the equation:

$$s_{20,w} = s \frac{\eta}{\eta_{20,w}} \frac{(1-\bar{v}\rho)_{20,w}}{(1-\bar{v}\rho)} \tag{3}$$

where η is the viscosity, $\bar{v}$ the partial specific volume of the macromolecule and ρ the solvent density. The subscripts indicate that these parameters are those for water at 20°C, while parameters with no subscript are those for the experimental conditions. This equation involves the term $(1-\bar{v}\rho)$, which is used as an approximation for the density increment, the increase in solution density for a unit increase in macromolecule concentration, from the equation:

$$\left(\frac{\partial \rho}{\partial c_2}\right)_{\mu} = (1-\bar{v}\rho) \tag{4}$$

where the partial derivative indicates that all components other than the macromolecule are at constant chemical potential (*i.e.* at dialysis equilibrium).

The sedimentation coefficient was classically interpreted into the molecular mass using the Svedberg equation:

$$M = RTs / D(1-\bar{v}\rho) \tag{5}$$

where R and T are the molar Gas constant and temperature (in °K), s is the sedimentation coefficient and D is the molar diffusion coefficient.

Moreover, the sedimentation coefficient itself can be a useful value to characterise a macromolecule (*e.g.* the typical reference to prokaryotic ribosomes as 70S, with 30 and 50S subunits) and the sedimentation coefficients of similar materials do correspond well with their relative sizes (due to the involvement of D as well as s, a dimerisation will typically result in an increase in s by $\sqrt{2}$).

While such relatively simple analysis was all that could be readily performed when photographs had to be measured by hand (*e.g.* to determine the radius of the midpoint of a sedimenting boundary at various times), the availability of computer-readable data has allowed significant improvements in the methods of analysis. For sedimentation velocity, a major step forward came with the introduction of the analysis of time-derivatives of the concentration[2, 3], leading to plots of $g(s^*)$ against s^*, where $g(s^*)$ is the amount of material sedimenting between s^* and $(s^*+\delta s^*)$, with the asterisk indicating an apparent value. Such plots are essentially a graph of the amount of material sedimenting at any given apparent sedimentation coefficient and, due to diffusion, will have a finite, ideally bell-shaped, width. This broadening of the boundary can be used to estimate the diffusion coefficient and thus to obtain the apparent molecular mass from Equation [5].

A very convenient program for this analysis has been developed by Philo[4], and this forms the basis for all the examples of sedimentation velocity which I give below. This program also uses model fitting to the raw $(\delta c/\delta t)$ data to estimate the diffusion (thereby avoiding any distortion introduced by transformation of this data into any derived form). The models used can have a number of components, and it is quite possible to obtain good fits and estimated parameters for more than one species, provided that they are sufficiently different that the cross-correlations during the fitting are not too great.

One inherent problem with all such analyses arises from any polydispersity in the sample, where this produces relatively small changes in the sedimentation coefficient. Thus a monomer/dimer mixture will usually give no problem, as it will satisfy the condition just given above, but size heterogeneity of a "single" component will cause the boundary to be broadened (other than by diffusion), leading to an overestimation of the diffusion coefficient and, hence, too low a value for the apparent molecular mass.

2.2 Sedimentation equilibrium

An alternative mode of analytical ultracentrifugation involves running the centrifuge at a relatively low speed, so that the macromolecules do not pellet, even though they do move down the sample cell under the influence of the sedimentation. Rather, as the concentration rises towards the bottom of the cell, the molecules diffuse back down the concentration gradient until finally an equilibrium distribution is established, when at any given radius the average sedimentation downwards and diffusion back upwards are equal, and no further bulk redistribution of sample is occurring. For a single macromolecular component, this distribution is described by the equation:

$$c_r = c_0 \exp\left(\sigma\left(r^2 - r_0^2\right)\right) \tag{6}$$

where σ is defined as: $\sigma = \dfrac{M_{app}(1-\bar{v}\rho)\omega^2}{2RT}$ (7)

with M_{app} being the apparent molecular mass of the component, c_r the concentration at radius r, and with the subscript 0 defining a reference concentration and radius.

This can be converted into the generally more useful form:

$$\overline{M}_{w,app} = \frac{d\ln(c)}{dr^2}\,\frac{2RT}{(1-\bar{v}\rho)\omega^2} \tag{8}$$

where $\overline{M}_{w,app}$ is the weight average, apparent molecular mass. By calculating $\left(d\ln(c)/dr^2\right)$ for a sliding window of datum points, it is possible to obtain estimates of $\overline{M}_{w,app}$, at essentially the mid-point concentration, and to plot these values. From the appearance of the curve, it is usually easy to assess the aggregation *etc.* shown by the macromolecule.

Besides obvious possibilities, such as a monomer/dimer equilibrium, a macromolecule may show non-ideal behaviour. The similarity to osmotic pressure is shown in the nature of the equation defining this behaviour:

$$\frac{1}{\overline{M}_{w,app}} = \frac{1}{M_0} + 2Bc \tag{9}$$

where M_0 is the "true" molecular mass (*i.e.* at concentration, c, equals zero, so with no effect of non-ideality) and B is the second osmotic virial coefficient.

A less convenient possibility is the occurrence of non-equilibrating aggregation, where the oligomers do not dissociate upon dilution. This can be readily seen by either loading cells at different initial concentrations, or running the centrifuge at different speeds, and then plotting $\overline{M}_{w,app}$ against c for these different conditions on the same graph. While the curves for an equilibrating system will essentially lie over each other, as the mixture of aggregates will always be the same at a given total concentration (depending upon K_d for the aggregation), for a non-equilibrating system the curves do not overlie, but in this case the results from the cell with lower loading concentration will show a higher apparent mass at any given total concentration. We found an example of this in some work on proteins with poly-glutamine tails[5], in which stably folded monomers, dimers and trimers co-exist, without equilibrating between themselves. In this case, each species can be isolated and remains stable over a period of months.

One problem with this type of analysis is that it is not reliable to fit parameters, such as K_d or B, to derived values (*e.g.* $\overline{M}_{w,app}$), since the errors will not be normally distributed (*i.e.* if the errors in measurements of c are normal, those in $\ln(c)$ will not be). This problem is now readily avoided by the expedient of fitting the original data directly. This has only become possible due to the power of modern computers, together with the computer-readable form of the data, and Johnson and Straume[6] have shown that it is possible to use Equation [6] to derive the following:

$$A = \varepsilon c_0 \exp\left(\sigma\left(r^2 - r_0^2\right)\right) + \Delta \qquad (10)$$

where A is the absorbance of the macromolecule, with extinction coefficient ε at the relevant wavelength, σ is as defined in Equation [7] and Δ is the error in the baseline.

Equation [10] can be adapted to allow for appropriate variations in M_{app}, for example in the case of a monomer/dimer equilibrium it becomes:

$$A_t = \varepsilon_1 \left(c_0 \exp\left(\sigma\left(r^2 - r_0^2\right)\right) + \frac{2\left(c_0 \exp\left(\sigma\left(r^2 - r_0^2\right)\right)\right)^2}{K_d} \right) + \Delta \qquad (11)$$

where M_{app} is now equivalent to the monomer mass, A_t is the total absorbance, ε_1 is the molar extinction coefficient for monomer and K_d is the dissociation constant. The equations for non-ideal macromolecules can be derived in a similar fashion, *e.g.* for a non-ideal monomer one gets the equation:

$$A = \varepsilon c_r = \varepsilon c_0 \exp\left(\sigma\left(r^2 - r_0^2\right) - 2B\left(c_r - c_0\right)\right) + \Delta \qquad (12)$$

where c_r is the concentration at radius r. Although Equation [12] is transcendental, *i.e.* the concentration c occurs on both sides of the equation and there is no analytical solution, it is again possible to fit numerically, with modern computers.

Each of these equations results in an expression for the total absorbance at radius, r (*i.e.* what has been measured in the data), which is calculated from the reference concentration, c_0, at the reference radius, r_0. Since the errors in the data will be Gaussian, and, moreover, the software on the Beckman instrument yields estimates of the standard deviation for each datum point (when several readings are averaged at each radius), it is appropriate to use least squares fitting to refine the parameters[6]. Typically these will be either K_d or B, an error in the measured baseline (Δ) and the monomer mass (M_1). Although this latter is in fact known to very high precision, it occurs in the product with $(1 - \bar{v}\rho)$ (Equation [7]), where the estimates for both $\bar{v}$ and ρ will contain error, and it is simplest to allow the single parameter to vary during the fitting.

The equations to fit other, more complicated models can be derived, but it is important to realise that fitting more parameters, while probably giving an apparently better fit to the data, is likely to be dangerous unless based upon a model known to be appropriate. Moreover, in many cases there will be very high co-variance between some of the parameters and so the estimates for them will be inaccurate. One problem with all fitting is in deciding which is the most appropriate model. Often this has been obvious from an initial plot of $\overline{M}_{w,app}$ against c for the data, but a further check is provided by a plot of the residuals between the calculated and measured absorbance values, against each radius value. For the appropriate model, these residuals should be small, of the order of magnitude of the standard deviations of the datum points, and randomly distributed around zero. Any systematic differences are good evidence that the model is not appropriate[6].

2.3 Isopycnic sedimentation

Although not used at all frequently now in the analytical ultracentrifuge, one early application of the analytical ultracentrifuge, which was very important, was the use of caesium chloride (CsCl) in the sample. This technique was originally developed by Meselson, Stahl and Vinograd[7], using the Spinco Model E centrifuge. It was then applied by Meselson and Stahl[8] in their classic demonstration of the semi-conservative replication of DNA, by isolating DNA from *E. coli* grown for successive generations in media with different isotopes, to produce density differences. They showed that if the DNA from the first generation was entirely heavy, the next generation would show intermediate density and all subsequent generations a mixture of the intermediate density and increasing amounts of light.

Important as this work was, in practice it is uncommon to use this technique in an analytical ultracentrifuge now, rather it is employed as a preparative technique, particularly for plasmid purification using ethidium bromide intercalation to generate a density difference between the closed, circular DNA of the plasmids and the linear, sheared cellular DNA.

3. IMPORTANCE OF THE CORRECT DENSITY INCREMENT

As already mentioned above, the density increment (usually substituted by the equivalent term $(1-\bar{v}\rho)$ – see Equation [4]) occurs in determining the apparent molecular mass. This term is essentially the remaining mass, when the contribution of the Archemedian buoyancy, as given by the product of the partial specific volume (the change in volume of the solution upon addition of a unit mass of the macromolecule) and the solution density, has been subtracted. Historically, most biochemists have used the partial specific volume for a protein, as defined as that of a unit mass *in vacuo*, for $\bar{v}$ and the density of the solvent (rather than the solution) for ρ. This has been convenient, especially as many proteins approximate to $\bar{v} \approx 0.72 - 0.73 \text{ ml/g}$ and solvents to $\rho \approx 1.01 \text{ g/ml}$, allowing these assumed values to be used in such cases. A more sophisticated approach is to calculate these parameters. The partial specific volume can be calculated using the method of Edsall[9], which can be reduced to the equation:

$$\bar{v} = \frac{\sum n_i M_i \bar{v}_i}{\sum n_i M_i} \tag{13}$$

where n_i is the number of moles, M_i the molecular mass and $\bar{v}_i$ the partial specific volume, of the ith component of the macromolecule. This calculation, and also the calculation of the solvent density and viscosity, has been greatly simplified by the development of the program Sednterp[10], and this program is now frequently used to obtain these essential parameters for the analysis of ultracentrifuge data.

The problem with all such calculations is that they inherently do not take into account solvation effects upon the macromolecule. This was addressed by Casassa and Eisenberg[11], who formulated the equation:

$$\left(\frac{\partial \rho}{\partial c_2}\right)_{\mu} = \left(1 - \phi' \rho_0\right) \tag{14}$$

where ϕ' is the "apparent partial specific volume" and ρ_0 the *solvent* density. It should be noted that ϕ' does not have any absolute meaning – it is just an apparent parameter, allowing the use of a formulation similar to that most commonly employed in the equations for molecular mass. These authors showed that, provided the condition of the partial derivative is fulfilled, *i.e.* that all components other than the macromolecule (component 2) are at constant chemical potential (dialysis equilibrium), the use of this density increment will yield the molecular mass of the species as measured for c_2. In other words, if the concentration measured is that of the protein alone, the molecular mass will be that of the protein component, irrespective of any potential solvation – for example by formation of detergent micelles with a membrane protein solubilised in detergent. This applies equally well for the binding of water, to the exclusion of other buffer components, or even for binding of Cs^- ions to DNA in density separations[12].

Eisenberg has revisited the question of 3 component systems more recently[13], showing that:

$$\left(\frac{\partial \rho}{\partial c_2}\right)_{\mu} = \left(1 - \rho_0 \bar{v}_2\right) + \xi_3\left(1 - \rho_0 \bar{v}_3\right) \tag{15}$$

where $\xi_3 = \left(\partial w_3 / \partial w_2\right)_{\mu}$, *i.e.* ξ_3 is an interaction parameter indicating the change in gram molality w_3 with the change in gram molality w_2, at constant chemical potential of components 1 and 3, due to binding of component 3 to component 2, and $\bar{v}_2$ and $\bar{v}_3$ are the partial specific volumes of the isolated components 2 and 3 (in the latter case it can be convenient to remember that $\bar{v}_3 = 1/\rho_3$). What is significant about Equation [15] is that, provided component 3 binds to 2, the density increment will always be larger than that of component 2 alone, and therefore ϕ' smaller than $\bar{v}_2$.

The density increment can be calculated from:

$$\delta\rho = \rho_{solution} - \rho_{solvent}; \tag{16}$$

and δc_2, which is readily measured for the protein alone (*i.e.* not including any bound buffer components) using amino acid analysis. The important condition is the dialysis equilibrium for the solution and solvent, which can be obtained either by extensive dialysis or by gel permeation chromatography into the relevant buffer. The one case where this can pose significant difficulty is with a detergent which is non-dialysable and where the detergent micelles are too similar in size to the protein/detergent micelles, so that these cannot be separated on a gel column. In practice, densities are readily measured to 10^{-6} g/ml, using the Paar density meter[14] which requires <1ml total volume. In view of the difference between the densities, it is necessary to have solution protein concentrations >2 mg/ml in order to obtain a reliable value for the density increment.

4. EXAMPLES OF STUDIES IN THE ANALYTICAL CENTRIFUGE

4.1 Mode of action of the Bovine ATPase-F_1 inhibitor, IF_1

Metabolism in the cell relies upon the presence of ATP as a source of energy for many processes. In eukaryotes, this ATP is synthesised using the

Table 1. Solvent densities, density increments and apparent partial specific volumes for IF_1 at various pH's.

pH	ρ_0 (g/ml)	$(\partial\rho/\partial c_2)_\mu$	ϕ' (ml/g)
5.0	1.002128	0.239	0.759
8.0	1.000957	0.239	0.760

proton-motive force generated across the inner mitochondrial membrane by transport of protons out of the matrix during oxidative electron transport. The enzyme involved is the mitochondrial F_0/F_1 ATP synthase, which couples inorganic phosphate to ADP, while allowing protons to re-enter the mitochondrion. The F_0 subunit is embedded in the membrane and appears to transport protons freely, while the F_1 subunit displays the catalytic activity. However under some adverse circumstances, for example anoxia, the proton-motive gradient can be lost, with the matrix falling to below pH 6.5, and F_1 will then act in reverse, *i.e.* as an ATPase, and thus destroy the cellular ATP. This is prevented by an inhibitor of F_1, called IF_1, which is activated at or below pH 6.5, then showing its inhibitory activity and preventing breakdown of ATP.

We were interested to characterise IF_1 and to study its interaction with F_1.

4.1.1 Sedimentation equilibrium studies of IF_1

In view of the importance of pH for the activity of IF_1, we looked to see whether this might be influencing its aggregation, by investigating this at pH's 5.0 and 8.0[15], at which pH's IF_1 is either active or inactive. Because of the somewhat unusual amino acid composition of this protein, we started by measuring its density increments at these pH's and then calculating the apparent partial specific volumes, obtaining values as shown in Table 1. It is perhaps worth noting that, despite the slight change in solvent density with pH, the density increment remains constant and the apparent partial specific volume alters to make up the difference.

As well as the wild type IF_1, this study also included a mutant, IF_1-M49K, with the methionine residue in position 49 mutated to histidine. This mutation abolishes the pH-dependent activation-inactivation, with the mutant protein active even at the higher pH. Sedimentation equilibrium experiments were performed on both proteins at each pH, and the resulting plots of $\overline{M}_{w,app}$ against concentration are shown in Figure 1. From the amino acid compositions of the proteins, the monomer masses are ~9.6 kDa and since the lowest values of $\overline{M}_{w,app}$, seen for the wild type protein at pH 5.0 (Fig. 1A), and for mutant M49K at both pH's (Fig. 1B and D), is ~kDa, it must be present as a minimum aggregate of dimmer under all conditions. (In practice, the structure of IF_1, determined at a later date[16], shows that the monomer chains form an anti-parallel dimer, held together by a coiled-coil, and this will be extremely stable and highly unlikely to dissociate at the concentrations used in any experiments.) Moreover, at pH 8.0 the wild type IF_1 shows a minimum mass of ~40 kDa (Fig. 1C), *i.e.* a tetramer (again as seen in the crystal structure).

The obvious model for fitting the data is that of a non-ideal monomer, but with the "monomer" actually a dimer for wild type at pH 5.0, and for mutant at both pH's, and a tetramer for wild type at pH 8.0. This model was found to fit well, as shown by the residual plots in each, although it was found that there was essentially no non-ideality in any case except the wild type at pH 5.0. The resulting dependence of $\overline{M}_{w,app}$ upon concentration has been plotted into each graph in Fig. 1, to show how the datum points fit to these curves.

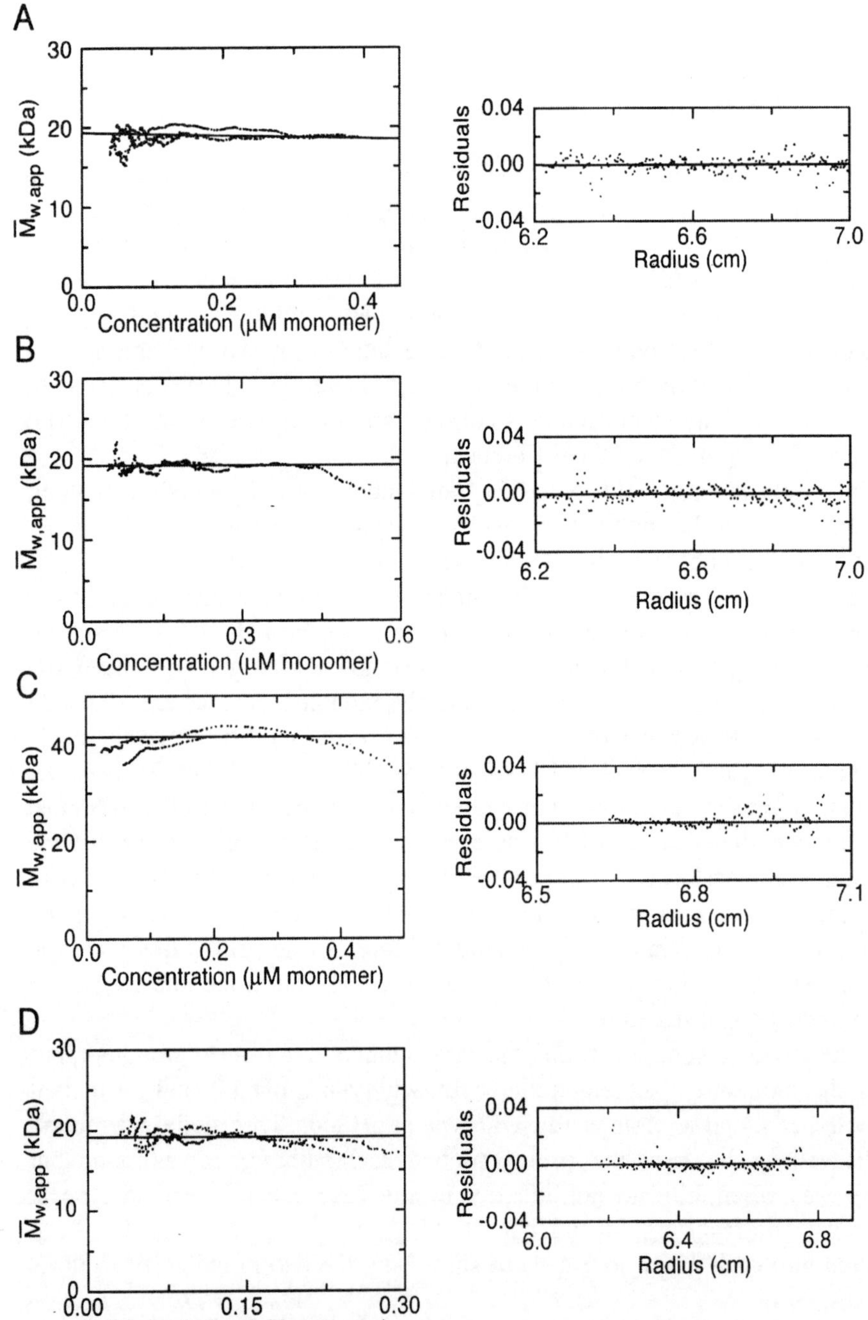

Figure 1. Plots of the results from sedimentation equilibrium analysis of protein IF_1 (A and (and mutant IF_1-M49K (B and D), at pH's 5.0 (A and B) and 8.0 (C and D).
Plots of $\overline{M}_{w,app}$ against c are shown at the left, with individual datum points plotted and also fitted lines for the optimum model – that of a non-ideal "monomer" (actually dimer) for plots A, B and D, and again a "monomer" (but now actually tetramer) for plot C. Plots of the residuals, against radius, for the fitting of the original data are shown on the right in each case.

These data strongly suggest that IF_1 is active as a dimer (*i.e.* at lower pH or as the mutant), but becomes inactivated when the dimers dimerise further, to give the tetramer. We were interested to study the effect upon F_1, but this could not be done using sedimentation equilibrium as the purified F_1 is not stable for long enough, so we used sedimentation velocity instead.

4.1.2 Sedimentation velocity studies with F_1 and IF_1

The effect of adding IF_1 to F_1 was studied at pH 6.5[17]. This pH is low enough for the IF_1 to be dissociated into dimers (data not shown) and gives greater stability to F_1 than lower pH's. Figure 2 shows sedimentation velocity data for F_1 and F_1+IF_1 (Fig. 2A and D), together with plots of $g(s^*)$ against s^* (Fig. 2B and E) and of the residuals for fitting the raw data with a model for a single component (Fig. 2C and F). The plots of the residuals show that there is essentially only a single component, and the fitting shows that F_1 alone has $s_{20,w} = 11.6\,S$, while together with IF_1 these change to $s_{20,w} = 16.0\,S$.

4.2 EmrE; a membrane-bound bacterial multi-drug transporter in detergent solution

Given the problems of drug-resistant bacteria, there is considerable interest in the resistance factors, many of which are in fact transporters which expel the antibiotics from the cell. One of these is EmrE, from *E. coli*, which will transport many antibiotics *in vivo* and can be overexpressed and isolated by detergent solubilisation of membranes with dodecyl maltoside (DDM). Such solutions show substrate or inhibitor binding, and from the stoichiometry of inhibitor binding and hetero-oligomerisation experiments with inactive mutants, it had been concluded that EmrE formed trimers or tetramers[18-20]. However, cryo-electronmicroscopy shows a dimer in the membrane[21] and measurements of inhibitor-binding to detergent solubilised EmrE gave ratios of 1:2, compatible with an active dimer structure (Chris Tate, personal communication). The structure is currently being studied intensively both by electron microscopy of membrane-bound EmrE and by X-ray crystallography. It was therefore of interest to look at the aggregation state of EmrE in DDM solution (Chris Tate and PJGB; unpublished work), to seek to resolve the conflicting, indirect claims about the oligomerisation.

In view of the detergent solution, the density increment was measured for the solution of EmrE in DDM, and the results are shown in Table 2. One initial problem was the difficulty in ensuring that a DDM solution is at

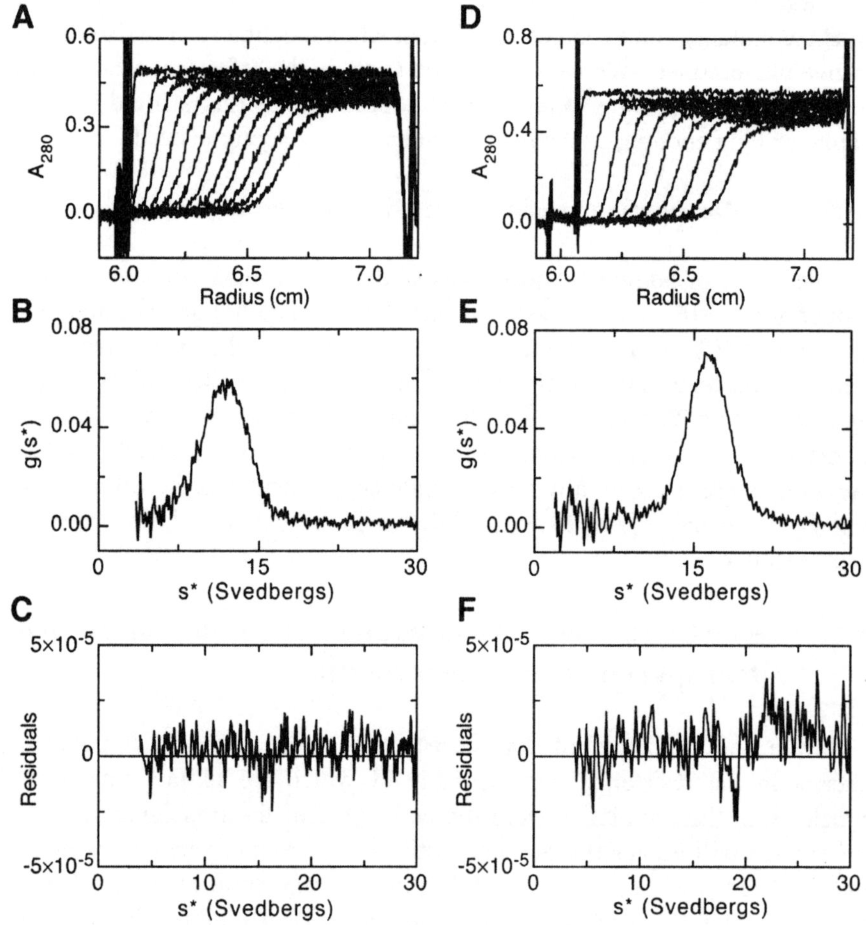

Figure 2. Sedimentation velocity analysis of F_1 without and with added IF_1.
A, B and C are F_1 alone, D, E and F are $F_1 + IF_1$.
A and D show the raw data, with scans at equal time intervals in each plot.
B and E show plots of g(s*) against s*.
C and F show plots of the residuals from model fitting against s*.

Table 2. Solvent density, density increment and apparent partial specific volume for EmrE solubilised in DDM solution.

ρ_0 (g/ml)	$\left(\partial\rho/\partial c_2\right)_\mu$	ϕ' (ml/g)
1.003936	0.960	0.040

dialysis equilibrium, since the detergent is non-dialysable and also EmrE has such a small monomer mass (15.2 kDa) that there is no resolution between the protein/detergent micelles and those of detergent alone in a gel filtration column. We used an alternative method to prepare the solution, by eluting EmrE from an ion-exchange column, equilibrated with the desired DDM concentration, using high salt and then dialysed the eluted peak against a buffer with the desired salt concentration and also the same DDM concentration.

While the solvent density is not unusual, the density increment is, initially at least, unexpectedly high resulting in an unusually low apparent partial specific volume. We therefore decided to confirm the measured value for the density increment by calculating it from the components, using Equation [15][13]. In making this calculation, we used the value for ρ_0 measured in determining the experimental density increment, and calculated $\bar{v}_2$ (*i.e.* the partial specific volume for EmrE) from the amino acid composition, using Sednterp (see above), giving a value of 0.7552 ml/g. Measurement of the binding of DDM to EmrE gave a mean value of 1:3.52 (±0.21) g:g, giving $\xi_3 (= w_3 / w_2) = 0.284$. The partial specific volume for DDM was calculated from the measured density of 1.225 g/ml, leading to $\bar{v}_3 = 0.816$ ml/g. These gave a calculated density increment for EmrE of 0.878, in reasonable agreement with that measured. Furthermore, it is known that EmrE, solubilised in DDM, still binds significant amounts of lipid. Although Equation [15] has been formulated in terms of a single "diffusible" component which binds to the macromolecular component (component 2), it is extendable to multiple components, each of which will contribute towards increasing the value of the density increment. The measured value is therefore very reasonable and, as pointed out by Casassa and Eisenberg[11], as the protein had been measured for the concentration When determining $\left(\frac{\partial \rho}{\partial c_2}\right)_\mu$, the molecular mass obtained from sedimentation equilibrium analysis will be that of the protein alone, without having to worry about the contributions from either the (known) amount of detergent or the (unknown) amount of lipid.

Sedimentation equilibrium experiments were therefore performed and analysed using the values in Table 2. The resulting plots are shown in Figure 3. All of the data fitted well to a monomer/dimer aggregation, with $K_d < 1\mu M$, although this value was not well determined as the concentrations used for the sedimentation were too high to allow accurate measurement. There was no evidence for higher aggregation in solution.

In view of the controversy about the aggregation state of EmrE, with the suggestions of trimers or tetramers in solution, we thought that it was desirable to investigate further, to see whether solvent conditions, or even time since solubilisation from the membrane, might be affecting the

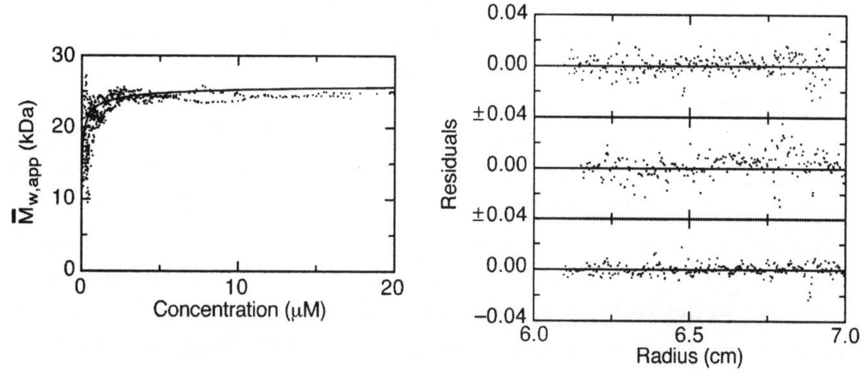

Figure 3. Sedimentation equilibrium analysis of EmrE, solubilised with DDM.
Left-hand plot shows individual points from separate cells, with line drawn for $M_1 = 13.25$ kDa, $K_d = 0.2$ μM.
Right-hand plots are residuals from fitting raw data for individual cells with model of ideal monomer/dimer aggregation..

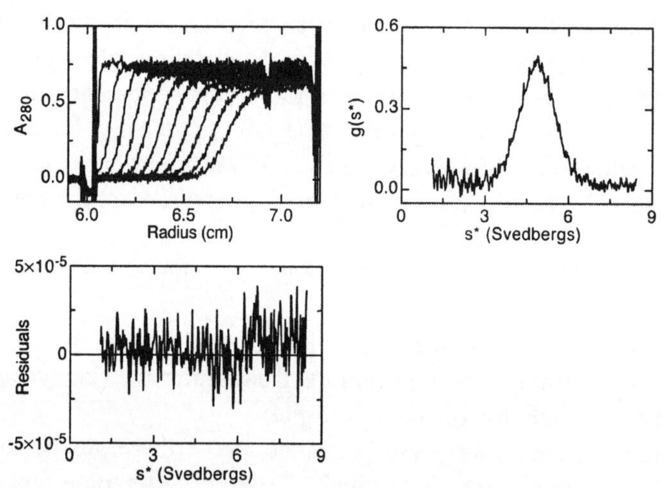

Figure 4. Sedimentation velocity analysis of EmrE. Scans are shown at equal time inervals, also the plots of g(s*) and the residuals from the model fitting, both against s*. These calculations were for scans towards the centre of cell.

aggregation state. In order to be able to obtain rapid analyses of the size, we turned to sedimentation velocity. An example of such an analysis is shown in Figure 4, with examples of the individual scans, a plot of g(s*) against s* for a set of 12 scans taken in the middle of the run and also the residual plot for model fitting to these scans. This model was for a single component, and gave $s_{20,w} = 4.90(\pm 0.003)\mathrm{S}$ and a molecular mass of 26.15 (±0.25) kDa, again compatible only with a dimer

We repeated the sedimentation velocity analyses with various samples, but in each case the results were essentially similar – the single boundary was running very close to the behaviour expected for a dimer. We therefore conclude that EmrE, as solubilised in DDM solution, is dimeric. This is in good agreement with our measurements of inhibitor binding, but does conflict with the conclusions drawn by others from indirect methods.

5. OVERALL REMARKS

I have discussed the various types of analysis available using the analytical centrifuge, including discussing at some length the importance of the accessory parameter which is essential for interpretation of the results into a molecular mass,.namely the density increment. I have dwelt upon this because the possible variations are frequently overlooked, with drastic consequences for the validity of any analysis.

My experimental examples have been chosen firstly to illustrate the usefulness of combining both sedimentation equilibrium and velocity, allowing one to look at even as complex a system as the interactions between, and control of, the mitochondrial F_1-ATP synthase and its physiological inhibitor IF_1, and secondly to show how the density increment matters, particularly for a membrane protein solubilised with detergent. While this latter may appear a fairly extreme example, in my experience people are turning to the analytical centrifuge more and more in cases which have proved difficult to resolve by other techniques. This means that many of the problems now being addressed with these techniques will not be straightforward, and so it is essential to be fully aware of the possible pitfalls, in order that any analysis should give the good results which are achievable.

REFERENCES

1. Svedberg, T. and Pedersen, K.O., 1940, *The Ultracentrifuge.* Oxford University Press, London and New York.
2. Stafford III, W.F., 1992, Boundary analysis in sedimentation transport experiments: a procedure for obtaining sedimentation coefficient distributions using the time derivative of the concentration profile. *Analyt. Biochem.* **203**: 295-301.
3. Stafford III, W.F., 1994, Methods for obtaining sedimentation coefficient distributions. In *Analytical Ultracentrifugation in Biochemistry and Polymer Sciencel* (Harding, S.E.,

Rowe, A.J. and Horton, J.C., eds), The Royal Society of Chemistry, Cambridge, UK, pp.359-393.

4. Philo, J.S., 2000, A method for directly fitting the time derivative of sedimentation velocity data and an alternative algorithm for calculating sedimentation coefficient distribution functions. *Analyt. Biochem.* **279**: 151-163.
5. Stott, K., Blackburn, J.M., Butler, P.J.G. and Perutz, M., 1995, Incorporation of glutamine repeats makes protein oligomerize: Implications for neurodegenerative diseases. *Proc. Nat. Acad. Sci. USA* **92**: 6509-6513.
6. Johnson, M.L. and Straume, M., 1994, Comments on the analysis of sedimentation equilibrium experiments. In *Modern Analytical UltracentrifugationI* (Schuster, T.M. and Laue, T.M., eds), Birkhäuser, Boston, USA, pp.37-65.
7. Meselson, M., Stahl, F.W. and Vinograd, J., 1957, Equilibrium sedimentation of macromolecules in density gradients. *Proc. Nat. Acad. Sci.* **43**: 581-588.
8. Meselson, M. and Stahl, F.W., 1958, The replication of DNA in *Escherichia coli. Proc. Nat. Acad. Sci.* **44**: 671-682.
9. Edsall, J.T., 1943, Apparent molal volume, heat capacity, compressibility and surface tension of dipolar ions in solutions. In *Proteins, amino acids and peptides as ions and dipolar ionsI* (Cohn, E.J. and Edsall, J.T., eds), Reinhold Publishing Corp., New York, pp.155-176.
10. Laue, T.M., Shah, B.D., Ridgeway, T.M. and Pelletier, S.L., 1992, Computer-aided interpretation of analytical sedimentation data for proteins. In *Analytical Ultracentrifugation in Biochemistry and Polymer ScienceI* (Harding, S.E., Rowe, A.J. and Horton, J.C., eds), Royal Society of Chemistry, Cambridge, UK, pp.90-125.
11. Casassa, E.F. and Eisenberg, H., 1964, Thermodynamic analysis of multicomponent solutions. *Adv. Prot. Chem.* **19**: 287-395.
12. Cohen, G. and Eisenberg, H., 1968, Deoxyribonucleate Solutions: Sedimentation in a Density Gradient, Partial Specific Volumes, Density and Refractive Index Increments, and Preferential Interactions. *Biopolymers* **6**: 1077-1100.
13. Eisenberg, H., 2000, Analytical ultracentrifugation in a Gibbsian perspective. *Biophys. Chem.* **88**: 1-9.
14. Kratky, O., Leopold, H. and Stabinger, H., 1973, The determination of the partial specific volume of proteins by the mechanical oscillator technique. *Methods In Enzymology* **27**: 98-110.
15. Cabezon, E., Butler, P.J.G., Runswick, M.J. and Walker, J.E., 2000, Modulation of the Oligomerization State of the Bovine F_1-ATPase Inhibitor Protein, IF_1, by pH. *J. Biol. Chem.* **275**: 25460-25464.
16. Cabezon, E., Runswick, M.J., Leslie, A.G. and Walker, J.E., 2001, The structure of bovine IF_1, the regulatory subunit of mitochondrial F-ATPase. *EMBO Journal* **20**: 6990-6996.
17. Cabezon, E., Arechaga, I., Butler, P.J.G. and Walker, J.E., 2000, Dimerization of bovine F_1-ATPase by binding the inhibitor protein, IF_1. *J. Biol. Chem.* **275**: 28353-28355.
18. Yerushalmi, H., Lebendiker, M. and Schuldiner, S., 1996, Negative dominance studies demonstrate the oligomeric structure of EmrE, a multidrug antiporter from Escherichia coli. *J. Biol. Chem.* **271**: 31044-31048.
19. Rotem, D., Sal-man, N. and Schuldiner, S., 2001, In vitro monomer swapping in EmrE, a multidrug transporter from Escherichia coli, reveals that the oligomer is the functional unit. *J. Biol. Chem.* **276**: 48243-48249.
20. Muth, T.R. and Schuldiner, S., 2000, A membrane-embedded glutamate is required for ligand binding to the multidrug transporter EmrE. *EMBO Journal* **19**: 234-240.

21. Tate, C.G., Kunji, E.R., Lebendiker, M. and Schuldiner, S., 2001, The projection structure of EmrE, a proton-linked multidrug transporter from Escherichia coli, at 7 A resolution. *EMBO J.* **20**: 77-81.

Time Resolved Protein Fluorescence. Application to Multi-Tryptophan Proteins

YVES ENGELBORGHS
Laboratory of Biomolecular Dynamics, University of Leuven, Celestijnenlaan 200D, B3001 Leuven, Belgium

1. INTRODUCTION

The analysis of tryptophan fluorescence of multi-Trp containing proteins is interesting in many respects. Trp is a very sensitive natural reporter of its local environment, and with several Trp-residues present, information can be obtained of several locations within the protein. If these natural locations are not of interest to us, site directed mutagenesis can be used to displace Trp-residues. Next to its sensitivity, protein fluorescence is also a very fast phenomenon, allowing to study time dependent phenomena in a very broad time window (from subnanosecond to hours), and potentially to collect structural information about kinetic intermediates. Finally tryptophan fluorescence anisotropy allows us to get a deeper insight into the dynamics of the protein on the nanosecond (ns) timescale, and the influence of e.g. ligand binding on it. Gregorio Weber, one of the pioneers of the use of fluorescence in biophysics has given a nice broad review on the possibilities of the technique[1].

The full richness of the information content of protein fluorescence only appears in a ns-time resolved study. Such an analysis is complicated for two reasons: (1) the fact that the fluorescence decay of a single Trp-residue usually has to be described by a sum of several exponentials, (2) the fact that many proteins contain more than one tryptophan. Despite these complications, many research groups have been able to obtain useful

Supramolecular Structure and Function 8, Edited by Pifat-Mrzljak
Kluwer Academic/Plenum Publishers, New York 2004

information out of a detailed time resolved analysis and to link this information to the function of the protein in question.

It is clear that the explanation of the origin of the different lifetimes for a single tryptophan is a necessary first step in the understanding of more complex situations. Therefore initially single Trp-containing proteins were studied. However, the number of these is limited. Fortunately much progress has been made in the analysis of complex proteins, by using site directed mutagenesis to get a specific replacement of one or more Trp-residues. In a number of cases a single tryptophan residue has also been introduced as a reporter group.

The early work on tryptophan fluorescence in proteins has been covered by the excellent review of J. Longworth[2]. In the recent years a lot of important progress has been made in instrumentation; the use of stable pulsed laser sources with narrow pulses and high repetition rates, and microchannel plates as very fast detectors. Also improvements have been made in data analysis with e.g. the method of maximal entropy which has, however, not yet become very wide spread[3]. These instrumental aspects have been described in nice reviews [4-6] and in an excellent textbook[7] and will not be covered in this chapter. This chapter elaborates on a previous review[8].

2. INDIVIDUAL TRYPTOPHANS

The ultimate goal of a lifetime analysis of a multi-Trp-protein is to be able to collect the parameters for each individual tryptophan. This be done by replacing all but one Trp -residues by site directed mutagenesis. Every mutation has, however, the danger of slightly changing the structure of the protein and when several Trp-residues have been mutated, the alterations accumulate. Sometimes compensating additional mutations can be made to restore the protein's stability and its correct conformation, together with a sufficient production upon expression in bacteria. A nice example is the case of colicin[9].

When only two Trp-residues are present chemical methods can be used to discriminate the contributions of individual Trp-residues and avoid the necessity for the construction of (-1W)-mutants. This is the case when only one of them is exposed. In this case one can obtain the lifetime parameters of the buried one, by totally quenching the fluorescence of the exposed one upon reaction with N-bromo-succinimide (NBS) [10]. We studied in this way with the thick anticoagulant peptide [11]. Upon titrating the protein with NBS, the lifetimes did not change, but the amplitude fraction of one lifetime decreased, the others increased. Therefore at least one lifetime can be attributed to the surface Trp. The reactivity with NBS was also used to demonstrate the existence of two conformations in the case of human serum albumine [12]. When the two Trp-residues show a marked difference in exposure, lifetime measurements in the presence of a collisional quencher immediately allow to attribute the lifetimes to the Trp- residues [13]. It should be noted, however, that even for a single tryptophan, different collisional quenching constants (kq) can be obtained for different lifetimes, as was observed in colicin [9] and many other proteins (see further).

Discrimination of two Trp residues has also been done in protein crystals, where the amplitude fractions depend on the crystal orientation [14]. This dependence is explained on the basis of a different orientation of the two Trp residues in the protein and therefore in the crystal.

When more than 2 Trp residues are present one cannot avoid site directed mutagenesis. Two ways of obtaining individual tryptophan data are open: (1) by subtraction of one Trp; (2) by leaving just a single Trp. By removing one Trp residue and subtracting the spectrum of the mutant from that of the wild type (WT), the spectrum of the Trp that has been removed can be obtained, provided the concentrations can be carefully controlled. When the lifetimes of the (-1W) mutant are compared to these of the WT, a lifetime that has disappeared from the collection clearly belongs to the Trp that was removed. The other method is to remove all but one Trp residue. Both methods clearly only work if perfect additivity is present. This implies that removing the Trp residues does not alter the properties of the other Trp's either through structural alterations, or to interactions in the excited state e.g. energy transfer. Pure additivity was observed in lactate dehydrogenase[15] and in chloramphenicol acetyltransferase[16] where three lifetimes were measured. In each case when one Trp was replaced, a particular lifetime disappeared. Such a simple situation is, however, not very general. Clearly, if a lifetime disappears upon replacing a Trp, it can be attributed to this residue, but one cannot exclude that the removed Trp has additional lifetimes similar to, and therefore hidden by, the others. This is the case in barnase with 3 Trp residues where (-1W) mutants where made initially, and later also (-2W) mutants. The results of the (-2W) mutants showed that small contributions of hidden lifetimes appeared in the double mutants [17,18].

In many cases the single tryptophan containing mutants display the same families of lifetimes as the wild type (short, middle, long). As it is known that the possibility of resolving exponentials is rather limited, the question can be asked whether the lifetime data of the WT protein can be considered to be linear combinations of these of the individual tryptophans within the lifetime families. This seems to be the case for a number of proteins e.g. lac repressor[19] glutamine binding protein[20], aspartate carbamylase[21] and the nuclear capsid protein (NCp7) from Human Immunodeficiency Virus

(HIV)[22] . In the case of the three-Trp containing proteins colicin [9] and barnase [17] the linear combination of the data of the individual Trp's does not fit the data of the WT. However, the linear combination works between the data for one particular single-Trp mutant and the data for the other mutant containing the two remaining tryptophan residues, indicating that interactions e.g. energy transfer occur between these two. A similar situation was observed in the 4-Trp containing protein Plasmin Activator Inhibitor type 1 (PAI1)[23] .

To check for the additivity we found it simplifying to use an amplitude average lifetime and an average family lifetime (short, middle, long)[24]. Salient features are more apparent in the average lifetime than in the full details of all the lifetimes and fractions. (In calculating the average lifetime it is usually accepted that the extinction coefficient is the same for the states that correspond to the different lifetimes. This is not necessarily true as we found a rather big variability of the extinction coefficient at 295 nm of a single tryptophan in the disulfide oxidoreductase from *Escherichia coli* [25] and in PAI1[23] . By calculating these average lifetimes obtained from individual tryptophans and comparing with wild type we could clearly demonstrate the presence of energy transfer within a particular tryptophan pair [9,17,23].

A typical example is Barnase, which is a small monomeric enzyme that has been extensively used as a model for studying the principles that rule protein stability and protein folding [26], protein-protein interactions [18,27,28] and electrostatics [29,30]. In barnase the tryptophans are found at positions 35, 71 and 94. The interaction of His18 with Trp94 causes a pH dependent quenching of its fluorescence, following the ionisation curve of the histidine residue ($pK_a = 7.75$) [17,31]. The fluorescence lifetimes of the proteins could be resolved and, in most cases, spectra and lifetimes could be attributed to single tryptophan residues using the method of subtraction [17]. Trp35 displays a pH independent lifetime of 4.3 to 4.8 ns. Trp71 and Trp94 behave as an energy transfer couple. When Trp71 and Trp94 are both present bidirectional energy transfer is deduced and calculated on the basis of the measured relaxation times and the formula of Porter [32] or Woolley [33]

The method of subtraction was applied to yeast actin[34] with 4 tryptophan residues and to the study of the nucleocapsid protein of HIV1[22] . In the latter case, the protein fluorescence is also used to study the binding of zinc ions to the finger motifs[35] and the sequence specific interactions with nucleic acids[36]

2.1 Spectral properties of single tryptophan residues

The difference between the wavelength of maximal absorbance and maximal emission, the so-called Stokes-shift is generally attributed to the rearrangement of the dipoles around the Trp-residue as a consequence of the altered dipole moment of tryptophan upon excitation. The rearrangement of solvent molecules is very fast at room temperature and leads to emission maxima between 340 and 350 nm for exposed tryptophan residues. Only at low temperatures this phenomenon occurs in the same time domain as the lifetime of the excited state and gives rise to spectra that shift with time in the nanosecond scale (Time Resolved Emission Spectra or TRES). It has been shown by theoretical calculations that the determining factor is the electric field in the direction of the transition moment 1La[37,38] and that this field can be influenced by the solvent but also by the protein environment. One can expect that the altered dipole moment upon excitation also leads to dynamic rearrangements in the protein matrix which could be a lot slower. Furthermore, exceptionally short wavelength maxima are observed when aromatic residues cluster arround the tryptophan as in the case of barstar[39] and the homeodomains[40] and lead to hydrogen bridge formation between the pyrrole nitrogen and the aromatic rings . In these situations the quantum yields are usually very low. The phenomenon of aromatic-aromatic residue interactions is often found in proteins, as shown by the analysis of group interactions in 33 refined high resolution crystal structures[41,42].

Fluorescence spectra of single tryptophan residue are nicely fitted by the log-normal function previously prescribed by Burstein and Emelyanenko[43].

$$I(\lambda) = I_{max}.\exp\left(-\frac{\ln 2}{\ln^2 \rho}.\ln^2\left\{\frac{a-1/\lambda}{a-1/\lambda_{max}}\right\}\right) \qquad (1)$$

Where λ_{max} is the wavelength of maximal fluorescence, a is the function-limiting point and ρ is the band asymmetry parameter.

$$\rho = (1/\lambda_{max} - 1/\lambda_-)/(1/\lambda_+ - 1/\lambda_{max}) \qquad (2)$$

$$a = 1/\lambda_{max} + \left(\rho.\left(1/\lambda_+ - 1/\lambda_-\right)\right)/\left(\rho^2 - 1\right) \qquad (3)$$

where for tryptophan λ_+ and λ_- are related to λ_{max} by the following relations:

$$1/\lambda_- = 0.831/\lambda_{max} + 7,07x10^{-4}(nm^{-1}) \quad (4)$$

and

$$1/\lambda_+ = 1,177/\lambda_{max} - 7,78x10^{-4}(nm^{-1}) \quad (5)$$

Spectra of proteins with more than one tryptophan can be decomposed into different components by using this log-normal function (Burstein et al., 2001)[44-46].

While doing the effort of getting good quality spectra, it is also worthwhile to determine the quantum yields, as will be shown further on. The quantum yield can only be determined in a relative manner and therefore at the same time the quantum yield of a known reference compound has to be determined as well [47] usually with the quantum yield of NATA as a reference.

$$\frac{Q}{Q_{ref}} = \frac{\int I_\lambda d\lambda}{O.D._{\lambda,ex}} / \frac{\int I_{ref,\lambda} d\lambda}{O.D_{ref,\lambda,ex}} \quad (6)$$

2.2. Factors determining the excited state lifetime of a single tryptophan

Even a single lifetime for a single tryptophan can vary a lot. The reason is that an excited tryptophan can return back to the ground state by a variety of different parallel pathways. Therefore (in the absence of fluorescence energy transfer) the global rate constant (the inverse of the lifetime) can be described as a sum of the different rate constants, as has been worked out by Chen and Barkley [48]:

$$1/\tau = k_r + k_{nr} \text{ with } k_{nr} = k_{isc} + k_{sol} + \Sigma k_{qi} \quad (7)$$

where k_r is the radiative rate constant, k_{nr} the nonradiative one, k_{isc} the rate constant for intersystem crossing, k_{sol} for solvent quenching, and k_{qi} for quenching by quenchers either present in solution or in the protein itself, using a variety of mechanisms i.e. proton and electron transfer. The different values for these rate constants are discussed in great detail by Chen and Barkley[48] and will only be presented briefly.

2.2.1. The radiative rate constant k_r

The average radiative rate constant can be determined by dividing the quantum yield by the wavelength independent amplitude-average lifetime [24,49,50]. This calculation can, however, give erroneous results since it does not take into account the possibility of static quenching or so-called Quasi-Static-Self-Quenching, which is very fast dynamic quenching that escapes lifetime measurements [49,51]. The possibility also exists to calculate k_r from spectral information based on the relation proposed by Strickler and Berg[52]. These authors relate the radiative rate constant with both the absorption and the emission band, and the molar absorption coefficient of a fluorophore

With an absorption coefficient of 5.540 $M^{-1}cm^{-1}$ at 280 nm for Trp [53]and using the log-normal function [43] for the shape of the emission spectrum of tryptophan we were able to replaced the complex formula of Strickler and Berg by a one parameter function $f(\lambda_{max})$[54] which was fitted to the different values of k_r obtained from 11 different mutants or states of a single protein i.e. the Sarcoplasmic Calcium-binding Protein of *Nereis Diversicolor* (NSCP). Our original 6th power series expansion [54] turned out to be very unstable[55] and can better be replaced by the following simple pragmatic expression for k_r:

$$k_r = -0.1138 + 54.0104/\lambda_{max} \ (ns^{-1}) \quad (\text{with } \lambda_{max} \text{ in nm}) \tag{8}$$

It should be noticed that this equation can only be applied to a tryptophan residue in the absence of any quenchers or charges in its immediate environment. A similar linear relation between k_r and ν_{max} ($\sim 1/\lambda_{max}$) was also observed by Privat et al.[56] for a number indole derivatives and a separate one for a number of proteins, with the exception of azurin. In the case of the different mutants of NSCP and the proteins mentioned by Privat et al., k_r shows rather limited variation between 0.055 ns^{-1} to 0.040 ns^{-1}. In a broad population of proteins the variation is much more pronounced[57] but part of this variation may be due to static quenching, and to variability in the absorption band. We find variability in the extinction coefficient at 295 for DsbA[25] and PAI1[23] which suggests the possibility of variation in this function of the absorption spectrum.

Next to azurin another exception to this rule is also the fact that a long lifetime component (25 ns, therefore $k_r < 0.04$ ns^{-1}) was observed for two membrane proteins, for which we expect a rather blue emission[58].

2.2.2. Intersystem crossing

The intersystem crossing rate constant k_{isc} is actually the rate constant for switching to the triplet state of indole, is temperature independent and is estimated[59,60] between 2 and 8 x 10^7 s^{-1} .

2.2.3. Quenching by water

The rate constant for solvent quenching k_{sol} is assumed to be responsible for the temperature dependence of the fluorescence lifetime, and is described in terms of the Arrhenius theory. With the frequency factor A = 10^{15} to 10^{17} s^{-1} and the activation energy around 44-52 kJ/mol the value of k_{sol} varies around 10^7 s^{-1} at 25 °C (ref 48 and references therein). For fully exposed 3-methylindole a value of 5.9 x 10^7 s^{-1} is found. In the protein, the rate constant will be more reduced dependent on the accessibility of the tryptophan considered, which can be calculated in the usual way[61].

2.2.4. Other quenching mechanisms

The rate constants for quenching in solution (k_{qi}) by different side chains have been determined and vary in magnitude: protonated amino groups (Lys or alpha-amino groups) and Tyr quench by excited-state proton transfer [62-64] with rate constants around 10^8 $M^{-1}s^{-1}$. The side chains of Gln, and Asn are weaker quenchers by electron transfer with rate constants around 4 x 10^7 M^{-1} s^{-1} and 2.4 x 10^7 $M^{-1}s^{-1}$ respectively[48]. The strongest quencher is S (and even more Se) in Cys [65-67] with k_q = 4 x 10^9 M^{-1} s^{-1} and in S-S bridges[48] with k_q> 10^{10} $M^{-1}s^{-1}$. Also protonated histidine is a good quencher [48,68-71] with k_q = 2.5 x 10^9 M^{-1} s^{-1} (unprotonated His quenches about 14 times less). In solution their quenching constant approaches that of diffusion controlled collisions and therefore their quenching efficiency is close to one i.e. almost every encounter leads to quenching. For $HisH^+$ we[71] found the efficiency in solution to be 0.32 . In the protein they cannot diffuse freely, and therefore their frequencies of collisions with the excited state tryptophan will depend on the dynamics of the protein matrix . In this way[71] intramolecular collisional frequencies of $HisH^+$ in anantin have been estimated to be 0.7 x 10^9 s^{-1} .

2.2.5. Electron transfer

Strong convincing evidence for electron transfer from the excited state is the recent observation of the reduction of SS-bridges that are found in the immediate environment of a Trp. The first example is present in the protein

Cutinase from *Fusarium solani pisi* [72-74]. The fluorescence of the single Trp69 is highly quenched in the native protein. This quenching is due to the presence of a nearby disulfide bridge between Cys31 and Cys109 and disrupting the disulfide bridge with DDT leads to a strong increase of the protein fluorescence just as in Dsba [25,72,73]. Very surprisingly, however, in Cutinase extensive UV irradiation of the protein leads to an almost tenfold increase of the quantum yield. This effect of UV-irradiation is due to the breaking of the disulfide bridge. The formation of a thiol can be demonstrated with the thiol reagent DTNB. The fact that the fluorescence increases proves that the tryptophan is not itself irreversibly oxidized in the process. The fluorescence properties of Trp69 in Cutinase are also influenced by an aromatic-amide hydrogen bond between the indole side chain of Trp69 and the peptide amide from Ala32, but NMR studies show that this hydrogen bond remains intact in the irradiated protein and cannot explain the observed effects.

Also in goat-α-lactalbumine reduction of the SS-bridge is observed upon illumination of the protein [75].

To explain the different lifetimes that can be observed, even in the absence of quenching side chains[25,76] electron transfer to the peptide carbonyl group has been suggested. This mechanism has been extensively studied [48,77-80]. Electron transfer is dependent on the distance between the donor and the acceptor, as well as on the medium between the two, therefore this mechanism should give rise to different lifetimes for different rotamers of tryptophan[54,78,79]. Also Pan et al.[81] describe the different lifetimes of different rotamers of tryptophan to electron transfer to the carbonyl of the peptide. A theoretical study of the quenching of tryptophan fluorescence by electron transfer to the carbonyl oxygen of the backbone was performed by Goldman et al.[79] and suggested that the $C^{\varepsilon 3}$ can be considered as the major donor.

Electron transfer can be described by the following equation derived by Marcus and Sutin [82]

$$k_{ET}(R) = k_0 \exp(-\beta (R - R_0)) \tag{9}$$

where $k_{ET}(R)$ is the rate constant for electron transfer, k_0 is the rate constant of electron transfer at the van der Waals contact distance (R_0 = 3Å), R is the distance between the donor and acceptor of electron transfer, and ß is a parameter dependent on the medium that expresses the sharpness of the distance dependence.

We have applied this relation to electron transfer in Trp in 20 different proteins, for which the lifetime data and the structure was available . With the distance between the carbonyl carbon of the peptide bond and the $C^{\varepsilon 3}$ atom of aromatic ring calculated from the known structure and assuming k_{nr} is essentially determined by k_{ET}, the constants k_0 and β could be fitted

These two constants were originally determined from two proteins[54] but now [83] we have 30 data pairs and fitting yields $k_0 = 5.046 \pm 0.3\ ns^{-1}$ and $\beta = 1.497 \pm 0.074\ Å^{-1}$.

These parameters have to be used very carefully: in view of the way the data have been selected, they can only be used to estimate k_{nr} in the absence of any quenchers, charges or aromatic rings in the immediate neighbourhood of the tryptophan.

2.2.6. Proton transfer

At pH values below 4, excited state protonation of the indole ring can occur (pK_a*= 2 to 3) and above pH 11 deprotonation can occur (pK_a* = 12 to 13) in both cases leading to quenching[84]. At intermediate pH values proton exchange can occur at ring positions in the presence of good proton donors e.g. ammonium groups. At room temperature the rate constants are estimated around $10^7\ s^{-1}$ and therefore are not the major determinant of short lifetimes[48].

2.2.7. Charge transfer

In a recent quantum mechanical study of Callis and Vivian[85] quenching by the pepide backbone is described as a charge transfer process, and the energy of the charge transfer band is strongly dependent on the electric field in the environment of the tryptophan[85]. This is probably a more general description of the excited state quenching than the electron transfer to the peptide carbonyl described above, and probably should be able to include the more complex situations that we have excluded by choosing tryptophan residues in a so-called "simple environment". Unfortunately these quantum mechanical calculations should be done for each case. So far no simple predictive tools have been derived.

With all the rate constants being defined, it is in principle possible to calculate a single lifetime for any tryptophan in a known protein environment. One of the problems is of course the frequency of collisions for the different quenchers which are in the environment of the tryptophan.

3. MULTIPLE LIFETIMES FOR A SINGLE TRYPTOPHAN

The quenching processes presented above are all assumed to be irreversible and they are able to explain the variation of the lifetime with variations of the environment of the tryptophan residue. However, they do not explain the existence of multiple lifetimes for a single residue, since all these pathways are parallel and consequently the rate constants have to be added to get the inverse of a single global lifetime.

In principle there are two basic mechanisms to explain multiple fluorescence lifetimes for a single tryptophan residue. One explanation is the existence of different states that do not interconvert on the time scale of the fluorescent phenomena, i.e. ground state heterogeneity. The most often suggested origin of these different states are the rotamers of tryptophan. For the second possibility, the origin has to be found in reversible excited state dynamics. Evidence for both mechanisms has been found, and therefore the situation for every tryptophan residue in a protein can be different.

3.1. Possibilities for microstates

Multiple lifetimes can originate form the presence of multiple microstates of tryptophan that do not interconvert on the time scale of ns. These microstates are distinguished from real conformational states, in that the overall protein fold does not change, but the local environment of the tryptophan residue does, either by different rotameric positions of the tryptophan rings, or by different details of the other side chains in the immediate environment. Evidence about protein dynamics comes from other sources as well[86-88] and very elegant experiments showing the penetration of the quencher acrylamide in the interior of a protein[89]. The big question is of course on which time scale this dynamics is manifested. Clear evidence for static heterogeneity in the lifetimes of tryptophan (interconversions between microstates are possible but are much slower than in ns scale) is presented in a number of cases: bacteriophage T4 lysozyme [90], phosphofructokinase of Bacillus stearothermophilus [91], and hemoglobin [92].

The most convincing evidence for the existence of different microstates for one tryptophan seems to be the fact that different lifetime components can have different quenching constants (k_q) for the same quencher, as is the case e.g. for colicin [9], and several other proteins [83].

The link between multi-exponential decay and the presence of rotamers about the C^{α}-C^{β} bond was made for Tyr by Gauduchon and Wahl[93] and for tryptophan by Szabo and Rainer [94]. The importance of rotations around the

C^β-C^γ bond in the amino acid tryptophan have been suggested by Engh et al. (1986)[95] because according to their molecular dynamics calculations C^α-C^β rotamers interconvert rapidly, while rotations about the C^β-C^γ are much more infrequent. In proteins one might expect that all these rotations are even more hindered. Nevertheless anisotropy studies in a variety of proteins yield rotational correlation times between 0.03 and 50 ns [96,97] indicating that no extrapolation from one protein to the other can be made. On top of that, X-ray structure analysis usually gives only one rotamer position for each tryptophan in a protein. The possibility to populate more than one rotamer which slowly interconvert is therefore not easy to prove. A correlation between the presence of different lifetimes and different rotamers by 1H-NMR data was done in Oxytocin for Tyr [98] and for Trp [99]. The results for Trp are surprising in the sense that the authors conclude that C^α-C^β rotations are more relevant than C^β-C^γ rotations, in contrast to what is expected from the molecular dynamics calculations mentioned before [95]. It should be noted, however, that these data are about a cyclic peptide and not about tryptophan itself.

The existence of different rotamers of tryptophan in different proteins present in the databases have been studied by several authors [100-102]. McGregor et al.[103] show a link with the secondary structure and the absence of g+ in the helical conformation due to steric clashes. Schrauber et al. (1993) [104] show by a statistical analysis of the data of 70 polypeptide chains that many deviations of the six standard rotamers occur in native proteins. Clayton and Sawyer [105,106] studied 5 different peptides of 18 amino acids long, which are able to form an amphipathic α-helix in the presence of lipid vesicles, but are unstructured in the absence. In the absence of vesicles the time dependence is tri-modal and the amplitude fractions correlate nicely with the fractions of occurrence of the different rotamers observed in the statistical study of the protein databank of McGregor et al.[103] resp. g-, g+ and t rotamers. Again a coupling with the secondary structure is observed, since for Trp in helices only a bimodal distribution of Trp- positions is found: t and g- and also the fluorescence lifetimes show only two lifetimes. In the helix structure, g+ does not occur. (Note that in their paper these authors use the opposite definition of g+ and g-). These results follow the same line as the evidence for constraints imposed by the secondary structure on the rotamer collection of tryptophan shown by Willis et al.[107]. However, since the relative amplitudes are wavelength dependent this correlation fits at their emission wavelength of 335 but may not necessarily fit at other wavelengths.

Pan et al.[81] studied rigid hexapeptides and also make a link between lifetimes and rotamers via the calculation of the lifetimes of the rotamers.

It should be noted that the statistical studies on rotamer values still assume one rotamer per tryptophan, as obtained form the crystal structures.

However, if rotamers are the source of lifetime heterogeneity of single tryptophan proteins, we have to assume that multiple rotamers appear even in the population of single-tryptophan proteins. We made an attempt to obtain these rotameric positions by MD calculations [54]. In a more recent study we have been able to calculated rotamer positions with the dead end elimination method[83] and correlate them successfully with lifetimes.

Using this correlation between rotamer and lifetime we have been able to estimate the parameters of electron transfer in the Trp-residue [83].

In the protein DSBA[5] we have observed a phenomenon that we attribute to the kinetics of microstate interconversion and that only occurs with a rate constant of 0.014 s^{-1} . The phenomenon consists in the recovery of some fluorescence (proven to be due to a long lifetime component) after NBS reaction at low concentrations and exhaustion of the reagent. On the basis of NMR and X-ray data we suggest that it is the p-/p+ transition that leads to a different exposure of the pyrrole ring, which is the site of reaction with NBS.

Evidence for slow exchange between discrete species was also shown to exist for thioredoxin[108] and probably serum albumin [12,138].

4. REVERSIBLE EXCITED STATE DYNAMIC PROCESSES

The processes of quenching and electron transfer described above are all considered irreversible parallel processes. Therefore the rate constants can be added up, and the more processes are taking place, the shorter the inverse lifetime of the tryptophan under consideration is.

Multi-exponential decay in single tryptophan proteins has been described as a coupling of internal motion and distance and angle dependent quenching efficiency of neighbouring quenchers by Tanaka and coworkers [109-112]. The same model was used to calculate lifetimes for Trp29 in erabutoxin b, where a Lys was the quencher, Trp59 in apocytochrome C where Tyr48 was the quencher of Trp59 and in Subtilisin (from streptomyces) where Trp86 is influenced by a neighbouring SS-bridge .

However, one can also imagine the existence of reversible excited state processes. Fluorescence energy transfer can be bidirectional [17,32,33] fast isomerization between different rotamers could take place, and even quenching by electron transfer has been suggested to be reversible [113,114].

These processes can be described by the following general scheme:

$$\begin{array}{ccc} & k_{12} & \\ \mathbf{A^*} & \leftrightarrow & \mathbf{B^*} \\ & k_{21} & \\ \downarrow k_1 & & \downarrow k_2 \\ A & \leftrightarrow & B \end{array} \tag{10}$$

Such a system gives rise to two relaxation times that are functions of all the four rate constants [115-117]:

$$\lambda\pm = S \pm \text{sqrt}\,\{S^2 - k_2k_{12} - k_1(k_2+k_{21})\,\} \tag{11}$$

With $S = (k_1 + k_2 + k_{12} + k_{21})/2$

This reversible mechanism can be experimentally tested since the amplitude ratio will change in the presence of external quenchers that increase k_1 and k_2 (eventually in a different way)[113,117], while heterogeneity of states will not lead to an alteration of the amplitude ratio unless preferential static or preferential quasi static quenching would occur.

5. EFFECTS OF LIGAND BINDING

The sensitivity of fluorescence and the extremely broad accessible time window makes it the technique of choice to study conformational changes of proteins. Almost every conformational change is accompanied by a fluorescence change, and the question is if we can get a more detailed interpretation of the observed changes. We suggest to analyse the change in fluorescence by calculating the ratio of the quantum yields (Q/Q_0) where Q is the quantum yield of the modified protein and Q_0 that of the reference (wildtype or protein in the absence of ligands). This ratio is split into three factors. This splitting is done by introducing an hypothetical intermediate for which the average lifetime is defined as $\Sigma\alpha_i\tau_{0i}$, where α_i are the amplitudes in the modified state and τ_{0i} are the lifetimes of the reference state. The three factors are: f_{kr} representing the change in the average radiative rate constant ($<k_r>$), the factor f_{PR} reflecting population reshuffling and a factor f_{DQ} representing pure dynamic quenching [24]:

$$\frac{Q}{Q_0} = \frac{<k_r> \Sigma \alpha_i \tau_i}{<k_{r0}> \Sigma \alpha_{0i} \tau_{0i}} = \frac{<k_r>}{<k_{r0}>} \frac{\Sigma \alpha_i \tau_{0i}}{\Sigma \alpha_{0i} \tau_{0i}} \frac{\Sigma \alpha_i \tau_i}{\Sigma \alpha_i \tau_{0i}}$$
$$= f_{kr} \, x \, f_{PR} \, x \, f_{DQ} \tag{12}$$

where τ_i is the fluorescence lifetime and α_i is the wavelength independent amplitude fraction, and the subscript (0) refers to the reference state. The wavelength independent amplitude fraction α_i can be calculated as described in the following equation :

$$\alpha_i = \frac{\int F_i(\lambda) / \tau_i \, d\lambda}{\Sigma_j \int F_j(\lambda) / \tau_j \, d\lambda} \tag{13}$$

where τ_i is the fluorescence lifetime of species i, F_i is the fluorescence intensity of species i obtained from the decay-associated spectra (DAS). Decay associated spectra are constructed by multiplying the intensity fraction with the intensity of the emission spectra at the respective wavelengths[118]. A log-normal function[43] can be fitted to the associated intensities to obtain the decay-associated spectra at every wavelength: This allows to integrate the fluorescence intensity of species i over all wavelengths. In our experience the fitting with a log-normal function works very well.

5.1. The change in the average radiative rate constant f_{kr}

The factor f_{kr} is determined by a change in the radiative rate constant, but also by a change in static quenching and even by a change in the molar absorption coefficient at the excitation wavelength. Generally the extinction coefficient of a single tryptophan is assumed to be the same for all microstates

Static quenching is also observed in proteins but it is possible that this is due to dynamic quenching of very high frequency that escapes detection by ns lifetime analysis [49,51]. In that case it is possible to calculate the boundaries of the lifetime and amplitude fraction of the non-detected component[24]. The molar absorption coefficient of proteins at 295 nm can change which leads directly to a change in the measured average radiative rate constant because the average radiative rate constant is the absorption fraction weighted radiative rate constant. But also there is an indirect change in the individual radiative rate constants as described in equation 2. If $f_{kr} = 1$ then none of these complications are in operation.

5.2. Heterogenous static quenching or population reshuffling (f_{PR})

The amplitude fraction can change due to heterogenous (selective) static quenching or due to the change in the microstates of tryptophan. If there is no change in the measured average radiative rate constant and the factor f_{PR} is different from 1 then static quenching can be excluded, and the changes can be attributed to population reshuffling. This is the case for Trp126 in the DsbA protein from E. Coli [25]. When mutations are made in the neighbourhood of this tryptophan, the balance of the amplitudes can be altered but the lifetimes don't change.

The binding of cyclosporin to cyclophilin also causes a strong fluorescence increase which was interpreted as due to population reshuffling towards a microstate with a longer lifetime[119].

5.3. The change in dynamic quenching (f_{DQ})

In the analysis of fluorescence changes upon oxidation or reduction of the protein DsbA this method seems to give clarifying results. DsbA is a monomeric, periplasmic 21.1 kDa protein (189 amino acids) that is required for efficient disulphide bond formation in secretory proteins in the bacterial periplasm. The enzyme contains a single, catalytic disulphide with the active-site sequence Cys30-Pro31-His32-Cys33. The X-ray structure of oxidized DsbA[120] as well as the X-ray[121] and NMR-structure[122] of reduced DsbA has revealed that the enzyme possesses a thioredoxin-like domain (residues 1-62 and 139-189), a motif found in all known structures of disulphide oxidoreductases. The sequence of the thioredoxin-like domain of DsbA is, however, only 10% identical with E. coli thioredoxin. DsbA possesses a second domain (residues 63-138) of unknown function, which is inserted into the thioredoxin motif and exclusively consists of α-helices. DsbA contains two tryptophans Trp76 and Trp126, which are not contained in the thioredoxin domain and are located in the α-helical domain. Trp76 is buried and about 12 Å apart from the disulphide, whereas Trp126 is even further away from the disulphide bridge (about 20 Å) and partially solvent-accessible.

Detailed lifetime studies showed that Trp76 is dynamically quenched by the disulfide ring but via the phenyl ring of Phe26. In the mutant F26L oxidation still leads to a more limited quenching but the lifetimes don't change, the amplitudes do. This analysis therefore suggests that this is due to a reshuffling of the populations. Also the lifetimes of Trp126 have been studied in great detail and a full scheme of excitation energy migration has been constructed [25].

6. TRYPTOPHAN DYNAMICS AND ANISOTROPY

Many authors have measured the rotational freedom of tryptophan using fluorescence anisotropy decay, both in time and frequency domain. Fluorescence depolarization is, however, not only induced by the dynamics of the fluorescent group, but also by energy transfer. Fluorescence anisotropy can also be used to collect information about associations of proteins. For a thorough discussion of tryptophan dynamics and anisotropy reference is made to the previous review [124].

7. INTRODUCING OR REMOVING TRYPTOPHAN

The increasing use of fluorescence spectroscopy is of course due to the high information content of the fluorescence signal, its broad time window, and the relative easiness of the measurements under a wide variety of conditions. However, not all interesting proteins have a tryptophan. Therefore a new tryptophan can be introduced, or placed on a more informative site. This has been done in many situations, but just to mention a few recent cases: the mutants F78W and F154W of chicken skeletal troponin C[125]. In the apo state Trp78 is buried and seems to be very rigid, while Trp154 is exposed. The situation inverts when Ca^{2+} is bound. The fluorescence changes allow the determination of the Ca^{2+}-binding constant at multiple sites and the results demonstrate interactions between the N- and C-end-domains. The role of Trp-5 in Annexin A3 was studied by engineering-out and was found to be important for membrane interactions [126]. The interaction of each of the 5 helices of apolipophorin-III from Locusta migratoria was studied by constructing a single tryptophan mutant in each of the 5 helices. In three helices the tryptophan interacted with the nonpolar domains of the lipids, in two others the tryptophan interacted with the polar head groups, indicating a different binding configuration for these helices [127]. A tryptophan was introduced in P21 close to the nucleotide binding sites and can be used to report on conformational changes and the binding of Berylliumfluoride [128-130]. The same proteins can also be studied by using fluorescent derivatives of the nucleotides and complementary information can be obtained [131,132].

When applying site directed mutagenesis, one can introduce tryptophan analogs with different spectral properties [133], but one can also engineer-in a Cys residue, which allows the selective positioning of other fluorescent molecules that bear SH-reactive groups. This method seems to be very promising, especially in combination with donor-donor energy migration [134] which avoids the problem of selectively putting in a donor and an acceptor [135]. The method was successfully applied to the study of protein-protein interactions [136] and conformational changes [137].

8. CONCLUSIONS

Recent work has clearly shown that it is possible to explain the different lifetimes of a single Trp-residues in terms of the possibility of having different rotameric states that interconvert only slowly if at all. The dead end elimination method has allowed us to explore these rotamers and in many cases links could be made with lifetimes on the basis of differences in accessibility. Experimental indications for the existence of different rotameric states are becoming available. The ultimate proof for the existence of different microstates that slowly interconvert would come from lifetime measurements on isolated (single) molecules of single-tryptophan proteins. But this is not an easy task, in view of the low photostability of Trp. Finally the analysis of the electron transfer as presented here can only be considered to be a pragmatic description and the parameters obtained have to used with great care in view of the selection of "simple" Trp-environments. For more complex environments a thorough quantum mechanical calculation, using the charge transfer model presented by Callis and Vivian might be necessary for each individual case.

REFERENCES

1. Weber, G., 1997, Fluorescence in biophysics: accomplishments and deficiencies, *Methods Enzym.* **278:** 1-15.
2. Longworth, J.W., 1983, Intrinsic fluorescence of proteins. In: *Time-Resolved Fluorescence Spectroscopy in Biochemistry and Biology* (R.B. Cundall and R.E. Dale, eds.), Nato ASI series. Series A: Life Sciences, Vol **69:** 651-725.
3. Brochon, J.C., 1994, Maximum entropy method of data analysis in time-resolved spectroscopy. *Methods Enzym.* **240:** 262-311.
4. Birch, D.J.S. and Imhof, R.E., 1991, Time-Domain Fluorescence Spectroscopy Using Time-Correlated Single-Photon Counting. *Topics in Fluorescence Spectroscopy* **1:** 1, 1-95 *in* (J.R. Lakowicz, ed.), Plenum Press, New York & London.

5. Lakowicz, J.R. and Gryczynski, I., 1991, Frequency-Domain Fluorescence Spectroscopy. In: *Topics in Fluorescence Spectroscopy* **1:** 5, 293-335, (J.R. Lakowicz, ed.), Plenum Press, New York & London.
6. Munro, I.H. and Martin, M.M., 1991, Time-Resolved Fluorescence Spectroscopy Using Synchrotron Radiation. *Topics in Fluorescence Spectroscopy* **1:** 4, 261-291 *in* Lakowicz, J.R. Ed. Plenum Press, New York & London.
7. Lakowicz, J.R., 1999, *Principles of Fluorescence Spectroscopy*, 2nd edition, Kluwer Academic/Plenum publishers, New York, Boston, Dordrecht, London, Moscow.
8. Engelborghs, Y., 2003, Correlating Protein Structure and Protein Fluorescence. *J. Fluorescence* **13:** 9-16.
9. Vos, R., Engelborghs, Y., Izard, J. and Baty, D., 1995, Fluorescence Study of the Three Tryptophan Residues of the Pore-Forming Domain of Colicin A Using Multifrequency Phase Fluorometry. *Biochemistry*, **34:** 1734-1743.
10. Spande, T.F., Green, N.M., Witkop, B., 1966, The reactivity toward N-bromosuccinimide of tryptophan in enzymes, zymogens, and inhibited enzymes. *Biochemistry*, **5:** 1926-1933.
11. Sillen, A., Vos, R. and Engelborghs, Y., 1996, Fluorescence Study of the Conformational Properties of Recombinant Tick Anticoagulant Peptide (*Ornithodorus moubata*) Using Multifrequency Phase Fluorometry. *Photochem. Photobiol.* **64:** 785-791.
12. Peterman, B.F. and Laidler, K.J., 1980, Study of Reactivity of Tryptophan Residues in Serum Albumins and Lysozyme by *N*-Bromosuccinamide Fluorescence Quenching. *Arch. Biochem. Biophys.* **199:** 158-164.
13. Eftink, M., 1983, Quenching-resolved emission anisotropy studies with single and multitryptophan-containing proteins. *Biophys.J.*, **43:** 323-334.
14. Willis, K.J., Szabo, A.G. and Krajcarski, D.T., 1991, Fluorescence decay kinetics of the tryptophyl residues of myoglobin single-crystals. *J. Am .Chem. Soc.*, **113:** 2000-2002.
15. Waldman, A.D., Clarke, A.R., Wigley, D.B., Hart, K.W., Chia, W.N., Barstow, D., Atkinson, T., Munro, I. and Holbrook, J.J., 1987, The use of site-directed mutagenesis and time-resolved fluorescence spectroscopy to assign the fluorescence contributions of individual tryptophan residues in Bacillus stearothermophilus lactate dehydrogenase. *Biochim. Biophys. Acta* **913:** 66-71.
16. Ellis, J., Bagshaw, C.R. and Shaw, W.V., 1995, Tryptophan fluorescence of chloramphenicol acetyltransferase: resolution of individual excited-state lifetimes by site-directed mutagenesis and multifrequency phase fluorometry. B*iochemistry*, **34:** 3513-3520.
17. Willaert, K., Loewenthal, R., Sancho, J., Froeyen, M., Fersht, A. and Engelborghs, Y., 1992, Determination of the excited-state lifetimes of the tryptophan residues in barnase, via multifrequency phase fluorometry of tryptophan mutants. *Biochemistry* **31:** 711-716.
18. De Beuckeleer, K, Volckaert, G. and Engelborghs, Y., 1999, Time Resolved Fluorescence and Phosphorescence Properties of the Individual Tryptophan Residues of Barnase: Evidence for Protein-Protein Interactions. *Proteins* **36:** 42-53.
19. Royer, C.A., Gardner, J.A., Beechem, J.M., Brochon, J.-C. and Matthews, K.S., 1990, Resolution of the fluorescence decay of the two tryptophan residues of lac repressor using single tryptophan mutants. *Biophys. J.*, **58:** 363-378.
20. Axelsen, P.H., Bajer, Z., Prendergast, F.G., Cottam, P.F. and Ho, C., 1991, Resolution of fluorescence intensity decays of the two tryptophan residues in glutamine-binding protein from Escherichia coli using single tryptophan mutants, *Biophys. J.*, **60:** 650-659.
21. Fetler, L., Tauc, P., Hervé, G., Ladjimi, M.M. and Brochon, J.-C., 1992, The tryptophan residues of aspartate transcarbamylase: site directed mutagenesis and time-resolved fluorescence spectroscopy . *Biochemistry*, **31:** 12504-12513.

22. Bombarda, E., Ababou, A., Vuillemier, C., Gérard, D., Roques, B.P., Piémont, E. and Mély, Y.,1999, Time-resolved fluorescence investigation of the human immunodeficiency virus type 1 nucleocapsid protein: influence of the binding of nucleic acids. *Biophys. J.*, **76:** 1561-1570.
23. Verheyden, S., Sillen, A., Gils, A., Declerck, P.J., and Engelborghs, Y., 2003, Tryptophan Properties in Fluorescence and Functional Stability of Plasminogen Activator Inhibitor 1. *Biophysical J.* **85:** 501-510.
24. Sillen, A. and Engelborghs, Y., 1998, The Correct Use of "Average" Fluorescence Parameters. *Photochem. Photobiol.*, **67:** 475-486.
25. Sillen, A., Hennecke, J., Roethlisberger, D., Glockshuber,R. and Engelborghs, Y., 1999, Fluorescence Quenching in the DsbA Protein from *Escherichia coli*: Complete Picture of the Excited-State Energy Pathway and Evidence for the Reshuffling Dynamics of the Microstates of Tryptophan. Proteins*: Struct. Funct.* Genet. **37:** 253-263.
26. Fersht, A.R., 1993, The sixth Datta Lecture. Protein folding and stability: the pathway of folding of barnase. *FEBS Letters* **325:** 5-16.
27. Schreiber, G. and Fersht, A.R., 1995, Energetics of protein - protein interactions: analysis of the barnase-barstar interface by single mutations and double mutant cycles. *J. Mol. Biol.* **248:** 478-486.
28. Zegers, I., Deswarte, J. and Wyns, L., 1999, Trimeric domain-swapped barnase. *Proc. Natl. Acad. Sci.* USA **96:** 818-822.
29. Loewenthal, R., Sancho, J, Reinikanen, T. and Fersht, A.R., 1993, Long-range surface charge-charge interactions in proteins. Comparison of experimental results with calculations from a theoretical method. *J.Mol. Biol.* **232:** 574-583.
30. Bastyns, K., Froeyen, M., Diaz, J.F., Volckaert, G. and Engelborghs, Y., 1996, Experimental and theoretical study of electrostatic effects on the isoelectric pH and the pKa of the catalytic residue His-102 of the recombinant ribonuclease from *Bacillus amyloliquefaciens* (barnase). *Proteins: Struct. Funct. Genet.* **24:** 370-378.
31. Loewenthal, R., Sancho, J. and Fersht, A.R., 1991, Fluorescence spectrum of barnase: contributions of three tryptophan residues and a histidine-related pH dependence. *Biochemistry* **30:** 6775-6779.
32. Porter, G.B., 1972, reversible energy-transfer. *Theor. Chim. Acta* **24:** 265-270.
33. Woolley, P., Steinhäuser, K.G. and Epe, B., 1987, Forster-type energy transfer. Simultaneous "forward" and "reverse" transfer between unlike fluorophores. *Biophys. Chem.*, **26:** 367-374.
34. Doyle, T.C., Hansen, J.E. and Reisler, E., 2001, Tryptophan fluorescence of yeast actin resolved via conserved mutations. *Biophys. J.* **80:** 427-434.
35. Mély, Y., De Rocquigny, H., Morellet, N., Roques, B.P. and Gérard, D., 1996, Zinc binding to the HIV-1 nucleocapsid protein: a thermodynamic investigation by fluorescence spectroscopy. *Biochemistry* **35:** 5175-5182.
36. Vuillemier, C., Bombarda, E., Morellet, N., Gérard, D., Roques, B.P. and Mély, Y., 1999, Nucleic acid sequence discrimination by the HIV-1 nucleocapsid protein NCp7: a fluorescence study. *Biochemistry* **38:** 16816-16825.
37. Callis, P.R. and Burgess, B.K., 1997, Tryptophan fluorescence shifts in proteins from hybrid simulations: an electrostatic approach *J. Phys. Chem. B*, **101:** 9429-9432.
38. Callis, P.R., 1997, 1La and 1Lb transitions of tryptophan: applications of theory and experimental observations to fluorescence of proteins, *Methods Enzymol.*, **278:** 113-150.
39. Gopalan, V., Golbik, R., Schreiber, G., Fersht, A.R. and Altman, S., 1997, Fluorescence Properties of a Tryptophan Residue in an Aromatic Core of the Protein Subunit of Ribonuclease P from *Escherichia coli. J. Mol. Biol.*, **267:** 765-769.

40. Nanda, V. and Brand, L., 2000, Aromatic Interactions in Homeodomains Contribute to the Low Quantum Yield of a Conserved, Buried Tryptophan. *Proteins* **40:** 112-125.
41. Burley, S.K. and Petsko, G.A., 1985, Aromatic-aromatic interaction: a mechanism of protein structure stabilization. *Science* **229:** 23-28.
42. Burley, S.K. and Petsko, G.A., 1986, Amino-aromatic interactions in proteins. *FEBS Lett.*, **203:** 139-143.
43. Burstein, E.A. and Emelyanenko, 1996, Log-normal description of fluorescence spectra of organic fluorophores. *Photochem. Photobiol.*, **64:** 316–320.
44. Burstein, E.A., Abornev, S.M., Reshetnyak, Y.K., 2001, Decomposition of protein tryptophan fluorescence spectra into log-normal components. I. Decomposition algorithms. *Biophys J.* **81:** 1699-1709.
45. Reshetnyak, Y.K., Burstein, E.A., 2001, Decomposition of protein tryptophan fluorescence spectra into log-normal components. II. The statistical proof of discreteness of tryptophan classes in proteins. *Biophys J.*, **81:** 1710-1734.
46. Reshetnyak, Y.K., Koshevnik, Y., Burstein, E.A., 2001, Decomposition of protein tryptophan fluorescence spectra into log-normal components. III. Correlation between fluorescence and microenvironment parameters of individual tryptophan residues. *Biophys J.*, **81:** 1735-1758.
47. Parker, C.A. and Rees, W.T., 1960, Correction of Fluorescence Spectra and Measurements of Fluorescence Quantum Efficiency. *Analyst.* **85:** 587-600.
48. Chen, Y. and Barkley, M.D., 1998, Toward Understanding Tryptophan Fluorescence in Proteins. *Biochemistry* **37:** 9976-9982.
49. Chen, R.F., Knutson, J.R., Ziffer, H. and Porter, D., 1991, Fluorescence of Tryptophan Dipeptides: Correlations with the Rotamer Model. *Biochemistry* **30:** 5184-5195.
50. Szabo, A.G. and Faerman, C., 1992, Dilemma of Correlating Fluorescence Quantum Yields and Intensity Decay Times in Single Tryptophan Mutant Proteins. *SPIE*, **1640:** 70-80.
51. Webber, S.E., 1997, The Role of Time-Dependent Measurements in Elucidating Static Versus Dynamic Quenching Processes. *Photochem. Photobiol.* **65:** 33-38.
52. Strickler, S.J. and Berg, R.A., 1962, *J. Chem. Phys.* **37:** 814.
53. Mach, H., Middaugh, C.R. and Lewis, R.V., 1992, Statistical Determination of the Average Values of the Extinction Coefficients of Tryptophan and Tyrosine in Native Proteins. *Anal. Biochem.* **200:** 74-80.
54. Sillen, A., Diaz, F. and Engelborghs, Y., 2000, A step toward the prediction of the fluorescence lifetimes of tryptophan residues in proteins based on structural and spectral data. *Protein Sci.* **9:** 158 -169.
55. Sillen, A., personal communication.
56. Privat, J.P., Wahl, P. and Auchet, J.-C., 1979, Rates of deactivation processes of indole derivatives in water-organic solvent mixtures. —Applications to tryptophyl fluorescence of proteins. *Biophys. Chem.* **9:** 223-233.
57. Eftink, M.R., 2001, Intrinsic Fluorescence of Proteins. In: *Topics in Fluorescence Spectroscopy* **6:** Lakowicz, J.R. Ed. pp. 1-16.
58. Döring, K., Konermann, L., Surrey, T. and Jähnig, F., 1995, A long lifetime component in the tryptophan fluorescence of some proteins. *Eur. Biophys. J.* **23:** 423-432.
59. Chen, Y., Liu, B., Yu, H.-T. and Barkley, M.D., 1996, The Peptide Bond Quenches Indole Fluorescence. *J.Am.Chem.Soc.* **118:** 9271-9278.
60. Klein, R., Tatischeff, I., Bazin, M. and Santus, R., 1981, Photophysics of indole - comparative-study of quenching, solvent, and temperature effects by laser flash-photolysis and fluorescence. *J. Phys. Chem.* **85:** 670–677.

61. Connolly, M.L., 1983, Solvent-accessible surfaces of proteins and nucleic acids. *Science* **221:** 709-713.
62. Yu, H.T., Colucci, W.J., McLaughlin, M.L. and Barkley, M.D., 1992, Fluorescence quenching in indoles by excited-state proton-transfer. *Am. Chem. Soc.* **114:** 8449-8454.
63. Eftink, M.R., Jia, Y., Hu, D. and Ghiron, C.A., 1995, Fluorescence studies with tryptophan analogs - excited-state interactions involving the side-chain amino group. *J. Phys. Chem.* **99:** 5713–5723.
64. Bushueva, T.L., Busel, E.P. and Burstein, E.A., 1975, Interaction of protein functional groups with indole chromophore .3. amine, amide, and thiol-groups. *Stud. Biophysica* **52:** 41-52.
65. Cowgill, R.W., 1967, Fluorescence and protein structure. XI. Fluorescence quenching by disulfide and sulfhydryl groups. *Biochim. Biophys. Acta* **140:** 37-44.
66. Steiner, R.F. and Kirby, E.P., 1969, The Interaction of the Ground and Excited States of Indole Derivatives with Electron Scavengers. *J. Phys. Chem.* **73:** 4130-4135.
67. Yuan, T., Weljie, A.M. and Vogel, H.J., 1998, Tryptophan fluorescence quenching by methionine and selenomethionine residues of calmodulin: orientation of peptide and protein binding. *Biochemistry* **37:** 3187-3195.
68. Shinitzky, M. and Goldman, R.,˙1967, Fluorometric detection of histiine-tryptophan complexes in peptides and proteins. *Eur. J. Biochem.*, **3:** 139-144.
69. Bushueva, T.L., Busel, E.P., Bushuev, V.N. and Burstein, E.A., 1974, Interaction of protein functional groups with indole chromophore .1. imidazole group. *Studia Biophys. Berlin*, **44:** 129-140.
70. Van Gilst, M. and Hudson, B.S., 1996, Histidine-tryptophan interactions in T4 lysozyme: 'anomalous' pH dependence of fluorescence. *Biophys. Chem.* **63:** 17-25.
71. Vos, R. and Engelborghs, Y., 1994, A fluorescence study of tryptophan-histidine interactions in the peptide anantin and in solution. *Photochem. Photobiol.*, **60:** 24-32.
72. Weisenborn, P.C.M., Meder, H., Egmond, M.R., Visser, T.J.W.G. and van Hoek, 1996, Photophysics of the single tryptophan residue in fusarium solani cutinase: evidence for the occurrence of conformational substates with unusual fluorescence behaviour A., *Biophys. Chem.* **58:** 281-288.
73. Prompers, J.J., Hilbers, C.W. and Pepermans, H.A.M., 1999, Tryptophan mediated photoreduction of disulfide bond causes unusual fluorescence behaviour of *Fusarium solani pisi* cutinase. *FEBS Letters* **456**, 409-416.
74. Neves-Petersen, M.T., Gryczynski, Z., Lakowicz, J., Fojan, P., Pedersen, S., Petersen, E. and Petersen, S.B., 2002, High probability of disrupting a disulphide bridge mediated by an endogenous excited tryptophan residue. *Protein Science* **11:** 588-600.
75. Vanhooren, A., Devreese, B., Vanhee, K., Van Beeumen, J. and Hanssens, I., 2002, Photoexcitation of Tryptophan Groups Induces Reduction of Two Disulfide Bonds in Goat α-lactalbumine. *Biochemistry* **41:** 11035-11043.
76. Hennecke, J., Sillen, A., Huber-Wunderlich, M., Engelborghs, Y. and Glockshuber, R., 1997, Quenching of tryptophan fluorescence by the active-site disulfide bridge in the DsbA protein from *Escherichia coli. Biochemistry* **36:** 6391-6400.
77. Ricci, R.W. and Nesta, J.M., 1976, Inter- and Intramolecular Quenching of Indole Fluorescence by Carbonyl Compounds, *J. Phys. Chem.*, **80:** 974-980.
78. Chang, M.C., Petrich, J.W., McDonald, D.B. and Fleming, G.R., 1983, Non-exponential fluorescence decay of tryptophan, tryptophylglycine, and glycyltryptophan. *J.Am.Chem.Soc.* **105:** 3819–3824.
79. Goldman, C., Pascutti, P.G., Piquini, P., Ito, A.S., 1995, On the contribution of electron transfer reactions to the quenching of tryptophan fluorescence. *J. Chem. Phys.* **103:**10614-10620.

80. Antonini, P.S., Hillen, W., Ettner, N., Hinichs, W., Fantucci, P., Doglia, S.M., Bousquet, J.-A. and Chabbert, M., 1997, Molecular mechanics analysis of Tet repressor TRP-43 fluorescence. *Biophys. J.* **72:** 1800-1811.
81. Pan, C.-P., Adams, P.D. and Barkley, M.D., 2000, Effect of backbone conformation on tryptophan fluorescence in rigid cyclic hexapeptides. *Biophys. J.* **78**: (part 2) pos 746.
82. Marcus, R.A. and Suttin, N., 1985, Electron transfers in chemistry and biology. *Biochim. Biophys. Acta* **811:** 265–322.
83. Hellings, M., De Maeyer, M., Verheyden, S., Hao, Q., Van Damme, E.J.M., Peumans, W.J. and Engelborghs, Y. 2003, The Dead-End Elimination Method, Tryptophan Rotamers, and Fluorescence Lifetimes, *Biophys. J.* **85:** 1894-1902.
84. Vander Donckt, E., *Bull. Soc. Chim. Belges* **78:** (1969) 69.
85. Callis, P. R., and Vivian, J.T., 2003, Understanding the variable fluorescence quantum yield of tryptophan in proteins using qm-mm simulations. quenching by charge transfer to the peptide backbone. *Chem. Phys. Let.*, **369:** 409-414
86. Frauenfelder, H., Parak, F. and Young, R.D., 1988, Conformational substates of proteins: a molecular dynamics analysis of myoglobin. *Ann. Rev. Biophys. Chem.* **17:** 451-579.
87. Elber, R. and Karplus, M., 1987, Multiple conformational states of proteins: a molecular dynamics analysis of myoglobin. *Science* **235**: 318–321.
88. Noguti, T. and Gō, N., 1989, Structural basis of hierarchical multiple substates of a protein. III: Side chain and main chain local conformations. *Proteins* **5:** 113-124.
89. Eftink, M. and Hagaman, X., 1985, Fluorescence quenching of the buried tryptophan residue of cod parvalbumin. *Biophys Chem.*, **22:** 173-180.
90. Harris, D.L. and Hudson, B.S., 1990, Photophysics of tryptophan in bacteriophage T4 lysozymes. *Biochemistry* **29:** 5276-5285.
91. Kim, S.J., Chowdhury, F.N., Stryjewski, W., Younathan, E.S., Russo, P.S. and Barkley, M.D., 1993, Time-resolved fluorescence of the single tryptophan of Bacillus stearothermophilus phosphofructokinase. *Biophys. J.* **65:** 215-226.
92. Szabo, A.G., Krajcarski, D., Zuker, M. and Alpert, B., 1984, Conformational heterogeneity in hemoglobin as determined by picosecond fluorescence decay. *Chem. Phys. Lettters* **108:**145–149.
93. Gauduchon, P. and Wahl, Ph., 1978, Pulsefluorimetry of tyrosyl peptides. *Biophys. Chem.* **8:** 87–104.
94. Szabo, A.G. and Rayner, D.M., 1980, Fluorescence Decay of Tryptophan Conformers in Aqueous Solution, *J. Am. Chem. Soc.* USA, **102:** 554-563.
95. Engh, R.A., Chen, L.X.-Q. and Fleming, G.R., 1986, Conformational dynamics of tryptophan - a proposal for the origin of the nonexponential fluorescence decay. *Chem. Phys. Letters* **126:** 365-371.
96. Lakowicz, J.R., Maliwil, B.P., Cherek, H. and Balter, A., 1983, Rotational freedom of tryptophan residues in proteins and peptides. *Biochemistry* **22:** 1741-1752.
97. Lakowicz, J.R., 2000, On spectral relaxation in proteins. *Photochem. Photobiol.*, **72:** 421–437.
98. Laws, W.R., Ross, J.B.A., Wyssbrod, H.R., Beechem, J.-M., Brand, L. and Sutherland, J.C., 1986, Time-resolved fluorescence and 1H NMR studies of tyrosine and tyrosine analogues: correlation of NMR-determined rotamer populations and fluorescence kinetics. *Biochemistry* **25:** 599-607.
99. Ross, J.B.A., Wyssbrod, H.R., Porter, R.A., Schwartz, G.P., Michaels, C.A. and Laws, W.R., 1992, Correlation of tryptophan fluorescence intensity decay parameters with 1H NMR-determined rotamer conformations: [tryptophan2]oxytocin. *Biochmistry* **31:** 1585-1594.

100. Janin, J., Wodak, J., Levitt, M. and Maigret, B., 1978, Conformation of amino acid side-chains in proteins. *J. Mol. Biol.* **125:** 357-386.
101. Bath, T.N., Sasisekharan, V. and Vijayan, M., 1979, An analysis of side-chain conformation in proteins. *Int. J. Protein Res.* **13:** 170-184.
102. Summers, N.L., Carlson, W.D. and Karplus, M., 1987, Analysis of side-chain orientations in homologous proteins. *J. Mol. Biol.* **196:**175-198.
103. McGregor, M.J., Islam, S.A. and Sternberg, M.J.E., 1987,Analysis of the Relationship Between Side-chain Conformation and Secondary Structure in Globular Proteins, *J. Mol. Biol.* **198:** 295-310.
104. Schrauber, H., Eisenhaber, F. and Argos, P., 1993, Rotamers: Tobe or not to be? An Analysis of Amino Acid Side-chain Conformations in Globular Proteins. *J. Mol. Biol.* **230:** 592-612.
105. Clayton, A.H.A. and Sawyer, W.H., 1999, The structure and orientation of class-A amphipathic peptides on a phospholipid bilayer surface, *Eur. Biophys. J.* **28:** 133-141.
106. Clayton, A.H.A. and Sawyer, W.H., 1999, Tryptophan Rotamer Distributions in Amphipathic Peptides at a Lipid Surface, *Biophys. J.* **76:** 3235-3242.
107. Willis, K.J., Neugebauer, W., Sikorska, M. and Szabo, A.G., 1994, Probing alpha-helical secondary structure at a specific site in model peptides via restriction of tryptophan side-chain rotamer conformation. *Biophys. J,* **66:** 1623-1630.
108. Mérola, F., Rigler, R., Holmgren, A. and Brochon, J.-C., 1989, Picosecond tryptophan fluorescence of thioredoxin: evidence for discrete species in slow exchange. *Biochemistry* **28:** 3383-3398.
109. Tanaka, F. and Mataga, N., 1982, Dynamic depolarization of interacting fluorophores. Effect of internal rotation and energy transfer. *Biophys. J.* **39:** 129-140
110. Tanaka, F. and Mataga, N., 1987, Fluorescence quenching dynamics of tryptophan in proteins. Effect of internal rotation under potential barrier. *Biophys. J.* **51:** 487-495.
111. Tanaka, F., Kaneda, N., Mataga, N., Tamai, N., Yamazaki, I. and Hayashi, K., 1987, Analyses of nonexponential fluorescence decay functions of a single tryptophan residue in erabutoxin-b. *J. Phys. Chem.* **91:** 6344-6346.
112. Tanaka, F., Tamai, N., Mataga, N., Tonomura, B. and Hiromi, K., 1994, Analysis of internal motion of single tryptophan in Streptomyces subtilisin inhibitor from its picosecond time-resolved fluorescence. *Biophys. J.* **67:** 874-880.
113. Van Gilst, M., Tang, C., Roth, A. and Hudson, B., 1994, Quenching Interactions and Nonexponential Decay: Tryptophan 138 of Bacteriophage T4 Lysozyme, *J. Fluorescence* **4:** 203-207.
114. Hudson, B.S., 1999, An ionization/recombination mechanism for complexity of the fluorescence of tryptophan in proteins. *Acc. Chem. Res.* **32:** 297-300.
115. Bernasconi, C.F., *Relaxation Kinetics*, Academic Press, New York & San Francisco, 1976.
116. McMahon, L.P., Yu, H.-T., Vela, M.A., Morales, G.A., Shui, L., Fronczek, F.R., McLaughlin, M.L. and Barkley, M.D., 1997, Conformer interconversion in the excited state of constrained tryptophan derivatives. *J. Phys. Chem. B* **101:** 3269-3280.
117. Boens, N., Szubiakowski, J. and Novikov, E., Ameloot, M., 2000, Testing the identifiability of a model for reversible intermolecular two-state excited state processes. *J. Chem. Phys.* **112:** 8260-8266.
118. Brochon, J.-C., Wahl, P., Charlier, M., Maurizot, J.C. and Hélene, C., 1977, Time resolved spectroscopy of the tryptophyl fluorescence of the E. coli LAC repressor. *Biochem. Biophys. Res. Commun.* **79:** 1261-1271.

119. Gastmans, M., Volckaert, G. and Engelborghs, Y., Tryptophan Microstate reshuffling Upon the Binding of Cyclosporin A to Human Cyclophilin A, 1999, *Proteins: Struct. Funct. Genet.* **35:** 464-474.
120. Guddat, L.W., Bardwell, J.C., Zander, T. and Martin, J.L., 1997, The uncharged surface features surrounding the active site of Escherichia coli DsbA are conserved and are implicated in peptide binding. *Protein Sci* **6:** 1148- 1156.
121. Guddat, L.W., Bardwell, J. and Martin, J.L., 1998, Crystal structures of reduced and oxidized DsbA: investigation of domain motion and thiolate stabilization. *Structure* **6:** 757-767.
122. Schirra, H.J., Renner, C., Czisch, M., Huber-Wunderlich, M., Holak, T.A. and Glockshuber, R., 1998, Structure of reduced DsbA from Escherichia coli in solution. *Biochemistry* **37:** 6263-6276.
123. Martin, J.L., 1995, Thioredoxin-A fold for all reasons. *Structure* **3:** 245-250.
124. Engelborghs, Y., 2001, The analysis of time resolved protein fluorescence in multi-tryptophan proteins. *Spectrachim. Acta Part A* **57:** 2255-2270.
125. Moncrieffe, M.C., Venyaminov, S.Y., Miller, T.E., Guzman, G., Potter, J.D. and Prendergast, F.G.,1999, Optical spectroscopic characterization of single tryptophan mutants of chicken skeletal troponin C: evidence for interdomain interaction. *Biochemistry* **39:** 11973 –11983.
126. Hofmann, A., Raguénes-Nicol, C., Favier-Perron, B., Mesonero, J., Huber, R., Russo-Marie, F. and Lewit-Bentley, A.,2000, The annexin A3-membrane interaction is modulated by an N-terminal tryptophan. *Biochemistry* **39**: 7712 –7721.
127. Soulages, J.L. and Arrese, E.L., 2000, Fluorescence spectroscopy of single tryptophan mutants of apolipophorin-III in discoidal lipoproteins of dimyristoylphosphatidylcholine. *Biochemistry* **39:** 10574-10580.
128. Rensland, H., John, J., Linke, R., Simon, I., Schlichting, I., Wittinghofer, A. and Goody, R.,1995, Substrate and product structural requirements for binding of nucleotides to H-ras p21: the mechanism of discrimination between guanosine and adenosine nucleotides. *Biochemistry* **34:** 593-599.
129. Díaz, J.F., Sillen, A. and Engelborghs, Y., 1997, Equilibrium and kinetic study of the conformational transition toward the active state of p21Ha-ras, induced by the binding of BeF3- to the GDP-bound state, in the absence of GTPase-activating proteins. *J. Biol. Chem.* **272:** 23138-23143.
130. Kuppens, S., Díaz, J.F. and Engelborghs, Y., 1999, Characterization of the hinges of the effector loop in the reaction pathway of the activation of *ras*-proteins. Kinetics of binding of beryllium trifluoride to V29G and I36G mutants of HA-*ras*-p21, *Protein Sci.* **8:**1860-1866.
131. Hazlett, T.L., Moore, K.J., Lowe, P.N., Jameson, D.M. and Eccleston, J.F.,1993, Solution dynamics of p21ras proteins bound with fluorescent nucleotides. *Biochemistry* **32:**13575-13583.
132. Jameson, D.M. and Eccleston, J.F., 1997, Fluorescent nucleotide analogs: synthesis and applications. *Methods Enzymol.* **278:** 363-390.
133. Ross JB, Szabo AG, Hogue CW., 1997, Enhancement of protein spectra with tryptophan analogs: fluorescence spectroscopy of protein-protein and protein-nucleic acid interactions. *Methods Enzymol.*, **278:** 151-190.
134. Bergstrom F, Hagglof P, Karolin J, Ny T, Johansson LB., 1999, The use of site-directed fluorophore labeling and donor-donor energy migration to investigate solution structure and dynamics in proteins. *Proc. Natl. Acad. Sci USA*, **96:** 12477-12481.

135. Deprez, E., Tauc, P., Leh, H., Mouscadet, J.-F., Auclair, C. and Brochon, J.-C., 2000, Oligomeric states of the HIV-1 integrase as measured by time-resolved fluorescence anisotropy. *Biochemistry* **39:** 9275-9284.
136. Fa, M., Bergström, F., Hägglöf, P., Wilczynska, M., Johansson, L.B. and Ny, T., 2000, The structure of a serpin-protease complex revealed by intramolecular distance measurements using donor-donor energy migration and mapping of interaction sites. *Structure Fold. Des.* **8:** 397- 405.
137. Fa, M., Bergström, F., Karolin, J., Johansson, L.B. and Ny, T., 2000, Conformational studies of plasminogen activator inhibitor type 1 by fluorescence spectroscopy. Analysis of the reactive centre of inhibitory and substrate forms, and of their respective reactive-centre cleaved forms. *Eur. J. Biochem.* **267:** 3729-3734.
138. Chadborn, N., Bryant, J., Bain, A.J. and O'Shea, P., 1999, Ligand-Dependent Conformational Equilibria of Serum Albumin Revealed by Tryptophan Fluorescence Quenching. *Biophys. J.* **76:** 2198-2207.

Novel (Bio)chemical and (Photo)physical Probes for Imaging Living Cells

ELIZABETH A. JARES-ERIJMAN[1], CARLA SPAGNUOLO[1], LUCIANA GIORDANO[1], MARIA ETCHEHON[1], JENNIFER KAWIOR[1], MARIA V. MAÑALICH-ARANA[1], MARIANO BOSSI[1], DIANE S. LIDKE[2], JANINE N. POST[2], RUDOLF J. VERMEIJ[2], RAINER HEINTZMANN[2], KEITH A. LIDKE[2], DONNA J. ARNDT-JOVIN[2], THOMAS M. JOVIN[2]
[1]Departamento de Química Orgánica, Facultad de Ciencias Exactas y Naturales, Universidad de Buenos Aires, Buenos Aires, Argentina; [2]Department of Molecular Biology, Max Planck Institute for Biophysical Chemistry, Göttingen, Germany

1. INTRODUCTION

The living cell mediates its internal state and the exchange of substances and information with its environment primarily via protein-protein interactions. The spatio-temporal disposition of structural, catalytic, and regulatory proteins defines the nature and functional state of the cell. Signaling mechanisms, as a prominent example, occupy a central role in this process, leading to a set of canonical questions, challenges and strategies (Table 1).

In applying fluorescence microscopy in cell biology to a particular system, one is faced with a multiplicity of molecules at every level of organization (external, membrane, cytoplasm). The elucidation of such an extensive degree of vertical and horizontal networking, extending into the downstream signaling cascades, requires imaging technology in addition to the classical biochemical

Supramolecular Structure and Function 8, Edited by Pifat-Mrzljak
Kluwer Academic/Plenum Publishers, New York 2004

Table 1. Issues and challenges in the field of signal transduction

molecules	which, where, when, how?
biosynthesis	"right-on-time" coordination
intermolecular interactions	conformational and association states
	stoichiometry, equilibria, kinetics
activation mechanisms(s)	binding, conformations, covalent modifications
signal propagation	cascades, "kiss and run" reactions
networking	redundancy, control points
spatial and temporal relationships	IN the cell
microscopy	resolution, contrast, speed
biochemistry	new probes (expression, photophysical)
image processing	analysis and reconstruction

and molecular biological methods based largely on classical "divide (separate) and conquer" protocols (Table 2). For example, the "orphan" (ligand-less) erbB2/HER2 receptor tyrosine kinase (RTK) is overexpressed and highly activated in a large fraction of breast tumors, forming characteristic homo- and heterodimers with three other members of this RTK family[1]. These are targets for the only anti-tumor immunotherapies in present clinical use, exemplified by the antibody specific for HER2, Herceptin[2]. Unfortunately, the modes of action

Table 2. Fishing for protein-protein interactions in the cell

Biochemical
(co)immunoprecipitation (IP)+gel electrophoresis
affinity chromatography
chem.,ical, ligand and photo(1-2 photon)-mediated cross-linking
Phenotypic complementation (yeast)
Yeast two-hybrid (YTHS): transcriptional readout (nucleus)
SOS-recruitment system (SRS): Ras-dependent growth (cytoplasm)
RNA interference: post-transcriptional gene
mRNA targeting, e.g. with ds siRNAs (19-mer)
Fluorescence (any cell) – microscopy & flow cytometry
VFPs (intein, paVFP)
Targeted external probes (FlAsHEDT$_2$, scFv Abs, Quantum Dots)
Fluorescence resonance energy transfer (FRET, EmFRET)
Correlation spectroscopy (FCS, ICS, ICCS, DICS) and diffusion
Magnetic microspheres with covareltnly bound ligands
Transfer labeling: tyramide signal amplification (TSA); other
Single-particle tracking; microarrays of reagents and cells

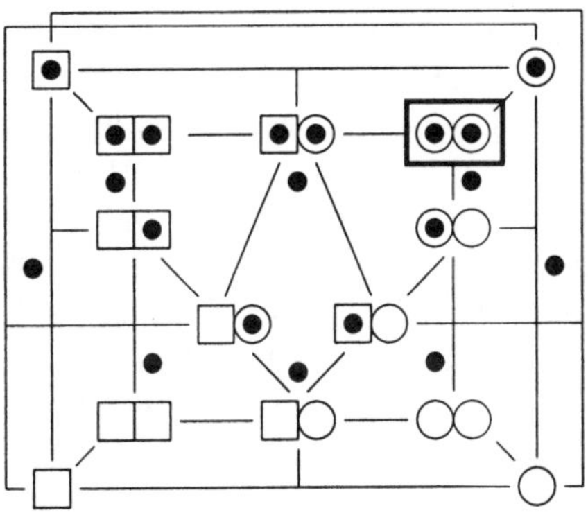

Figure 1. Binding, conformational, and association states for a putative ligand-dependent receptor. The receptor can adopt two alternative conformational states (square, circles) and exist as a monomeric or dimeric species. The ligand (dark circles) can bind via parallel sequential pathways. If there is high cooperativity, the system will progress from the isolated unliganded receptor (empty single square, bottom left) to the fully liganded dimeric "circular" receptor (enclosed by the rectangle, upper right). The biophysical resolution of this scheme, particularly in the cellular context, is difficult!

of such agents are poorly understood. Thus, in order to elucidate the repertoire of the RTKs under normal and pathological conditions, one must evaluate their localization and molecular structural and functional state(s) in defined cell populations, be it cell culture lines or primary patient-derived cells.

The thermodynamic and kinetic complexity is evident from the minimal scheme defining the interplay between ligand binding, conformational states (2), and association states (2) for a prototypic growth factor receptor (Figure 1).

Although Table 2 cannot be regarded as comprehensive, it emphasizes that in addition to established biochemical and genetic approaches, physico-chemical techniques offer the versatility required for assessing molecular interactions in the cell. In particular, fluorescence unites the features of great sensitivity and selectivity with high contrast, even under conditions of low local molecular density, i.e. concentration.

2. PROBES

New probes for new strategies. Emerging technologies, utilizing a combination of chemistry, physics and molecular biology, are creating an increasing interest in smart materials serving as reporters and sensors in micro- and nano-systems. Such devices constitute unique tools for a plethora of biological applications and therapies, allowing *in vivo* and real time monitoring of biomolecular structure and function. Imaging live cells poses a number of problems, not the least of which is to minimize the perturbation of the physiological state and viability of the cells with the probes and "tools". One seeks to achieve single molecule sensitivity (when appropriate and desirable), monitor fast kinetic processes, and observe molecular interactions occurring on distance scales far beyond the optical resolution of light microscopes. In the remainder of this contribution we discuss a few methods that have already attained some of these goals and present some model systems that hold the promise for achieving others.

In vivo expressed fluorescent molecules are revolutionizing cell biology. The advent of expression vectors for visible fluorescent proteins (VFPs) such as the green fluorescent protein from the jellyfish *Aequorea-victoria*[3] provided biologists with a tool of immense utility. Fusions of spectral variants of this protein and related molecules of coral origin to selected target proteins constitute expression probes for determining molecular concentration (including of ions), structure, distribution, and "age" *in vivo* (reviewed in[4-6]); other VFP chimeras are also available[7-10]. This technology and that of RNA interference (RNAi) arguably represent the most significant technical advances for cell biological studies in the past decade. The most recent developments in the VFP field include mutants with increased spectral ranges[11], photoconversion capabilities[12,13], improved photostability and brightness[14], reduced environmental (pH, chloride) sensitivity[15], faster maturation rate[14], and a suppressed tendency to oligomerize[8,16]. The advent of a photoactivatable VFP[12] is of particular significance. Localized application of blue light to cells (we have utilized two-photon activation as well) permits the activation of a fluorescence signal at an arbitrary location and time, an invaluable feature for studies of protein translocation and association. Fusion with particular protein binding elements has extended the use of these VFPs to probe molecular interactions and concentrations[17].

Molecular interactions can be quantitated in the light microscope by exploiting fluorescence resonance energy transfer (FRET; see Section 3 and reviews[18-20]). The VFPs are particularly well suited for FRET measurements due to the spectral overlap properties of their excitation and emission spectra leading to classical FRET[21] or homoFRET (discussed in Section 3.3). By linking the donor and acceptor VFPs via reporter peptides that undergo conformational change and/or covalent modification upon binding or enzymatic reaction, a number of groups have generated "smart tools" for measuring ions[6,17], phosphorylation states[7] and the concentrations of other ligands[4-6]. The resulting perturbation of the FRET signal serves as a monitor of the underlying time-dependent process.

A recent addition to the VFP toolbox is the method denoted as Bimolecular Fluorescence Complementation (BiFC), conceived as a means for assessing protein-protein interactions *in vivo*[22]. Non-fluorescent fragments of spectrally distinct VFP half-molecules are fused to different proteins of interest. If the latter are brought into proximity by the association of their respective fusion partners, the coupled VFP fragments join such as to generate an intrinsic VFP chromophore exhibiting fluorescence after a maturation period of several hours. Related complementation systems are available for studying protein-protein interactions: the fragmentation assay (PCA) and the intein-mediated protein splicing of fragmented VFP and luciferase[23].

In all the VFP-based techniques, two fundamental problems must be addressed: the possible functional (e.g. unphysiological) consequences of overexpression and the need to distinguish between VFP-fused proteins in their natural cellular compartment from nascent, reclaimed, and degraded molecules elsewhere in the cell. Total internal reflection microscopy[24,25] and other superresolution techniques provide the means for solving the latter problem.

Integrated strategies and dyes. An alternative strategy to VFP fusions is to combine exogenous probes with the expression of small peptides incorporated in the proteins of interest and serving as highly selective and affine binding sites. This approach offers the advantage of a target with a greatly reduced size compared to VFPs and the versatility afforded by the ligand in terms of lifetime, large Stokes shift, or some other property. One such system is based on specific hexapeptide sequences with 4 cysteines that are rarely found in genomically coded proteins. Application of an exogeneous, membrane permeable, non-fluorescent probe, e.g. biarsenylated fluorescein ($FlAsH \cdot EDT_2$) or resorufin ($ReAsH \cdot EDT_2$) derivatives, leads to binding to the peptide target and the generation of a specific fluorescent signal[26,27]. These are very promising reagents, but it has become apparent that additional chemical modifications are required to reduce the background in most cellular applications (see also[28,29]).

3. FRET

Fluorescence is an emission process competing with a variety of other pathways for deactivating the excited state (Figure 2). The reciprocal of the composite rate constants is the fluorescence lifetime. FRET is one of the many mechanisms competing with radiative decay and is thereby manifested by a reduction in the fluorescence lifetime and quantum yield (Figure 3). According to the concept introduced elsewhere[18] of a fluorophore as a photophysical "enzyme", FRET serves to modulate the turnover rate (Figure 4).

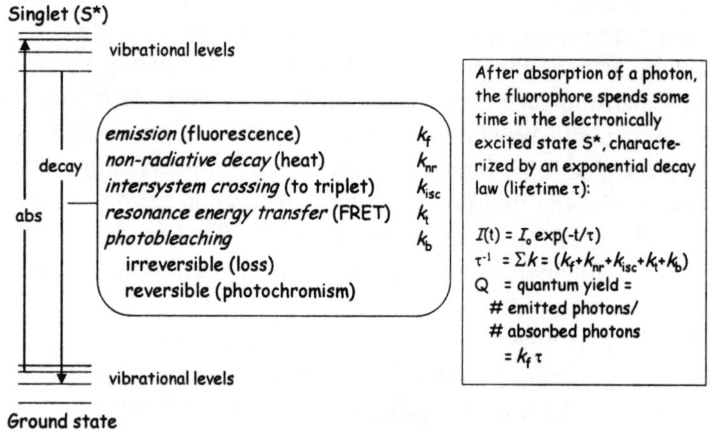

Figure 2. Jablonski energy diagram indicating the different decay pathways from the singlet excited state of a fluorophore

In FRET microscopy, one must adopt a formalism appropriate for quantitating the transfer efficiency under conditions of arbitrary, generally unknown, intramolecular and/or intermolecular stoichiometries, distributions, and microenvironments of donor and acceptor. Furthermore, continuous, i.e. non-destructive methods of observation (by FRET) are desirable in studies of live cells. We have characterized extensively the available methods for performing FRET measurements in imaging systems[18]. The reader is referred to that publication for background material and extensive references.

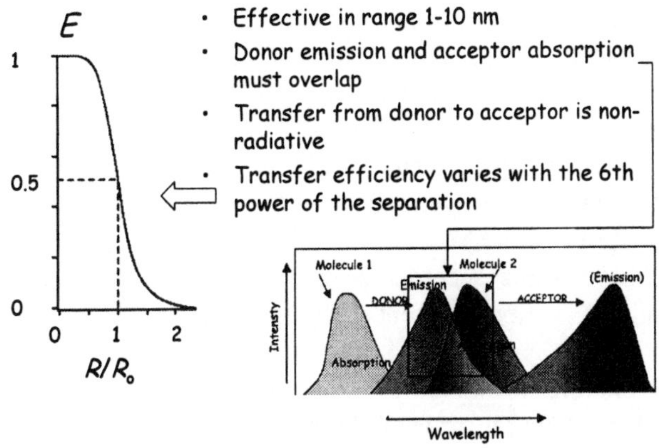

Figure 3. Two major determinants of FRET efficiency (E): (i) the distance (R) between donor and acceptor (left), normalized by the 50% transfer distance R_o, and (ii) the spectral overlap between donor emission and acceptor absorption (bottom right).

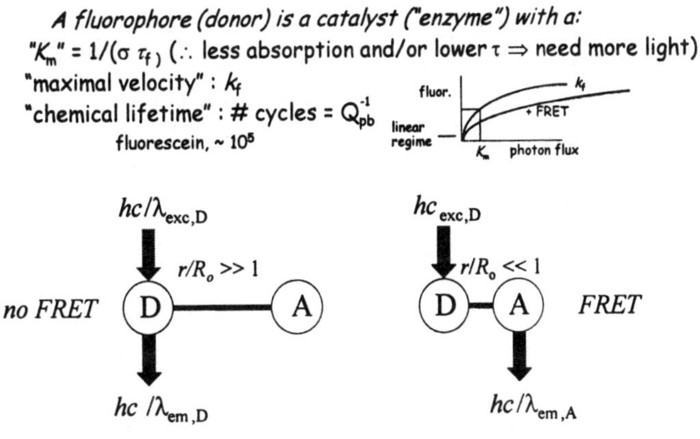

Figure 4. A FRET donor-acceptor pair, functioning as a photophysical catalyst with two alternative "products": donor and acceptor emission (photons) with ensemble mean quantum efficiencies, $Q_d(1\text{-}E)$ and $Q_a\ E$, respectively.

3 1. Photochromic FRET (pcFRET)

Among the different methods available for the quantitative determination of FRET in the fluorescence microscope the so-called acceptor depletion techniques[18] are of fundamental importance. This class permits the *in situ* generation of the reference state in which the donor is un-quenched, required for the calculation of FRET at any given sample position. However, acceptor depletion by photobleaching is irreversible, in contrast to reversible depletion using photochromic FRET (pcFRET).

In pcFRET, a photochromic acceptor is cycled repeatedly between FRET "competent (*on*)" and FRET "incompetent (*off*)" states by alternative exposures to UV and visible light[30,31] (Figure 5). Besides being reversible, pcFRET requires only a few photons to induce the interconversions between states, in contrast to $\sim 10^4$ to 10^6 photons for irreversible photobleaching.

3.2 Biarsenical dyes for pcFRET

To obtain switchable acceptors for pcFRET which could be introduced into living cells, we designed a series of photochromic compounds combined with biarsenical ligands. As explained in Section 2, these compounds target specific genetically encoded peptides such as the CCPGCC. The pcFRET strategy was tested with a series of model compounds incorporating a fluorescent donor (without arsenic) bound through a tether to a spiropyran or to a diheteroarylethene (DAE) acceptor. The DAE family of photochromic compounds proved superior as acceptors for FRET in biological media due to their greater stability, the lack of fatigue processes and their lack of spontaneous conversion to the FRET competent form in polar media, characteristic of the spiropyran family. The reversibility of the process was observed both in organic solvents and in aqueous media for the diheteroarylethene family of acceptors.

The FlAsH ligand functions as a donor for pcFRET. One important issue and objective in the design of a biarsenical pcFRET dye was to demonstrate that the resulting dye-peptide complex would indeed demonstrate resonance energy transfer between the donor and acceptor units. A titration of FlAsH·EDT_2 with the synthetic peptide CCPGCC is shown in Figure 6. The increase in fluorescence emission of the probe upon addition of the peptide reflects the formation of the FlAsH-CCPGCC complex.

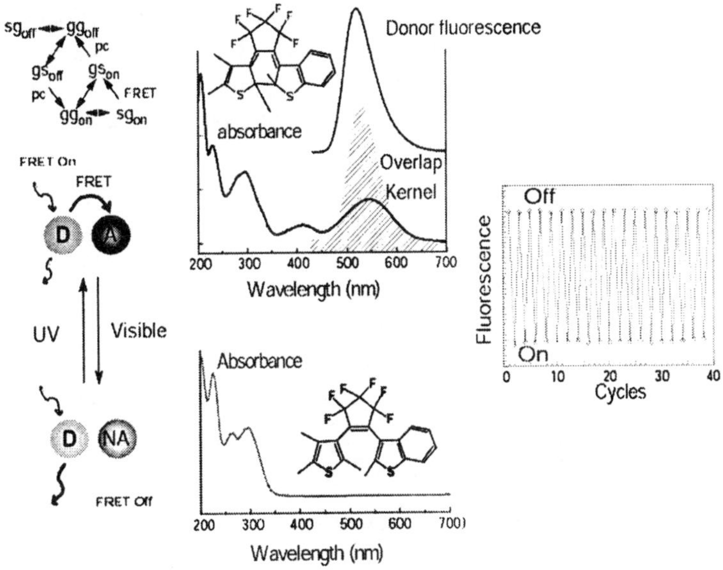

Figure 5. pcFRET. Resonance energy transfer takes place only when the accepting quencher is in its closed, absorbing conformation. The transitions between the closed (*on*) and open (*off*) conformations are promoted by UV (*off* → *on*) or visible (*on* → *off*) light. The switching can be performed due to the lack of absorbance in the visible spectrum of the open conformation. Figure adapted from Ref. 18

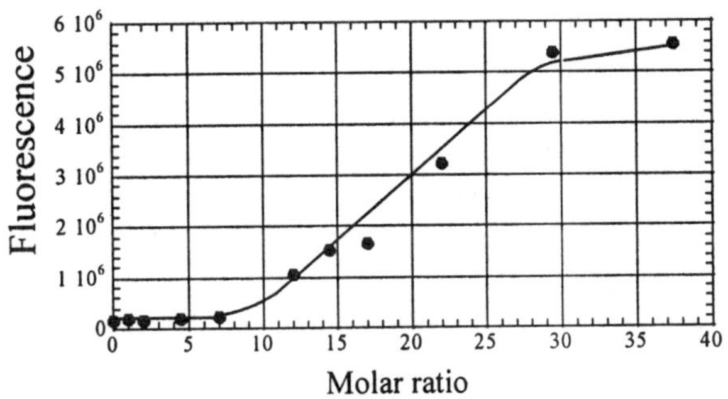

Figure 6. Fluorogenic behavior of FlAsH·EDT$_2$ as a function of target peptide concentration. 50 mM solution of FlAsH·EDT$_2$ in EtOH:Tris buffer (4:1 v/v) titrated with the hexapeptide CCPGCC. The excitation and emission peaks were 508 and 524 nm, respectively.

Quenching of FlAsH-peptide emission. A solution of FlAsH-CCPGCC complex was titrated with a suitable diheteroarylethene in its closed form. A decrease in the emission was observed throughout the titration (Figure 7). A change in slope was observed at a concentration consistent with intermolecular resonance energy transfer.

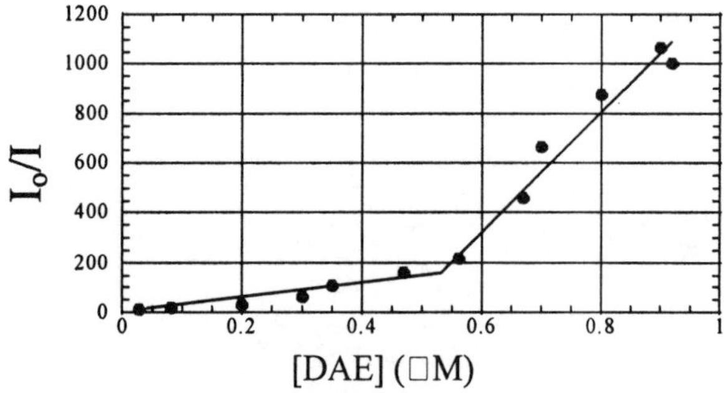

Figure 7. Front face fluorescence of a FlAsH-CCPGCC complex titrated with a free diheteroarylethene. 50 mM solution of FlAsH-CCPGCC complex in EtOH:Tris buffer (4:1 v/v). The change in slope occurred at 0.54 mM of the photochromic compound.

Modulation of FlAsH-peptide emission by cyclical photoconversion. FRET between the FlAsH-peptide complex and the diheteroarylethene was inhibited by photoswitching the acceptor from its closed to its open form by irradiation with visible light (520 nm). The FRET process was enabled again by photoconverting the photochromic open form by UV irradiation at 340 nm. The on/off cycling was repeated to show the reversibility of the process (Figure 8).

Additional design considerations. In order to apply the photochromic FlAsH·EDT_2 system in pcFRET, a covalent linkage between the biarsenical dye and a photochromic acceptor is required. The resulting system ensures proximity of the two units as well as a constant donor-acceptor ratio. Modulation of fluorescence by pcFRET[30,31] allows for the subtraction of autofluorescence noise and greater stability of the dye. UV irradiation is used to convert the diheteroarylethene to its closed "FRET competent" form, which quenches the emission of the dye, thus creating the reference signal that is required for background subtraction.

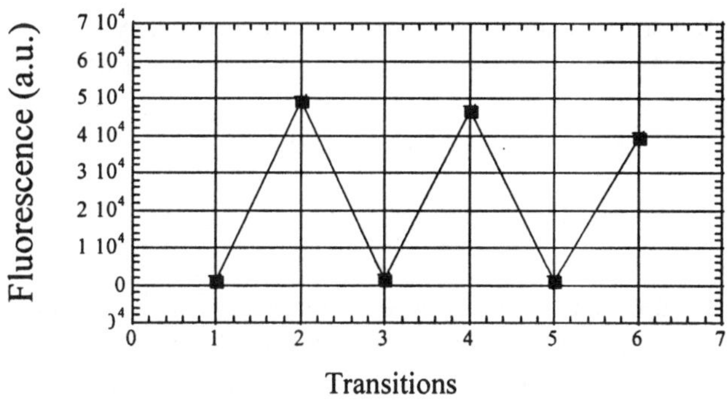

Figure 8. Fluorescence monitoring of cyclical deactivation and activation of FRET between FlAsH-peptide and a diheteroarylethene by photoconversion of the diheteroarylethene via irradiation with VIS and UV light at 520 and 340 nm, respectively.

Diheteroarylethenes with spectral properties in the red were prepared and are presently being evaluated as candidates for pcFRET acceptors. In particular, acceptors with absorption in the red and near infrared spectral regions are of special interest; combined with appropriate donors, these acceptors can be used to take advantage of sensitive regions of CCD cameras and reduce cell damage. Another group of photochromic compounds synthesized for the same purpose belongs to the family of the fulgimides. Fulgimides are potentially useful photochromic compounds for pcFRET in biological systems because they can be converted to the FRET competent form of the acceptor at longer wavelengths (400-500 nm) than the diheteroarylethenes implemented so far.

3.3 Monitoring homoassociation with homo FRET (emFRET)

Classical FRET occurs between two fluorophores of different color when the emission spectrum of an excited donor chromophore overlaps the excitation spectrum of the acceptor. In the same way, energy migration (emFRET, homoFRET) can occur between like fluorescent molecules when the separation between emission and excitation peaks is small. Since the transfer is between like molecules, the excited state lifetime and intensity are not changed. The observable effect of emFRET is a depolarization of the donor fluorescence due to the secondarily excited molecules that have an orientation uncorrelated with the original distribution established by photoselection. As a result, the proximity of two like fluorophores can be detected by measuring the depolarization, or loss in anisotropy, of the emitted signal. The steady-state anisotropy (r) of the system is determined by selectively exciting the cell with linearly polarized light and comparing the intensities of the parallel ($I_{\parallel}$) and perpendicular ($I_{\perp}$) polarized emission signals with respect to the exciting polarization:

$$r = (I_{\parallel} - I_{\perp})/(I_{\parallel} + 2I_{\perp})$$

EmFRET has the distinct advantage over classical FRET in that it requires only a single fluorophore, rendering it an ideal method for detecting homoassociation. The implication for studies of VFP-fusion proteins in live cells is that only a single molecular species needs to be expressed in a cell to evaluate intermolecular association. If the cellular population displays a range of expression level, the functionally important dependence on molecular density (e.g. surface density for membrane proteins) can be readily evaluated.

EmFRET is manifested in solution as a process of concentration-dependent depolarization. As the concentration increases, the intermolecular separation decreases, leading to a greater efficiency of emFRET. In addition, any binding equilibrium involving monomeric and oligomeric species favors the latter, such that the depolarization effects are further enhanced. The resulting decrease in steady-state anisotropy as a function of concentration can be seen in Figure 9 for the VFP Citrine. The fit is to a monomer-dimer equilibrium, yielding a K_d = 0.24±0.11 mM, a monomer $r = 0.31±0.02$ and a dimer $r = 0.16±0.05$.
We have also adapted emFRET measurements to the widefield[32] and confocal microscopes[33] to study receptor tyrosine kinases (RTK) homoaggregation in response to external regulatory ligands such as growth factors. A central dogma of RTK signal transduction is that dimerization of such proteins is a result of ligand binding and activation of the latent protein kinase in the cytoplasmic domain[34]. We previously reported evidence based on FRET-FLIM imaging of the EGF receptor (EGFR/erbB1), suggesting that a substantial fraction of such receptors might be pre-associated (e.g. dimerized) in the absence of ligand[35]. These experiments have been extended to confocal laser scanning microscopy (CLSM) using VFP fusions with various members of the erbB RTK family. Anisotropy measurements were performed using a Zeiss Model 310 CLSM adapted with a polarizing beam splitter simultaneously directing the two polarized emission components to dual photomultipliers. Images were corrected for background intensities from a cell-free region of the same frame and the anisotropies calculated on a pixel-by-pixel basis.

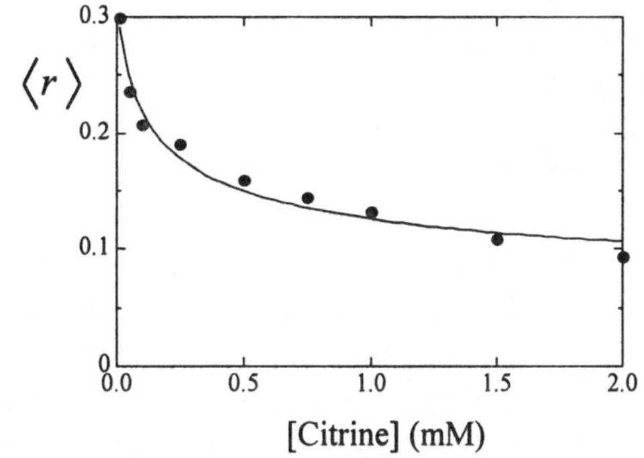

Figure 9. Average anisotropy as a function of protein concentration for purified Citrine (a variant of YFP[15]). The anisotropy of VFP in solution was measured using a Cary Eclipse spectrofluorometer equipped with motorized excitation and emission polarizers. Inner filter effects were minimized by use of a 0.25 mm square cross-section cuvette. The mean anisotropy was calculated as a weighted average over the emission band. See text for a discussion of the fit to a monomer-dimer equilibrium (solid line).

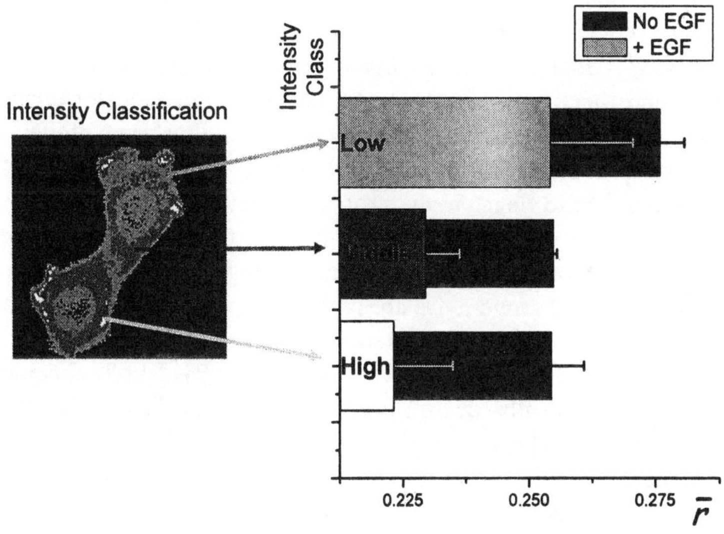

Figure 10. CHO cell expressing erbB1-eGFP. Left panel: image separated into low (white), middle (dark grey) and high (grey) intensity classes. The plot on the right indicates the measured anisotropy in each region without EGF (black) and after 2 min of EGF stimulation at 37 °C (gray). Cells were fixed in MeOH before measurement.

In Figure 10 (left panel), the pixels are sorted into three categories of signal intensity (low, medium, high). Higher concentrations (higher intensity regions) were generally correlated with lower anisotropy, indicative of concentration depolarization (emFRET) and a closer proximity of molecules in these areas. One can also compare the mean anisotropies of the three intensity classes under different physiological conditions. An example of this is shown in Figure 10 for the EGFR-eGFP construct stably expressed at high levels in Chinese hamster ovary (CHO) cell lines with and without EGF stimulation. We have obtained trend lines characteristic for different states of the RTKs during activation and internalization stages of signal transduction using this procedure. Pixel-by-pixel analysis indicated that the highest intensity pixels, i.e. of highest local receptor concentrations, exhibited the lowest anisotropies, reflecting a higher extent of association in these regions, and the values decrease further upon addition of EGF, thus indicating an increased clustering.

4. NANO AND MICROPARTICULATE PROBES

Other probes and particles are of relevance for fluorescence *in vivo* imaging. Among them, microspheres, nanocrystals and phosphors can be evaluated on and in cells by virtue of attached ligands, morphology, spectroscopy, and/or localization and functional effects. Bioconjugated semiconductor quantum dots (QDs) offer an alternative to organic molecules as fluorescence probes. QDs are finding widespread application as labeling reagents for cells and macromolecules due to their unique properties and commercial availability[36,37]. Appropriately designed QDs are very photostable and non-toxic, can be excited with one or more photons over a wide spectral range[38], yet emit in a narrow and programmable spectral range. QDs are typically capped with a polymer bearing specific binding moieties such as (strept)avidin, protein G or A, biotin or conjugatable chemical groups (Quantum Dot Corp., Evident Technologies, and others). We have recently demonstrated that growth factor ligands can be coupled to QDs and behave similarly to the ligand itself and furthermore that single ligands on a QD are sufficient to induce the biological effect[39]. An example of such a study is featured below. QDs can serve as FRET donors in aqueous systems[40,41], a further property of these extraordinary materials that will undoubtedly lead to many applications.

4.1. Endosomal trafficking of EGF-QDs

We have demonstrated that loading QDs with protein ligands provides a reliable method for tracking membrane protein internalization and trafficking, as well as to detect protein-protein interactions[39]. In those experiments, EGF (the ligand for EGFR/erbB1) was coupled to QDs (EGF-QD) through a streptavidin-biotin complex. The EGF-QD was able to bind EGFR, activating and inducing the internalization of the complex. After the internalization of the EGF-QD-EGFR complex, the endosomal trafficking could be observed. QDs are highly photostable and do not photobleach, such that they can be tracked for long periods of time. Figure 11 shows a CHO cell expressing EGFR-eGFP, 30 minutes after addition of EGF-QD. Endocytic vesicles containing EGFR-eGFP surrounding the EGF-QD (white) underwent Brownian movement, directed linear motion, and vesicular fusion. The left panel shows specific vesicles at the beginning of their motion and the panel on the right shows the destination. The lines in the images track the path taken by the vesicles. The vesicles moved at a rate ~0.4 μm/sec, presumably transported by motor proteins along the microtubules.

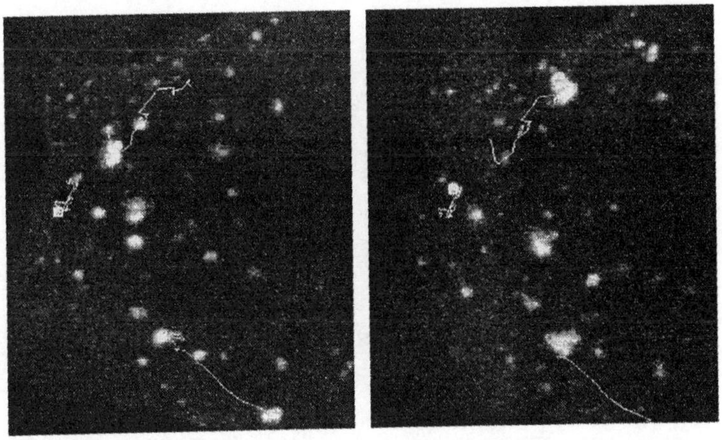

Figure 11. ErbB-eGFP construct stably expressed at high levels in Chinese hamster ovary (CHO) cell lines. See text for details.

4.2 Focal probing with magnetic ligand-bearing microspheres

Magnetic microspheres with attached ligands for cell surface receptors are powerful tools for investigating local binding reactions, activation and downstream signaling[42,43]. Figure 12 shows two CHO cells stably expressing about $6{\cdot}10^5$ human erbB1- eGFP fusion proteins on the cell surface[42]. One cell has bound magnetic microspheres to which EGF was coupled covalently. The microspheres were added to the cells placed over a magnet for 5 minutes at room temperature, after which the cells were subsequently washed and fixed in -20°C MeOH. Activated erbB1 can be monitored via the specific phosphorylation of tyrosine residues in the cytoplasmic domain of the receptor. Staining with a monoclonal antibody to phosphotyrosine and Cy3 labelled GAMIG $(Fab)_2$ revealed localized activation (left panel of Figure 12) and the recruitment and clustering of the activated erbB1 receptors around the microspheres (eGFP signal, right panel). Adaptor proteins of the downstream signaling pathway have been shown to be recruited to such sites in a time dependent manner[42,43]. Unliganded microspheres neither bind specifically nor cause activation.

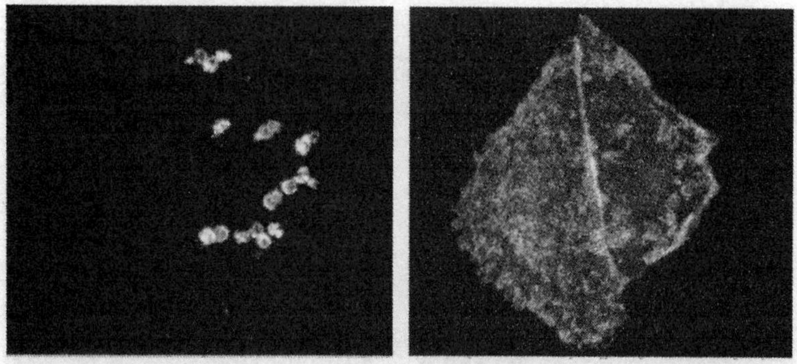

Figure 12. CHO-erbB1-eGFP cells were grown in Dulbecco's DMEM + 10% FCS. Cells were starved for 24 h before reacting with EGF-coupled microspheres in Ringer's physiological buffer containing 20 mM glucose, 0.1 % BSA. After fixation and immunostaining against activated erbB1 with anti-phosphotyrosine (pY), imaging was perfomed on a Zeiss 310 CLSM with a 63×, 1.4 NA Apochromat objective. eGFP excitation at 488 nm and emission at 520±25 nm; Cy3: excitation at 532 nm, emission at 585 nm. Confocal Z sections at 0.5 μm; displayed are extended focus images of 16 sections; left: anti-PY; right: eGFP.

This form of localized activation has been demonstrated on human, mouse and hamster cells. However, our results are in contrast to studies using non-magnetic beads from another group[44] reporting extensive propagation of the signal throughout the plasma membrane. The use of liganded magnetic microspheres to investigate localized activation, internalization or inhibition has been extended in our laboratory to include the use of coupled inhibitors and antibodies to cell surface receptors.

ACKNOWLEDGMENTS

E.A.J.-E. and T.M.J. are recipients of a joint grant (I/77 897) from the Volkswagen Foundation. E.A.J.-E. is indebted for financial support to the Agencia Nacional de Promoción de la Ciencia y Tecnología (ANPCyT), Fundación Antorchas, Consejo Nacional de Investigaciones Científicas y Técnicas (CONICET), Secretaría de Ciencia, Tecnología e Innovación Productiva (SECyT) and the Universidad de Buenos Aires (UBA). The laboratory of T.M.J. was supported by the Max Planck Society, European Union FP5 Projects QLG1-2000-01260 and QLG2-CT-2001-02278, and the Center of the Molecular Physiology of the Brain funded by the German Research Council (DFG). J.N.P. was supported by a grant from the German Research Council to D.J.A.-J.

REFERENCES

1. Yarden, Y., and Slikowski, M.X., 2001, Untangling the ErbB signalling network. *Nat. Rev. Mol. Cell Biol.* **2**: 127-137.
2. Clynes, R.A., Towers, T.L., Presta, L.G., and Ravetch, J.V., 2000, Inhibitory Fc receptors modulate *in vivo* cytotoxicity against tumor targets. *Nature Med.* **6**: 443-446.
3. Prasher, D.C., Eckenrode, V.K., Ward, W.W., Prendergast, F.G., and Cormier, M.J., 1992, Primary structure of the *Aequorea-victoria* green fluorescent protein. *Gene* **111**: 229-233.
4. Zhang, J., Campbell, R.E., Ting, A.Y., and Tsien, R.Y., 2002, Creating new fluorescent probes for cell biology. *Nat. Rev. Mol. Cell Biol.* **3**: 906-918.
5. Lippincott-Schwartz, J., and Patterson, G.H., 2003, Development and use of fluorescent protein markers in living cells. *Science* **300**: 87-91.
6. Miyawaki, A., 2003, Visualization of the spatial and temporal dynamics of intracellular signaling. *Dev. Cell* **4**: 295-305.
7. Sato, M., Ozawa, T., Inukai, K., Asano, T., and Umezawa, Y., 2002, Fluorescent indicators for imaging protein phosphorylation in single living cells. *Nat. Biotechnol.* **20**: 287-294.

8. Zacharias, D.A., Violin, J.D., Newton, A.C., and Tsien, R.Y., 2002, Partitioning of lipid-modified monomeric GFPs into membrane microdomains of live cells. *Science* **296**: 913-916.
9. Zeytun, A., Jeromin, Scalettar, B.A., Waldo, G.S., and Bradbury, A.R.M., 2003, Fluorobodies combine GFP fluorescence with the binding characteristics of antibodies. *Nat. Biotechnol.* **21**: 1473-1479.
10. Kurokawa, K., Mochizuki, N., Ohba, Y., Mizuno, H., Miyawaki, A., and Matsuda, M., 2001, A pair of fluorescent resonance energy transfer-based probes for tyrosine phosphorylation of the CrkII adaptor protein *in vivo*. *J. Biol. Chem.* **276**: 31305-31310.
11. Wiedenmann, J., Schenk, A., Röcker, C., Girod, A., Spindler, K.-D., and Nienhaus, G.U., 2002, A far-red fluorescent protein with fast maturation and reduced oligomerization tendency from *Entacmea quadricolor* (Anthoza, Actinaria). *Proc. Nat. Acad. Sci. U.S.A.* **99**: 11646-11651.
12. Patterson, G.H., and Lippincott-Schwartz, J., 2002, A photoactivatable GFP for selective photolabeling of proteins and cells. *Science* **297**: 1873-1877.
13. Haker, A., Hendriks, J., van Stokkum, I.H.M., Heberle, J., Hellingwerf, K.J., Crielaard, W., and Genach, T., 2003, Two photocycles of photoactive yellow protein from *Rhodobacter sphaeroides*. *J. Biol. Chem.* **278**: 8442-8451.
14. Nagai, T., Ibata, K., Park, E.S., Kubota, M., Mikoshiba, K., and Miyawaki, A., 2002, A variant of yellow fluorescent protein with fast and efficient maturation for cell-biological applications. *Nat. Biotechnol.* **20**: 87-90.
15. Griesbeck, O., Baird, G.S., Campbell, R.E., Zacharias, D.A., and Tsien, R.Y., 2001, Reducing the environmental sensitivity of yellow fluorescent protein. *J. Biol. Chem.* **276**: 29188-29194.
16. Campbell, R.E., Tour, O., Palmer, A.E., Steinbach, P.A., Baird, G.S., Zacharias, D.A., and Tsien, R.Y., 2002, A monomeric red fluorescent protein. *Proc. Nat. Acad. Sci. U.S.A.* **99**: 7877-7882.
17. Miyawaki, A., Llopis, J., Heim, R., McCaffery, J.M., Adams, J.A., Ikura, M., and Tsien, R.Y., 1997, Fluorescent indicators for Ca^{2+} based on green fluorescent proteins and calmodulin. *Nature* **388**: 882-887.
18. Jares-Erijman, E.A., and Jovin, T.M., 2003, FRET imaging. *Nat. Biotechnol.* **21**: 1387-1395.
19. Clegg, R.M., 1995, Fluorescence resonance energy transfer. *Curr. Opin. Biotechn.* **6**: 103-110.
20. Clegg, R.M., Gadella Jr., T.W.J., and Jovin, T.M., 1994, Lifetime-resolved fluorescence imaging. *Proc. SPIE* **2137**: 105-118.
21. Patterson, G.H., Piston, D.W., and Barisas, B.G., 2000, Forster distances between green fluorescent protein pairs. *Anal. Biochem.* **284**: 438-440.
22. Hu, C.D., and Kerppola, T.K., 2003, Simultaneous visualization of multiple protein interactions in living cells using multicolor fluorescence complementation analysis. *Nat. Biotechnol.* **21**: 539-545.
23. Ozawa, T., and Umezawa, Y., 2002, Peptide assemblies in living cells. Methods for detecting protein-protein interactions. *Supramol. Chem.* **14**: 271-280.
24. Marriott, G., and Parker, I., eds., 2003, *Methods Enzymol. Biophotonics, Part B*. **361,** Academic Press, San Diego, CA.
25. Riven, I., Kalmanzon, E., Segev, L., and Reuveny, E., 2003, Conformational rearrangements associated with the gating of the G protein-coupled potassium channel revealed. *Neuron* **38**: 225-235.
26. Griffin, B.A., Adams, S.R., Jones, J., and Tsien, R.Y., 2000, Fluorescent labeling of recombinant proteins in living cells with FlAsH. *Methods Enzymol.* **327**: 565-578.

27. Gaietta, G., Deerinck, T.J., Adams, S.R., Bouwer, J., Tour, O., Laird, D.W., Sosinsky, G.E., Tsien, R.Y., and Ellisman, M.H., 2002, Multicolor and electron microscopic imaging of connexin trafficking. *Science* **296**: 503-507.
28. Falk, M.M., 2002, Genetic tags for labelling live cells: gap junctions and beyond. *Trends Cell Biol.* **12**: 399-404.
29. Stroffekova, K., Proenza, C., and Beam, K.G., 2001, The protein-labelling reagent FlAsH·EDT_2 binds not only to CCXXCC motifs but also non-specifically to endogenous cysteine-rich proteins. *Eur. J. Physiol.* **442**: 859-866.
30. Giordano, L., Jovin, T.M., Irie, M., and Jares-Erijman, E.A., 2002, Diheteroarylethenes as thermally stable photoswitchable acceptors in photochromic fluorescence resonance energy transfer (pcFRET). *J. Am. Chem. Soc.* **124**: 7481-7489.
31. Song, L., Jares-Erijman, E.A., and Jovin, T.M., 2002, A photochromic acceptor as a reversible light-driven switch in fluorescence resonance energy transfer (FRET). *J. Photochem. Photobiol. A* **150**: 177-185.
32. Clayton, A.H.A., Hanley, Q.S., Arndt-Jovin, D.J., Subramaniam, V., and Jovin, T.M., 2002, Dynamic fluorescence anisotropy imaging microscopy in the frequency domain (rFLIM). *Biophys. J.* **83**: 1631-1649.
33. Lidke, D.S., Nagy, P., Barisas, B.G., Heintzmann, R., Post, J.N., Lidke, K.A., Clayton, A.H.A., Arndt-Jovin, D.J., and Jovin, T.M., 2003, Imaging molecular interactions in cells by dynamic and static fluorescence anisotropy (rFLIM and emFRET). *Biochem. Soc. Trans.* **31**: 1020-1027.
34. Schlessinger, J., 2002, Ligand-induced, receptor-mediated dimerization and activation of EGF receptor. *Cell* **110**: 669-672.
35. Gadella Jr., T.W.J., and Jovin, T.M., 1995, Oligomerization of epidermal growth factor receptors on A431 cells studied by time-resolved fluorescence imaging microscopy: A stereochemical model for tyrosine kinase receptor activation. *J. Cell Biol.* **129**: 1543-1558.
36. Wu, X., Liu, H., Liu, J., Haley, K.N., Treadway, J.A., Larson, J.P., Ge, N., Peale, F., and Bruchez, M.P., 2003, Immunofluorescent labeling of cancer marker Her2 and other cellular targets with semiconductor quantum dots. *Nat. Biotechnol.* **21**: 41-46.
37. Jaiswal, J.K., Mattoussi, H., Mauro, J.M., and Simon, S.M., 2003, Long-term multiple color imaging of live cells using quantum dot bioconjugates. *Nat. Biotechnol.* **21**: 47-51.
38. Larson, D.R., Zipfel, W.R., Williams, R.M., Clark, S.W., Bruchez, M.P., Wise, F.W., and Webb, W.W., 2003, Water-soluble quantum dots for multiphoton fluorescence imaging *in vivo*. *Science* **300**: 1434-1436.
39. Lidke, D.S., Nagy, P., Heintzmann, R., Arndt-Jovin, D.J., Post, J.N., Grecco, H., Jares-Erijman, E.A., and Jovin, T.M., 2004, Quantum dot ligands provide new insights into erbB/HER receptor-mediated signal transduction. *Nat. Biotechnol.*, in press (DOI: 10.1038/Nbt1929).
40. Medintz, I.L., Trammell, S.A., Mattoussi, H., and Mauro, J.M., 2004, Reversible modulation of quantum dot photoluminescence using a protein-bound photochromic fluorescence resonance energy transfer acceptor. *J. Am. Chem. Soc.* **126**: 30-31.
41. Clapp, A.R., Medintz, I.L., Mauro, J.M., Fisher, B.R., Bawendi, M.G., and Mattoussi, H., 2004, Fluorescence Resonance Energy Transfer between Quantum Dot donors and dye-labeled protein acceptors. *J. Am. Chem. Soc.* **126**: 301-310.
42. Brock, R., and Jovin, T., 2003, Quantitative image analysis of cellular protein translocation induced by magnetic microspheres: application to the EGF receptor. *Cytometry A* **52A**: 1-11.

43. Brock, R., and Jovin, T.M., 2001, Heterogeneity of signal transduction at the subcellular level: microsphere-based focal EGF receptor activation and stimulation of Shc translocation. *J. Cell Sci.* **114**: 2437-2447.

44. Verveer, P.J., Wouters, F.S., Reynolds, A.R., and Bastiaens, P.I.H., 2000, Quantitative imaging of lateral ErbB1 receptor signal propagation in the plasma membrane. *Science* **290**: 1567-1570.

Observing Structure and Dynamics of Membrane Proteins by High-resolution Microscopy

ANDREAS ENGEL
M.E. Müller Institute for Structural Biology, Biozentrum, University of Basel
Klingelbergstrasse 50-70, CH - 4056 Basel, Switzerland

1. INTRODUCTION

Biological membranes fulfill many vital functions as interfaces to the outside world, as interfaces between cells, and as boundaries of intracellular compartments. Thus, biological membranes are related to numerous diseases such as hyperinsulinemia, nephrogenic diabetes insipidus, congestive heart failure, liver cirrhosis, cystic fibrosis, hyper- and hypotension, lung edema, epilepsy, retinitis pigmentosa and cataract. About 40% of the sequenced genes appear to code for membrane proteins, of which only 50 crystal structures are available and among them less than ten of mammalian proteins, compared to at least 5,000 unique structures of soluble proteins.

Progress in 3D crystallization of membrane proteins has recently improved[1,2], leading to several new membrane protein structures[3-13]. The strength of X-ray crystallography is the established technology that allows data to be collected and structures to be solved to high resolution with enormous efficiency. Nevertheless, this route to establish the atomic structure of a membrane protein is still risky as result of the crystallization bottleneck. Solution NMR is the other established method for atomic structure determination. It does not require 3D crystals and allows the dynamics of a protein to be measured. Progress towards assessing the structure of large complexes[14] or membrane proteins has been reported[15]. Difficulties with the stability of solubilized membrane proteins, however, can be a problem.

Supramolecular Structure and Function 8, Edited by Pifat-Mrzljak
Kluwer Academic/Plenum Publishers, New York 2004

To observe membrane proteins at work they must reside in their native environment, i.e. in a physiological buffer and embedded in a lipid bilayer. Therefore, a powerful alternative to determine the structure of a membrane protein is its reconstitution into two-dimensional (2D) crystals in the presence of lipids[16]. This approach restores the native environment of membrane proteins and thus their biological activity. Cryo-electron microscopy then allows the assessment of the static 3D structure of 2D crystals at atomic resolution[17]. The atomic force microscope (AFM) on the other hand is the only instrument that provides sub-nanometer spatial resolution when operated in buffer solution[18]. The outstanding signal-to-noise (S/N) ratio provided by the AFM makes the observation of single protein conformations and their dynamics possible[19]. In addition, the AFM stylus is a nanotool that allows single molecules to be manipulated[20].

This review discusses the methods to assemble 2D crystals and to determine their 3D structure.

2. 2D CRYSTALLIZATION OF MEMBRANE PROTEINS BY RECONSTITUTION INTO LIPID BILAYERS

In contrast to 3D crystallization the membrane proteins are reconstituted into a native-like environment during 2D crystallization[21-23]. To produce high-quality 2D crystals, the protein of interest must be purified away from other proteins and contaminants as in 3D crystallization. This involves solubilization of the original membrane with detergent and commonly requires several subsequent separation steps. Finally the pure protein is obtained in a detergent solution, often with residual lipids. In fact, the latter often contribute to the stability of the membrane protein and may be essential for successful 2D crystallization[24]. The nature of the detergent in both, protein purification and subsequent crystallization trials is a critical determinant of success.

Mixing lipids and protein, both solubilized in detergents and decreasing the detergent concentration by (i) dialysis, (ii) Biobeads, or (iii) dilution achieves the reconstitution of membrane proteins into bilayers (Fig. 1). During this process, the small micellar structures coalesce into larger structures leading to the formation of vesicles and sheets. Both structures consist of lipid bilayers containing varying amounts of protein. At the start of a typical reconstitution experiment, an excess of detergent ensures a homogenous distribution of protein and lipid in micelles. As the detergent concentration is decreased, lipid and protein interact due to the exposure of their hydrophobic surfaces. With an excess of lipid over protein, the protein is mainly incorporated into lipid bilayers, similar to its native state. In an excess of protein over lipid, some of the protein aggregates, likely in a denatured form. An important parameter is therefore the lipid-to-protein ratio (LPR), which should be low enough to promote crystal contacts between protein molecules, but not so low that the protein is lost to aggregation.

When the membrane protein is reconstituted from a mixture of solubilized components, crystal ordering of proteins is facilitated by many factors. As mentioned above, the LPR of the reconstitution experiment is critical for crystal packing during reconstitution. As this parameter is quite unpredictable, it has to be optimized by carefully designed reconstitution series. While the lipid content of the reconstitution mixture is in general a well-controlled parameter, the content of monodisperse protein is sometimes unknown, because protein assays do not indicate the amount of aggregates. Therefore, it is advisable to determine the fraction of aggregated protein in a given protein batch by, for example, negative stain electron microscopy, ultracentrifugation or light scattering. Alternatively, it has also been possible to reconstitute with an excess of lipids and then improve crystal packing by mild digestion of the lipids with phospholipase A2[25,26].

The lipid mixture used for reconstitution has an influence on the crystallization results. Crystallization is more likely to occur when the lipid bilayer is in the fluid phase and thus warrants some lateral mobility of the inserted membrane proteins. While saturated lipids are chemically more stable and preferred, unsaturated lipids such as those from *E. coli* have often been successfully used to produce highly ordered crystals[27]. A good compromise is dimyristoyl-phosphatidylcholine (DMPC), a lipid frequently employed with success, which has saturated fatty acids but has a phase transition temperature close to room temperature (23°C). Native lipids are often ideal in terms of stability and transition temperatures, and they also provide mixtures of head group charges and molecular geometries similar to membranes the protein originated from. Since synthetic lipids, *E. coli* lipids, soybean lecithin and egg lecithin have all been successfully used for 2D crystallization, no general recommendations can be made on which lipid or lipid mixture is most suitable for any one particular membrane protein.

The pertinent interactions depend on the shape and surface charges of the components. For a given protein, the lipid-detergent mixture, pH, counter ions and temperature must all be optimized. In addition, the concentration, the ratio of the respective components and the detergent removal rate are critical. The latter is dictated by the physico-chemical properties of the detergent and the method of detergent removal (see Fig. 1). Detergents with a high critical micellar concentration (cmc) can efficiently be removed by dialysis, the most successful crystallization method[27]. Low cmc detergents

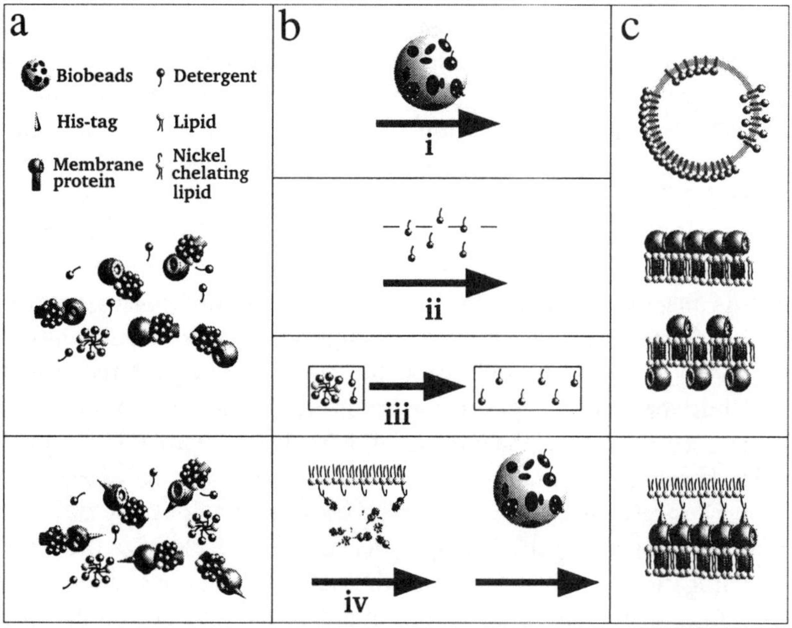

Figure 1. 2D crystallization by the detergent removal principle. (a) Starting conditions: detergent solubilized proteins are mixed with lipid/detergent micelles. (b) Detergent removal using (i) Biobeads[23], (ii) dialysis [21,22], (iii) dilution[28,29], and (iv) Biobeads combined with a monolayer at the air-water interface, which binds His-tagged proteins [30]. (c) The resulting 2D crystals can exist as a vesicular or a sheet-like assembly, or they are trapped at the air-water interface (by courtesy of Werten et al. [31]).

can be eliminated by adsorption to Biobeads, the other frequently used method[23]. More recently, experiments have shown that dilution of the protein, lipid, detergent mixture induces crystallization of membrane proteins[29]. Finally, 2D crystallization can also be promoted by concentrating the solubilized membrane protein at a lipid monolayer spread at the air-water interface, exploiting specific interactions between protein and lipid[30]. Altogether this gives a multidimensional parameter space that needs to be experimentally sampled, a similar task to that carried out in 3D crystallization screens. The difficulty of such experiments is the management of the screens and the assessment of results. With 2D crystallization the latter is particularly cumbersome because 2D crystals cannot be detected by light microscopy and screening by electron microscopy is time consuming.

3. ATOMIC FORCE MICROSCOPY

The AFM records topographs by raster scanning a sample below a sharp stylus that is attached to a flexible cantilever. During scanning a servo displaces the sample vertically to keep the cantilever deflection constant. An optical system resolves cantilever deflections of 0.1 nm, which corresponds to a force difference of typically 10-50 pN[19]. With modern instruments stable contact mode operation (see Fig. 2) is possible at forces of some 50 pN, provided the sample is in an aqueous solution. Various ways to exploit the deflection signal yield quite different types of images, as illustrated and explained in Fig. 2a. Figure 2b displays the experimental details of an AFM, which allows scanning a sample in buffer solution, and is equipped with an efficient system for exchanging the buffer.

3.1 Imaging and nanomanipulation

To achieve high resolution, topographs are recorded in buffer solution. In a simplified model, electrostatic and van der Waals forces govern the tip sample interactions in aqueous solutions. Hydrophilic surfaces are charged in water, leading to long-range electrostatic interactions. They can be attractive or repulsive, depending on the surface charges, which depend on the pH. Screening the surface charges between tip and sample with electrolytes allows the electrostatic interactions to be controlled. Since the stylus (silicon nitride, Si_3N_4) is negatively charged at neutral pH, and protein layers are often negatively charged as well, the electrostatic forces are frequently repulsive. In biological systems, van der Waals interactions do not depend on the ionic strength, decay rapidly, and are always attractive. The DLVO (Derjaguin, Landau, Verwey, Overbeek) theory describes these forces quantitatively[32] and allows the interactions between a spherical tip and a planar sample to be modeled, providing clues to optimize the recording conditions[33].

While suppliers specify tip radii of 10-50 nm, topographs of flat biological surfaces that exhibit a resolution of 1 nm or better have been acquired routinely. Therefore, the tips employed most likely had a single nm-sized asperity that protruded sufficiently to contour the finest surface structures. Such a small asperity exerts a prohibitively high pressure on the underlying structure, inducing its deformation. However, electrolytes can be used to adjust the tip-sample interactions, provided that the electrostatic forces are repulsive. The tip then surfs on a cushion of electrostatic repulsion while the small asperity is in contact with the sample. Fig. 3 illustrates this situation, and it shows the changes of repulsive forces

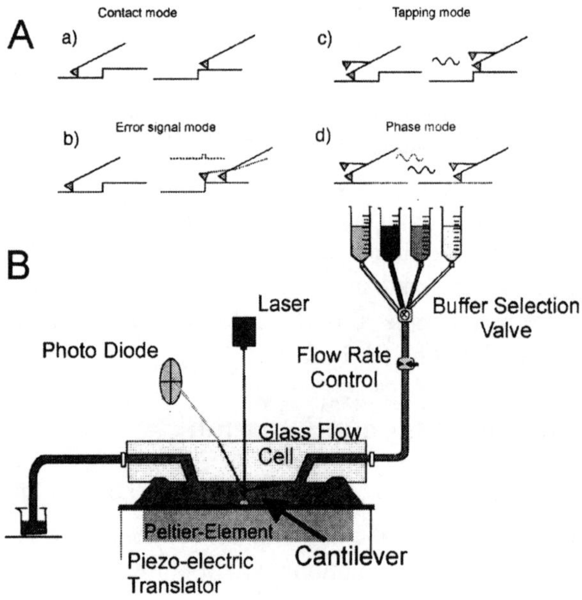

Figure 2. AFM Instrumentation and modes of operation. A) Imaging modes: In the contact mode (a) the servo system moves the sample down (with respect to the tip) over elevations and up over depressions while it is raster scanned below the fixed stylus. To react the servo needs a difference signal to keep the cantilever deflection constant at sharp edges (b). This error signal provides an image that reveals the edges of the surface topography. Approaching the sample with an oscillating cantilever results in tapping of the sample by the stylus (c). This reduces the oscillation amplitude providing the signal to activate the servo. Because the tip-sample contact is disrupted periodically, the friction forces are eliminated. The phase difference between the measured oscillation (solid wave) and the oscillation driving the cantilever depends on the mechanical properties of the sample (d). This phase signal thus produces a sensitive material contrast. B) Key elements of an AFM are the cantilever with a pyramidal stylus that touches the sample, an optical lever consisting of a laser and a photo diode to measure the cantilever's deflection, a piezo-electric translator to displace the sample in x, y, z, and a computer to control these movements and store the surface contours. The instrument is working in buffer solution under ambient condition. The springboard type cantilever (dimensions ~200 X 20 X 0.5 μm, spring constant 0.1-0.01 N/m) is deflected upwards when the tip is pushed towards the sample surface (repulsive forces) or downwards when the tip is retracted from the sample surface (attractive forces). Different liquids can be injected and a Peltier-element allows precise adjustment of the temperature (by courtesy of Hegner & Engel [34]).

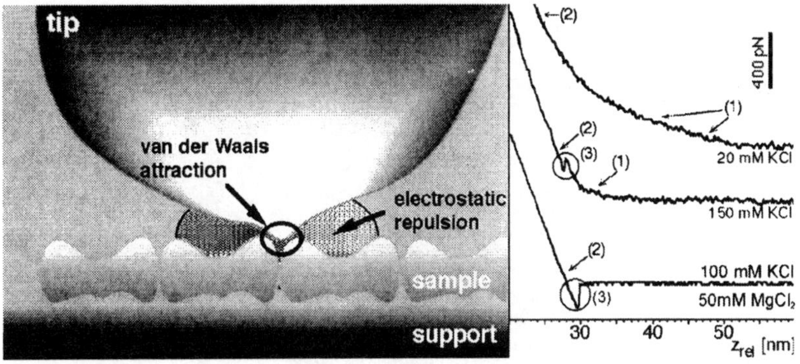

Figure 3. DLVO Forces: Force-distance-curves recorded on the extracellular purple membrane surface. The data was obtained for different electrolyte concentrations at constant pH (7.6). Force curves were recorded during the approach of sample and AFM tip. Arrows (1) mark the onset of measurable electrostatic repulsion, whereas arrows (2) indicate the point of contact between tip and sample. The snap-in (3) is the result of van der Waals attraction (by courtesy of D.J. Müller).

between the stylus and the sample resulting from changes of the ionic strength[33].

By operating commercial AFMs under such optimal recording conditions, the surfaces of bimolecular are contoured routinely at a lateral resolution better than 1 nm and a vertical resolution around 0.1 nm[16,35-45]. Forces need to be minimized to obtain high-resolution images, whether the AFM is operated in the contact mode or the tapping mode. However, the stylus may be used as nanoscalpel to dissect supramolecular assemblies. In this case, the force applied to the tip is increased to 1-10 nN, depending on the nanodissection to be achieved[20]. Quite small forces (typically 1 nN) are sufficient to separate stacked layers of membranes or 2D crystals[38,46]. Even smaller forces and repeated scanning at high magnification suffice to push away extrinsic proteins that are specifically complexed to an integral membrane protein[37] (Fig. 4). A major advantage of the AFM is the excellent S/N ratio, which has allowed high-resolution imaging of native membranes[47,48].

3.2 Single molecule force spectroscopy

The sensitivity of the cantilever deflection detector has promoted force measurements with the AFM. Unfolding forces of biomolecules that are tethered to support and tip can be monitored while retracting the tip[49-51]. To record such force-distance curves, the tip is approached vertically

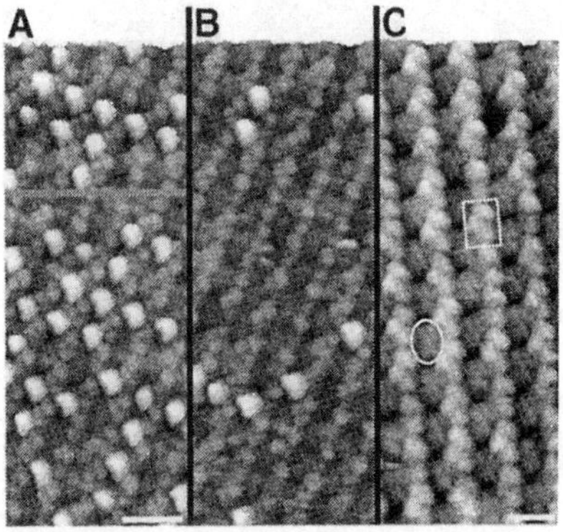

Figure 4. Topograph of a 2D photosystem I (PSI) crystal. The up-and-down orientation of the PSI complexes which give the crystal a p12$_1$2 symmetry is clearly visible. The high protrusions arise from the extrinsic subunits PsaC, PsaD and PsaE located on the stromal side. **B**, The water-soluble, extrinsic subunits disappear upon repetitive scanning of the same area. **C**, At higher magnification, the stromal surface depleted of the extrinsic subunits unveils the topography of the reaction center core (ellipse). The lumenal surface is marked by a rectangle. Scale bars represent 25 nm (**A** and **B**) and 15 nm (**C**). The contact mode topographs were recorded in buffer solution (300 mM KCl, 20 mM Tris-HCl, pH 7.8), exhibit vertical full gray level ranges of 4 nm (**A** and **B**) and 2 nm (**C**), and are displayed as reliefs tilted by 5° (by courtesy of Fotiadis et al. [20]).

towards the sample until a molecule is attached to the tip, and subsequently retracted ('fly-fishing'). Attachment is either due to an unspecific hydrophobic interaction between tip and several exposed residues of the protein or by specific covalent interaction of a cysteine with a gold coated tip. Acquisition of high-resolution images before and after 'unzipping' a biomolecule allows the defect produced to be directly visualized[52] (Fig. 5). Only specialized AFMs can resolve forces below 20 pN, which is still too large for the direct monitoring of molecular motors, whose forces are in the range of 1 – 10 pN. Optical tweezers are the tools of choice for such measurements [34].

4. ELECTRON CRYSTALLOGRAPHY

Transmission electron microscopy (EM) has progressed steadily, and modern instruments equipped with a field-emission gun are now available that efficiently transfer the atomic scale structural information from the sample to the image, which represents a projection of the 3D potential distribution of the object[17]. Meaningful information, however, can only be

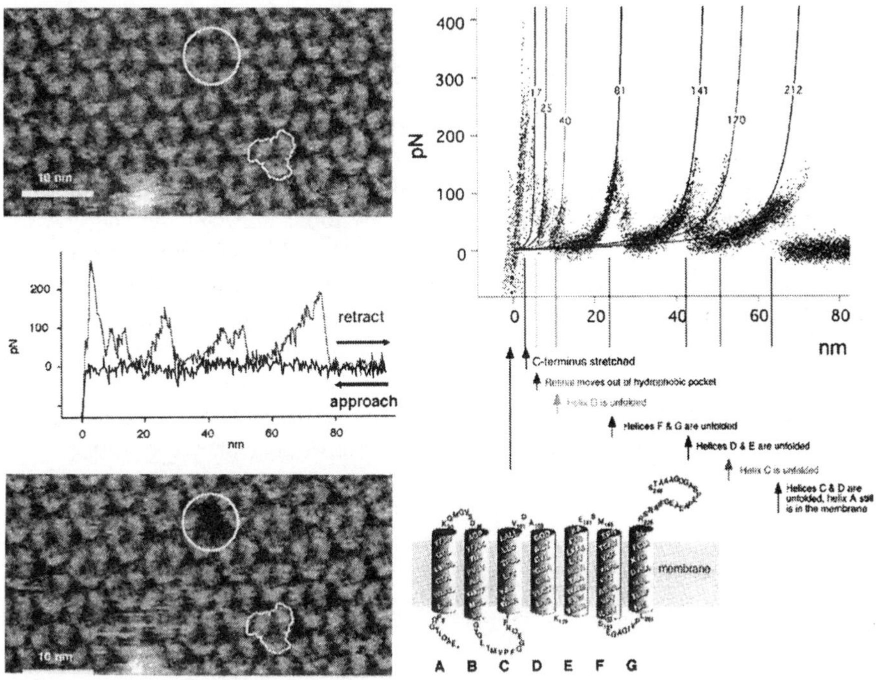

Figure 5. High-resolution topograph of the cytosolic purple membrane surface before (top left) and after (bottom left) extraction of a single bacteriorhodopsin molecule (circle). The forces required to unfold the membrane bound protein reflect its 7 transmembrane helical topology (middle left and diagram bottom right). A series of force-extension curves reveal overall reproducibility and fitting of the worm-like-chain model (top right) (by courtesy of D.J. Müller).

extracted when the sample is structurally preserved in spite of the vacuum within the electron optical system. Suitable methods produce biomolecules embedded in a thin vitrified water layer[53], or embedded in a layer of partially dried and frozen sugar[54]. Since all these samples are highly sensitive to the electron beam, images are recorded at low electron doses. High-resolution data of proteins are recorded at doses below 5 electrons/Å^2, with the sample kept at liquid nitrogen temperature (77 Kelvin) [53], or at doses below 20 electrons/Å^2 when the sample is kept at liquid helium temperature (4.3 Kelvin) [55]. Such recording doses produce inherently noisy images, whose information must be extracted by image processing.

2D crystals allow the signal of membrane proteins to be extracted by crystallographic methods. Using this approach, the atomic structure of bacteriorhodopsin (bR) was first determined[56], and subsequently refined using field emission gun microscopes while keeping the sample at 4.3 Kelvin[57]. Two other atomic structures of membrane proteins were solved using similar instruments: the light-harvesting complex of plants[58] and aquaporin-1 (AQP1), the first and only human membrane channel solved to atomic resolution up to now[59].

Interpreting a cryo-EM potential map in terms of an atomic protein structure typically still is a challenging task, even when relatively high-resolution data of better than 4 Å are available. The first step is to determine the macromolecular fold and to trace the backbone through the map. A number of approaches have been applied to tackle this problem. First, the backbone trace (and thereby the three-dimensional fold) can be derived by visual inspection. Further, structural clues can be obtained from bioinformatics methods: Extensive sequence alignments and the analysis of correlated mutations led to valuable insights in the case of AQP1[60]. Finally, unambiguously derived. Once the fold has been determined, an initial model of the backbone structure can be generated. Depending on the resolution, what follows are many rounds of manual model (re)building and

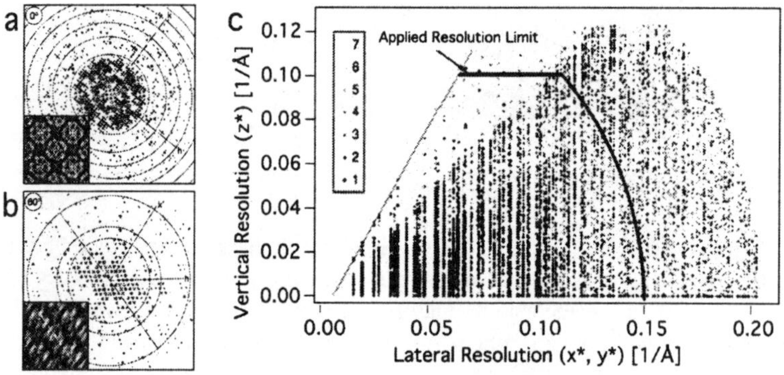

Figure 6. Data collection and processing to build the 6.9 Å potential map of the glycerol facilitator GlpF. a) The projection map of GlpF tetramers recorded at a sample-tilt of 0 degrees together with the calculated Fourier transform. b) The projection map of GlpF tetramers recorded at a sample-tilt of 60 degrees together with the calculated Fourier transform. c) The merged data set used to establish the potential map at 6.9 Å resolution. The data density is low between 45 and 60 degrees, but the information extends beyond the applied resolution limit (by courtesy of H. Stahlberg and T. Braun).

refinement. The structures of bR[62] and AQP1[59,63] have been solved by EM and were later confirmed by X-ray crystallography[13,64]. These successes demonstrate that even at a resolution of around 3.5 Å, a cryo-EM map

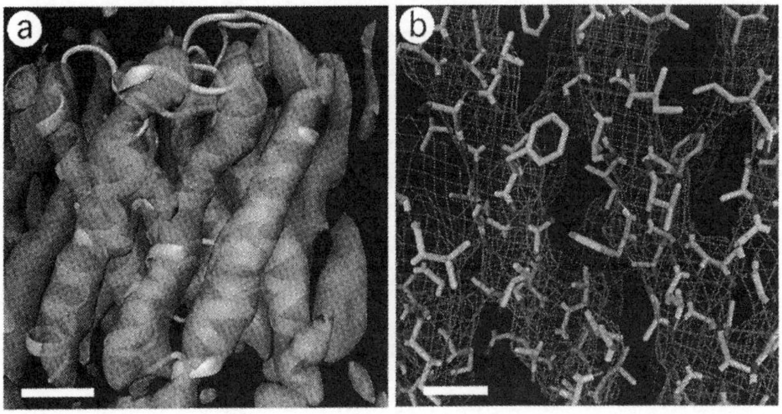

Figure 7. Progress in the determination of the AQP1 atomic model a) At 6 Å resolution the secondary structure becomes visible. b) At 4.5 Å resolution the peptide backbone can be placed, the side chains, however, are not discernible in the map. Scale bars represent 1 nm in a, and 0.5 nm in b.

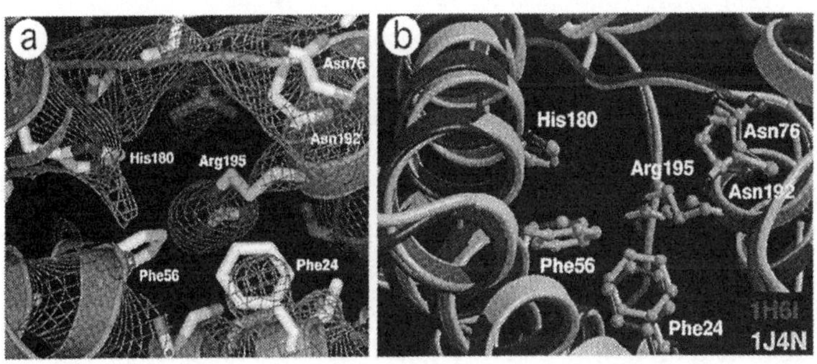

Figure 8. Atomic models of AQP1. a) The potential map of the channel region with the refined atomic model based on the 3.8 Å resolution data obtained by electron crystallography. At this resolution, the definition of the side chains becomes clear, but the detailed packing arrangement needs to be refined by statistical methods. b) Comparison of atomic models of AQP1 obtained by X-ray (yellow, bovine AQP1) and electron crystallography (red, human AQP1) reveals the close agreement of the independently determined structures.

contains sufficient structural information to uniquely define the atomic structure.

The practical challenge is to find this optimal structure in the high-dimensional search space. Since an exhaustive search is not feasible, there is no straightforward general method to arrive at the correct (or best possible) solution at this level of resolution. The available tools are for a substantial part the same as used in X-ray crystallography, but if used inappropriately, the usually lower resolution may lead to inaccurate interpretations of the data, slow convergence of the refinement, or poor model quality. In contrast to X-ray crystallography, experimental phase information is available in the case of EM. Unfortunately, this advantage is not as yet fully exploited. Therefore, the development of novel automated techniques specifically targeted at model building and refinement of EM data is mandatory.

5. CONCLUSIONS AND PERSPECTIVES

Microscopy techniques open important possibilities to assess the structure, dynamics and function of membrane proteins. The examples presented here intend to demonstrate that these techniques, although used and developed in relatively few laboratories, now reach a level that allows their routine application. Compared to the efforts invested for X-ray crystallography in sample preparation (i.e., high-throughput 3D crystallization using robots) and the costs of synchrotrons as well as NMR instruments, the investment in microscopic techniques is relatively small. Yet no other technique offers such a wide insight into the structure of living matter from the cellular down to the atomic level.

Here we have concentrated on discussing the high-resolution end, where significant progress has been made. The results indicate that further developments of microscopy-based methods are important and promising. The AFM will provide information about the native organization of biological membranes, the dynamics of membrane proteins and their molecular interaction with ligands. Electron crystallography is expected to provide the atomic structure of membrane proteins that are not amenable to 3D crystallization. Ultimately, however, the same bottle necks as found in X-ray crystallography will limit the progress of electron crystallography, namely, the large scale expression of functional membrane proteins and the problem of their crystallization.

ACKNOWLEDGEMENTS

The work presented here was supported by the Swiss National Research Foundation, the M.E. Müller Foundation, the Swiss National Center of

Competence in Research (NCCR) 'Structural Biology' and the NCCR 'Nanoscale Science.' The author is indebted to his creative and motivated collaborators and students, who have all contributed to the work summarized here. Particular thanks are given to Dr. Dimitrios Fotiadis for excellent and constructive discussions.

REFERENCES

1. Hunte, C., Michel, H., 2002. Crystallisation of membrane proteins mediated by antibody fragments. *Curr Opin Struct Biol.* 12: 503.
2. Nollert, P., Qiu, H., Caffrey, M., Rosenbusch, J.P., Landau, E.M., 2001, Molecular mechanism for the crystallization of bacteriorhodopsin in lipidic cubic phases. *FEBS Lett.* **504**: 179-186.
3. Abramson, J., Smirnova, I., Kasho, V., Verner, G., Kaback, H.R., Iwata, S., 2003, Structure and mechanism of the lactose permease of Escherichia coli. *Science* **301**: 610-615.
4. Dutzler, R., Campbell, E.B., Cadene, M., Chait, B.T., MacKinnon, R., 2002, X-ray structure of a ClC chloride channel at 3.0 A reveals the molecular basis of anion selectivity. *Nature* **415**: 287-294.
5. Dutzler, R., Campbell, E.B., MacKinnon, R., 2003, Gating the selectivity filter in ClC chloride channels. *Science* **300**: 108-112.
6. Fu, D., Libson, A., Miercke, L.J., Weitzman, C., Nollert, P., Krucinski, J., Stroud, R.M., 2000, Structure of a glycerol-conducting channel and the basis for its selectivity. *Science* **290**: 481-486.
7. Gordeliy, V.I., Labahn, J., Moukhametzianov, R., Efremov, R., Granzin, J., Schlesinger, R., Buldt, G., Savopol, T., Scheidig, A.J., Klare, J.P., et al., 2002, Molecular basis of transmembrane signalling by sensory rhodopsin II-transducer complex. *Nature* **419**: 484-487.
8. Huang, Y., Lemieux, M.J., Song, J., Auer, M., Wang, D.N., 2003, Structure and mechanism of the glycerol-3-phosphate transporter from Escherichia coli. *Science* **301**: 616-620.
9. Jiang, Y., Lee, A., Chen, J., Ruta, V., Cadene, M., Chait, B.T., MacKinnon, R., 2003, X-ray structure of a voltage-dependent K+ channel. *Nature* **423**: 33-41.
10. Kuo, A., Gulbis, J.M., Antcliff, J.F., Rahman, T., Lowe, E.D., Zimmer, J., Cuthbertson, J., Ashcroft, F.M., Ezaki, T., Doyle, D.A., 2003, Crystal structure of the potassium channel KirBac1.1 in the closed state. *Science* **300**: 1922-1926.
11. Locher, K.P., Lee, A.T., Rees, D.C., 2002, The E. coli BtuCD structure: a framework for ABC transporter architecture and mechanism. *Science* **296**: 1091-1098.
12. Palczewski, K., Kumasaka, T., Hori, T., Behnke, C.A., Motoshima, H., Fox, B.A., Le Trong, I., Teller, D.C., Okada, T., Stenkamp, R.E., et al., 2000, Crystal structure of rhodopsin: A G protein-coupled receptor. *Science* **289**: 739-745.
13. Sui, H., Han, B.G., Lee, J.K., Walian, P., Jap, B.K., 2001, Structural basis of water-specific transport through the AQP1 water channel. *Nature* **414**: 872-878.
14. Fiaux, J., Bertelsen, E.B., Horwich, A.L., Wüthrich, K., 2002, NMR analysis of a 900K GroEL GroES complex. *Nature* **418**: 207-211.
15. Fernandez, C., Wuthrich, K., 2003, NMR solution structure determination of membrane proteins reconstituted in detergent micelles. *FEBS Lett.* **555**: 144-150.
16. Stahlberg, H., Fotiadis, D., Scheuring, S., Remigy, H., Braun, T., Mitsuoka, K., Fujiyoshi, Y., Engel, A., 2001, Two-dimensional crystals: a powerful approach to assess structure, function and dynamics of membrane proteins. *FEBS Lett.* **504**: 166-172.

17. Fujiyoshi, Y., 1998. The structural study of membrane proteins by electron crystallography. *Adv Biophys.* **35**: 25-80.
18. Müller, D.J., Engel, A., 2002, Conformations, flexibility, and interactions observed on individual membrane proteins by atomic force microscopy. *Methods Cell Biol.* **68**: 257-299.
19. Engel, A., Muller, D.J., 2000, Observing single biomolecules at work with the atomic force microscope. *Nat Struct Biol.* **7**: 715-718.
20. Fotiadis, D., Scheuring, S., Muller, S.A., Engel, A., Muller, D.J., 2002, Imaging and manipulation of biological structures with the AFM. *Micron* **33**: 385-397.
21. Jap, B.K., Zulauf, M., Scheybani, T., Hefti, A., Baumeister, W., Aebi, U., Engel, A., 1992, 2D crystallization: from art to science. *Ultramicroscopy* **46**: 45-84.
22. Kühlbrandt, W., 1992, Two-dimensional crystallization of membrane proteins. *Q Rev Biophys.* **25**: 1-49.
23. Rigaud, J.L., Mosser, G., Lacapere, J.J., Olofsson, A., Levy, D., Ranck, J.L., 1997, Bio-Beads: an efficient strategy for two-dimensional crystallization of membrane proteins. *Journal of Structural Biology* **118**: 226-235.
24. Zhuang, J., Privé, G.G., Werner, G.E., Ringler, P., Kaback, H.R., Engel, A., 1999, Two-dimensional crystallization of Escherichia coli lactose permease. *J Struct Biol.* **125**: 63-75.
25. Mannella, C.A., 1984, Phospholipase-induced crystallization of channels in mitochondrial outer membranes. *Science* **224**: 165-166.
26. Jap, B.K., Li, H., 1995, Structure of the osmo-regulated H2O-channel, AQP-CHIP, in projection at 3.5 A resolution. *J Mol Biol.* **251**: 413-420.
27. Ringler, P., Heymann, B., Engel, A., 2000, *Two-dimensional crystallization of membrane proteins*. Edited by Baldwin SA: Oxford University Press.
28. Dolder, M., Engel, A., Zulauf, M., 1996, The micelle to vesicle transition of lipids and detergents in the presence of a membrane protein: towards a rationale for 2D crystallization. *FEBS Letters* **382**: 203-208.
29. Rémigy, H., Caujolle-Bert, D., Suda, K., Schenk, A., Chami, M., Engel, A., 2003, Membrane protein reconstitution and crystallization by controlled dilution. *FEBS Lett.* **555**: 160-169.
30. Levy, D., Chami, M., Rigaud, J.L., 2001, Two-dimensional crystallization of membrane proteins: the lipid layer strategy. *FEBS Lett.* **504**: 187-193.
31. Werten, P.J., Remigy, H.W., de Groot, B.L., Fotiadis, D., Philippsen, A., Stahlberg, H., Grubmuller, H., Engel, A., 2002, Progress in the analysis of membrane protein structure and function. *FEBS Lett.* **529**: 65-72.
32. Israelachvili, J., 1991, *Intermolecular & surface forces* edn Second Edition. London: Academic Press Limited.
33. Müller, D.J., Fotiadis, D., Scheuring, S., Müller, S.A., Engel, A., 1999, Electrostatically balanced subnanometer imaging of biological specimens by atomic force microscope. *Biophys J.* **76**: 1101-1111.
34. Hegner, M., Engel, A., 2002, Single Molecule Imaging and Manipulation. *Chimia* **56**: 506–514.
35. Czajkowsky, D.M., Iwamoto, H., Cover, T.L., Shao, Z., 1999, The vacuolating toxin from Helicobacter pylori forms hexameric pores in lipid bilayers at low pH. *Proc Natl Acad Sci U S A* **96**: 2001-2006.
36. Czajkowsky, D.M., Shao, Z., 1998, Submolecular resolution of single macromolecules with atomic force microscopy. *FEBS Lett.* **430**: 51-54.
37. Fotiadis, D., Muller, D.J., Tsiotis, G., Hasler, L., Tittmann, P., Mini, T., Jeno, P., Gross, H., Engel, A., 1998, Surface analysis of the photosystem I complex by electron and atomic force microscopy. *J Mol Biol.* **283**: 83-94.

38. Fotiadis, D., Hasler, L., Muller, D.J., Stahlberg, H., Kistler, J., Engel, A., 2000, Surface tongue-and-groove contours on lens MIP facilitate cell-to-cell adherence. *J Mol Biol.* **300**: 779-789.
39. Fotiadis, D., Jeno, P., Mini, T., Wirtz, S., Müller, S.A., Fraysse, L., Kjellbom, P., Engel, A., 2001, Structural characterization of two aquaporins isolated from native spinach leaf plasma membranes. *J Biol Chem.* **276**: 1707-1714.
40. Mou, J., Czajkowsky, D.M., Sheng, S.J., Ho, R., Shao, Z., 2996, High resolution surface structure of E. coli GroES oligomer by atomic force microscopy. *FEBS Lett.* **381**: 161-164.
41. Mou, J., Czajkowsky, D.M., Zhang, Y., Shao, Z., 1995, High-resolution atomic-force microscopy of DNA: the pitch of the double helix. *FEBS Lett.* **371**: 279-282.
42. Müller, D.J., Engel, A., Carrascosa, J.L., Velez, M., 1997, The bacteriophage phi29 head-tail connector imaged at high resolution with the atomic force microscope in buffer solution. *Embo J.* **16**: 2547-2553.
43. Müller, D.J., Engel, A., 1999, Voltage and pH-induced channel closure of porin OmpF visualized by atomic force microscopy. *J Mol Biol.* **285**: 1347-1351.
44. Scheuring, S., Stahlberg, H., Chami, M., Houssin, C., Rigaud, J.L., Engel, A., 2002, Charting and unzipping the surface layer of Corynebacterium glutamicum with the atomic force microscope. *Mol Microbiol.* **44**: 675-684.
45. Scheuring, S, Muller, D.J., Stahlberg, H., Engel, H.A., Engel, A., 2002, Sampling the conformational space of membrane protein surfaces with the AFM. *Eur Biophys J.* **31**: 172-178.
46. Fotiadis, D., Qian, P., Philippsen, A., Bullough, P.A., Engel, A., Hunter, C.N., 2003, Structural analysis of the RC-LH1 photosynthetic core complex of Rhodospirillum rubrum using atomic force microscopy. *J Biol Chem.* in press.
47. Fotiadis, D., Liang, Y., Filipek, S., Saperstein, D.A., Engel, A., Palczewski, K., 2003, Atomic-force microscopy: Rhodopsin dimers in native disc membranes. *Nature* **421**: 127-128.
48. Liang, Y., Fotiadis, D., Filipek, S., Saperstein, D.A., Palczewski, K., Engel, A., 2003, Organization of the G protein-coupled receptors rhodopsin and opsin in native membranes. *J Biol Chem* **278**: 21655-21662.
49. Rief, M., Gautel, M., Oesterhelt, F., Fernandez, J.M., Gaub, H.E., 1997, Reversible unfolding of individual titin immunoglobulin domains by AFM. *Science* **276**: 1109-1112.
50. Rief, M., Oesterhelt, F., Heymann, B., Gaub, H.E., 1997, Single molecule force spectroscopy on polysaccharides by AFM. *Science* **275**: 1295-1298.
51. Rief, M., Gautel, M., Gaub, H.E., 2000, Unfolding forces of titin and fibronectin domains directly measured by AFM. *Adv. Exp. Med. Biol.* **481**: 129-136.
52. Oesterhelt, F., Oesterhelt, D., Pfeiffer, M., Engel, A., Gaub, H.E., Muller, D.J., 2000, Unfolding pathways of individual bacteriorhodopsins. *Science* **288**: 143-146.
53. Dubochet, J., Adrian, M., Chang, J.-J., Homo, J.-C., Lepault, J., McDowall, A.W., Schultz, P., 1988, Cryo-electron microscopy of vitrified specimens. *Quaterly Review of Biophysics* **21**: 129-228.
54. Unwin, P.N., Henderson, R., 1975, Molecular structure determination by electron microscopy of unstained crystalline specimens. *J Mol Biol.* **94**: 425-440.
55. Fujiyoshi, Y., Mizusaki, T., Morikawa, K., Yamagishi, H., Aoki, Y., Kihara, H., Harada, Y., 1991, Development of a superfluid helium stage for high-resolution electron microscopy. *Ultramic.* **38**: 241-251.
56. Henderson, R., Baldwin, J.M., Ceska, T.A., Zemlin, F., Beckmann, E., Downing, K.H., 1990, Model for the structure of bacteriorhodopsin based on high-resolution electron cryo-microscopy. *J Mol Biol.* **213**: 899-929.

57. Kimura, Y., Vassylyev, D.G., Miyazawa, A., Kidera, A., Matsushima, M., Mitsuoka, K., Murata, K., Hirai, T., Fujiyoshi, Y., 1997, Surface of bacteriorhodopsin revealed by high-resolution electron crystallography. *Nature* **389**: 206-211.
58. Kühlbrandt, W., Wang, D.N., Fujiyoshi, Y., 1994, Atomic model of plant light-harvesting complex by electron crystallography. *Nature* **367**: 614-621.
59. Murata, K., Mitsuoka, K., Hirai, T., Walz, T., Agre, P., Heymann, J.B., Engel, A., Fujiyoshi, Y., 2000, Structural determinants of water permeation through aquaporin-1. *Nature* **407**: 599-605.
60. Heymann, J.B., Engel, A., 2000, Structural clues in the sequences of the aquaporins. *J Mol Biol.* **295**: 1039-1053.
61. de Groot, B.L., Heymann, J.B., Engel, A., Mitsuoka, K., Fujiyoshi, Y., Grubmuller, H., 2000, The fold of human aquaporin 1. *J Mol Biol.* **300**: 987-994.
62. Mitsuoka, K., Hirai, T., Murata, K., Miyazawa, A., Kidera, A., Kimura, Y., Fujiyoshi, Y., 1999, The structure of bacteriorhodopsin at 3.0 A resolution based on electron crystallography: implication of the charge distribution. *J Mol Biol.* **286**: 861-882.
63. de Groot, B.L., Engel, A., Grubmuller, H., 2001, A refined structure of human aquaporin-1. *FEBS Lett.* **504**: 206-211.
64. Luecke, H., Schobert, B., Richter, H.T., Cartailler, J.P., Lanyi, J.K., 1999, Structure of bacteriorhodopsin at 1.55 A resolution. *J Mol Biol.* **291**: 899-911.

A 2D-Infrared Study of Human Lipoproteins

XABIER COTO, IBÓN ILORO, and JOSÉ LUIS R. ARRONDO
Unidad de Biofísica (CSIC-UPV/EHU) and Departamento de Bioquímica, Universidad del País Vasco, P.O. Box 644, E-48080 Bilbao, Spain.

1. INTRODUCTION

Lipids are transported in human blood plasma by lipoproteins consisting of a nonpolar core where triacylglycerols and cholesteryl esters are hidden surrounded by a monolayer facing the water composed of phospholipid, cholesterol and proteins, giving these lipid-rich structures water solubility. Blood plasma lipoproteins are classified on the basis of their density, which in turn is a reflection of their lipid content. The greater their lipid contents the lower their density. In a previous work we have defined the major classes that we use in our infrared studies[1]. These are VLDL or very low density lipoprotein, LDL or low density lipoprotein and HDL or high density lipoprotein. The mean characteristics of the samples that are used in this work are depicted in Table 1.

Table I. Composition of lipoproteins from human blood serum in fasting subjects

	VLDL	LDL	HDL
Density	0.95-1.006	1.019-1.063	1.063-1.210
Proteins (%)	B-100, C-I, C-II, C-III, E (8)	B-100 (22)	A-I, A-II (40-50)
Triglycerides	55	6	8
Phospholipids	18	22	22
Cholesterol	7	8	3-5
Cholesteryl esters	12	42	14-17

Supramolecular Structure and Function 8, Edited by Pifat-Mrzljak
Kluwer Academic/Plenum Publishers, New York 2004

Biosynthesis of VLDL cholesterol, triglycerides and apolipoproteins takes part in liver hepatocites. These particles have apoB-100, apoC and apoE proteins. VLDL are converted into LDL, cholesteryl ester-rich particles that have a single molecule of apoB-100. LDL carries cholesterol from liver to peripheral tissues where it is used in membrane and steroid biosynthesis. HDL is synthesised from liver precursors and are matured in plasma. Apolipoproteins in HDL are apoA-I and apoA-II. HDL removes cholesterol from the cells carrying it to the liver in a "reverse transport".

1.1 Infrared Spectroscopy

Infrared spectroscopy has become a widely used tool in the study of protein structure. In principle, a structure as large as a protein would give rise to an enormous number of overlapping vibrational modes obscuring the information that could be obtained in practice, but because of the repeating patterns of the biological molecules, e.g. the secondary structure of the protein backbone, the spectra are much simpler and useful structural information can be obtained. Structural analysis usually implies a mathematical approach in order to extract the information contained in the composite bands, designated in IR spectroscopy as amide bands, obtained from proteins. Commonly used methods of analysis imply narrowing the intrinsic bandwidths to visualize the overlapping band components and then decomposing the original band contour into these components by means of an iterative process. The various components are finally assigned to protein or subunit structural features[2]. External perturbations such as temperature are commonly used to obtain a deeper insight in protein structure by means of infrared spectroscopy. Thermal profiles have been often used to study conformational changes in proteins[3].

2. TWO-DIMENSIONAL IR SPECTROSCOPY (2D-IR)

The use of 2D-IR has been proposed recently. In this procedure the spectra before and after an external perturbation are correlated, to increase the amount of information obtained from the infrared spectrum[4]. Proteins are a good target for this method, since changes induced by an external perturbation can be studied in more detail than with the conventional infrared.

This approach, essentially different from 2D-NMR spectroscopy, uses correlation analysis of the dynamic fluctuations caused by an external perturbation to enhance spectral resolution without assuming any lineshape models for the bands. The perturbation can be achieved through changes in temperature[5], by the presence of lipids[6,7], or other external ligands[8]. The power of the 2D correlation approach results primarily in an increase of the spectral resolution by a dispersal of the peaks along a second dimension that also reveals the time-course of the events induced by the perturbation[9]. Correlation between bands are found through the so-called synchronous and asynchronous contour maps that correspond to the real and imaginary parts of the cross-correlation of spectral intensity at two wavenumbers. In a synchronous 2D map, the peaks located along the diagonal (autopeaks) correspond to changes in intensity induced (in this case) by temperature and they are always positive. The cross-relation peaks indicate an in-phase relationship between the two bands involved, i.e. that two vibrations of the protein characterized by two different wavenumbers (ν1 and ν2), are being affected simultaneously. Asynchronous maps show not-in-phase cross-relation between the bands and give an idea of the sequential order of events produced by the perturbation, i.e. an asynchronous peak is produced if the vibrations of the functional groups corresponding to the varying wavenumbers change each at a different time. The asynchronous peak will be positive if the change in ν1 occurs previous in time to ν2, and negative in the opposite case.

Figure 1 shows the correlation maps corresponding to a protein being denatured by temperature. The synchronous plot shows autopeaks at 1617 and 1657 cm^{-1}. Positive cross-peaks are located at 1617/1682. Negative ones are at 1617/1657 and 1617/1665, indicating that α-helix and β-turns are contributing, to the aggregation/denaturation. The asynchronous map shows the main correlation peaks at 1617/1650 and 1630/1652 indicating that the major event taking place in region II is the rise of the 1617 cm^{-1} aggregation band, now at the expenses of the α and β components.

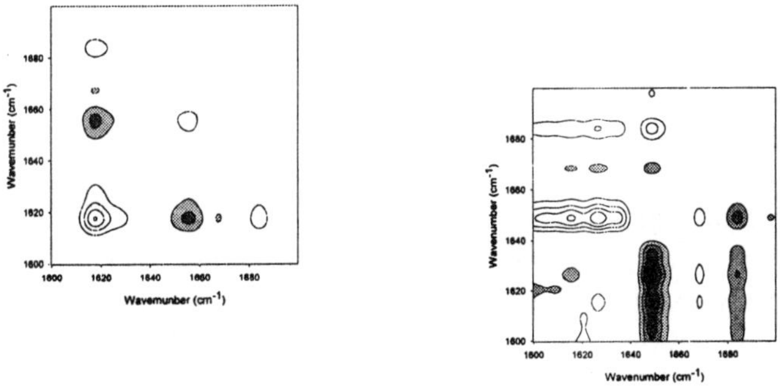

Figure 1. Synchronous (left) and asynchronous (right) correlation map contour in the region 1700-1600 cm^{-1} of chicken annexin in the interval 20-80 °C, in which the protein is denatured by temperature[10]. White peaks correspond to positive correlations and grey peaks to negative ones.

3. 2D-IR OF LIPOPROTEINS

The study of protein structural changes caused by an external perturbation such as temperature provides information on the conformation and organization of the protein during unfolding. In a previous work[1] we showed that the distinct functionality of lipoproteins gives rise to differences in structure that are reflected in the infrared spectra. Specially, it is clear that VLDL and LDL amide I spectra are dominated by the special characteristics of apoB-100. The difference between VLDL or LDL and HDL is also evident in the thermal profiles. Figure 2 shows the 20-80 °C interval of LDL and HDL, where the three dimensional profile of LDL is slightly different from a classical protein profile[2,11] because the low frequency band corresponding to aggregation is very close to the one arising from the β-strands. However, this does not mean that the bands attributed to β-strands have structural characteristics of an aggregated protein, since the high frequency component, located at 1695 cm^{-1} in the β-strands is not affected by the isotopic shift produced when the protein in measured in a D_2O medium[12]. This band shifts to 1682 cm^{-1} after protein denaturation, a similar behaviour to aggregated proteins. On the other hand, HDL thermal profile is similar to soluble or membrane proteins undergoing aggregation.

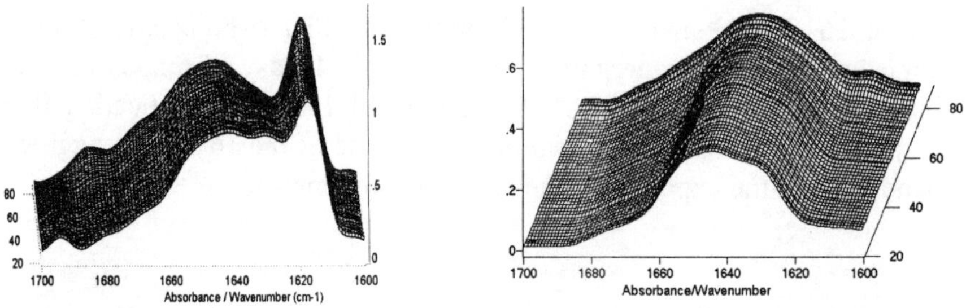

Figure 2. Thermal profile in a D_2O medium of LDL (left) and HDL (right) in the amide I region (1700-1600 cm^{-1}).

However, the information that can be extracted from these profiles is not exhaustive and more details on the effect of temperature can be obtained by using 2D-IR. As stated above, two contour maps are obtained for each series of spectra. Figure 3 shows the synchronous maps of VLDL, LDL and HDL corresponding to the amide I band in the interval 20-80 °C. The corresponding autopeak and cross-peak values are shown in Table 2.

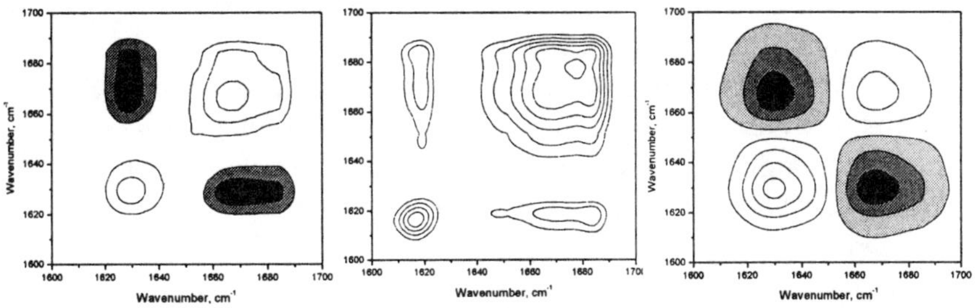

Figure 3. 2D-IR synchronous maps of VLDL (left), LDL (centre) and HDL (right) in the 1600-1700 cm^{-1} region corresponding to the amide I region and the 20-80 °C interval. The spectra were taken in a D_2O medium.

Table 2. Values of the autopeaks and cross-peaks corresponding to the synchronous maps shown in Fig. 3

20-80°C	Autopeaks	Cross-peaks
VLDL	1630,1664, 1680	(–) 1630-1664 (–) 1630-1680
LDL	1618, 1665 1682	(+) 1618-1664 (+) 1618-1682 (+) 1665-1682
HDL	1630,1670	(–) 1630-1670

From the maps of Fig.3 and the values of Table 2 several features can be observed. First point is that the three plots are distinct, that is, each lipoprotein map is different and therefore we can distinguish them by the values of the peaks in the maps. Second, VLDL and LDL are similar by shape and distinct from HDL. On the other hand, VLDL and HDL are closer by position. These differences would be a reflect of the unusual thermal behaviour of apoB100 described previously[13]. ApoB100 has two distinctive peaks at 1618 and 1695 cm^{-1}. These band components have been assigned to a protein structure penetrating the monolayer and establishing hydrophobic interactions deeper than the outer charged shell. As can be seen in Fig. 2, the LDL changes in bandshape after denaturation are related with the appereance of the 1682 cm^{1} band, what is reflected with a stronger cross-peak region. HDL does not have apoB100 and therefore there is not such shift and the peaks in the square delimited by the autopeaks and cross-peaks[4] are similar in intensity. In VLDL the map is conditioned by the abundance of apoB100, giving a picture similar in shape to LDL, but modulated by the presence of other proteins, with "classical" thermal profiles. It must be noted that the cross-peaks are negative in VLDL and positive in LDL. This is due to the fact that in LDL the synchronous map is dominated by the bands at

1618 and 1682 cm^{-1}, being the component at 1665 cm^{-1} obscured, whereas in VLDL the map is dominated by the bands at 1630 and 1665 cm^{-1}, and the obscured component is the 1680 cm^{-1} band.

The asynchronous maps, which give an idea of the sequential order of the events, provide more information on the events taking place during the thermal unfolding. The asynchronous maps and the cross-peaks of the different lipoproteins are shown in Figure 4 and Table 3.

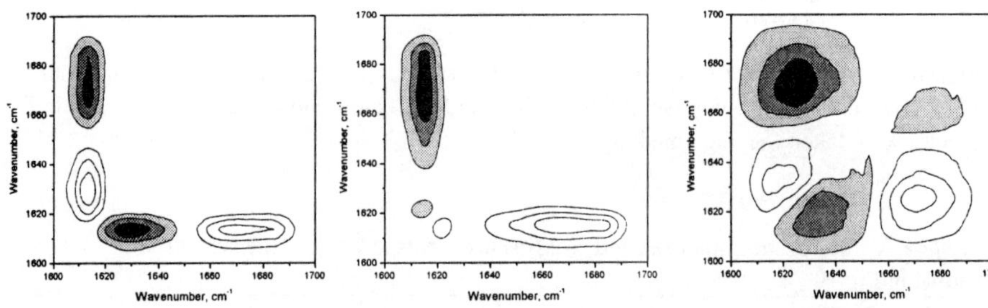

Figure 4. 2D-IR asynchronous maps of VLDL (left), LDL (centre) and HDL3 (right) in the 1600-1700 cm^{-1} region corresponding to the amide I region and the 20-80 °C interval. The spectra were taken in a D_2O medium

Table 3. Values of the cross-peaks corresponding to the asynchronous maps shown in Fig. 4.

20-80°C	Cross-peaks
VLDL	(+) 1617-1630 (–) 1617-1665 (–) 1617-1682
LDL	(–) 1618-1665 (–) 1618-1682 (–) 1618-1620
HDL	(+) 1622-1640 (–) 1625-1670

In this case, the difference in shape between VLDL or LDL and HDL is more evident. Based in simulation studies made with artificial curves[14], the shape of VLDL and LDL map contours corresponds to a process that is dominated by one event, whereas in the case of HDL the pattern is more common of processes that are changing not only in intensity but also in band position or bandwidth. The interpretation of Table 3 would tell us that in the positive pairs the event associated with the first wavenumber is previous to the second, e.g. in the case of VLDL the change associated with the band at 1617 cm^{-1} is previous to the one at 1630 cm^{-1}, but this change at 1617 cm^{-1} is subsequent to the one at 1665 cm^{-1}.

These studies, corresponding to the process in the interval 20-80 °C, give us an overall idea of what is happening in the apolipoproteins during thermal denaturation. A more careful study of the temperature-induced unfolding has been obtained in other proteins by looking at narrower temperature intervals[15].

One of the advantages of infrared spectroscopy in the study of membranes is the possibility of studying the bands corresponding to lipids in the same experiment. The 2D-IR maps of the region 1700-1800 cm^{-1} is shown in Figure 5. This region corresponds to lipid carbonyls that arise mostly from triglycerides in VLD;, cholesteryl esters in LDL and phospholipids and cholesteryl esters in HDL as deducted from Table 1.

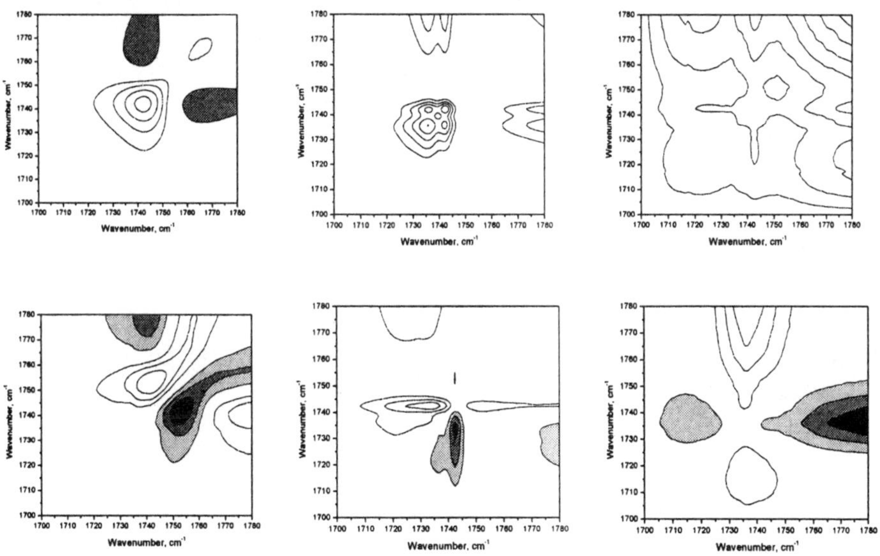

Figure 5. 2D-IR synchronous (top) and asynchronous (bottom) maps of VLDL (left), LDL (centre) and HDL3 (right) in the 1700-1800 cm-1 region corresponding to the carbonyl lipid region in the 20-80 °C interval. The spectra were taken in a D2O medium

From the lipid maps and taking into account simulation studies[14], it can be assumed that in VLDL, a change takes place in the lipid components during the protein thermal denaturation process associated with changes in band position (butterfly shape in the asynchronous map). .In LDL a single process, associated mainly with changes in band intensity is observed, whereas the differences in HDL can be associated with noise (no difference), with a very little intensity above noise seen in the asynchronous cross-peak.

4. CONCLUSION

The distinct functionality of lipoproteins gives rise to differences in structure that are reflected in the infrared spectra. 2D-IR is a data treatment based in the changes induced by an external perturbation, temperature in our study, which increases the sensibility of the infrared analysis. Thus, from the results presented above is clear that in VLDL and LDL, opposite to HDL, the amide I spectrum is dominated by the special characteristics of apoB-100 but the presence of other proteins in VLDL induces a different pattern in the synchronous and asynchronous maps producing a distinctive pattern for every lipoprotein. The lipid moiety of the different lipoproteins also produce 2D-IR plots, but with less features when compared with proteins, and without significant changes in HDL. The use of temperature as perturbing agent in protein thermal denaturation can be enhanced with studies in shorter temperature ranges which will increase the sensitivity of the technique during thermal unfolding.

ACKNOWLEDGEMENTS

This work was supported by grant BMC2002-01438 (Ministerio de Ciencia y Tecnologia) and 9/UPV00042.310-13552 from Universidad del Pais Vasco

REFERENCES

1. Coto, X. and Arrondo, J.L.R., 2001, Infrared spectroscopy of lipoproteins. In *Supramolecular Structure and Function vol.7* (G. Pifat-Mrzljak ed.), KluwerAcademic, London, pp. 75-87.
2. Arrondo, J.L.R. and Goñi, F.M., 1999, Structure and dynamics of membrane proteins as studied by infrared spectroscopy. *Prog.Biophys.Mol.Biol.* **72**:367-405.
3. Arrondo, J.L.R., Castresana, J., Valpuesta, J.M., and Goñi, F.M., 1994, The Structure and Thermal Denaturation of Crystalline and Non-Crystalline Cytochrome Oxidase as Studied by Infrared Spectroscopy. *Biochemistry* **33**:11650-11655.
4. Noda, I., Dowrey, A.E., Marcott, C., Story, G.M., and Ozaki, Y., 2000, Generalized Two-Dimensional Correlation Spectroscopy. *Appl.Spectrosc.* **54**:236A-248a.
5. Fabian, H., Mantsch, H.H., and Schultz, C.P., 1999, Two-dimensional IR correlation spectroscopy: Sequential events in the unfolding process of the lambda Cro-V55C repressor protein. *Proc.Natl.Acad.Sci.USA* **96**:13153-13158.
6. Shanmukh, S., Howell, P., Baatz, J.E., and Dluhy, R.A., 2002, Effect of hydrophobic surfactant proteins SP-B and SP-C on phospholipid monolayers. Protein structure studied using 2D IR and beta correlation analysis. *Biophys.J.* **83**:2126-2141.

7. Torrecillas, A., Corbalan-Garcia, S., and Gomez-Fernandez, J.C., 2003, Structural Study of the C2 Domains of the Classical PKC Isoenzymes Using Infrared Spectroscopy and Two-Dimensional Infrared Correlation Spectroscopy. *Biochemistry* **42**:11669-11681.
8. Pastrana-Rios, B., Ocana, W., Rios, M., Vargas, G.L., Ysa, G., Poynter, G., Tapia, J., and Salisbury, J.L., 2002, Centrin: its secondary structure in the presence and absence of cations. *Biochemistry* **41**:6911-6919.
9. Sasic, S., Muszynski, A., and Ozaki, Y., 2001, New Insight into the Mathematical Background of Generalized Two-Dimensional Correlation Spectroscopy and the Influence of Mean Normalization Pretreatment on Two-Dimensional Correlation Spectra. *Appl.Spectrosc.* **55**:343-349.
10. Turnay, J., Olmo, N., Gasset, M., Iloro, I., Arrondo, J.L.R., and Lizarbe, M.A., 2002, Calcium-dependent conformational rearrangements and protein stability in chicken annexin A5. *Biophys.J.* **83**:2280-2291.
11. Chehín, R., Iloro, I., and Arrondo, J.L.R., 1998, Thermal denaturation of membrane proteins: An infrared study. *Period.biol.* **100**:13-19.
12. Susi, H., 1969, Infrared Spectra of Biological Macromolecules and Related Systems. In *Structure and Stability of Biological Macromolecules* (S.N. Timasheff and L. Stevens, eds), Dekker, New York, pp. 575-663.
13. Bañuelos, S., Arrondo, J.L.R., Goñi, F.M., and Pifat, G., 1995, Surface-Core Relationships in Human Low Density Lipoprotein as Studied by Infrared Spectroscopy. *J.Biol.Chem.* **270**:9192-9196.
14. Arrondo, J.L.R., Iloro, I., Aguirre, J., and Goñi, F.M., 2003, A Two-dimensional IR Spectroscopic (2D-IR) Simulation of Protein Conformational Changes. *Spectroscopy* (In press)
15. Paquet, M.J., Laviolette, M., Pezolet, M., and Auger, M., 2001, Two-dimensional infrared correlation spectroscopy study of the aggregation of cytochrome c in the presence of dimyristoylphosphatidylglycerol. *Biophys J* **81**:305-312.

An Introduction to Biological Solid State NMR

ANDREW DODD AND FRANCES SEPAROVIC
School of Chemistry, University of Melbourne, Melbourne, VIC 3010, Australia

1. INTRODUCTION

With its obvious links to better understanding of living systems, the field of structural biology is strongly driven by the need to increase our knowledge of the nature and function of important biological molecules. Peptides, proteins and their complexes are of key interest due to their specific functional roles in the life cycle of the cell, yet the diversity of their structural form and biophysical properties makes them difficult to fully characterize using any one technique.

In the solution state, NMR spectroscopy has proved to be a powerful and reliable tool in the structural elucidation of biological systems, making it, along with X-ray crystallography, a generous contributor to the protein data bank. However, the scope of crystallography is somewhat hampered by the prerequisite of reliable quality single crystals while solution NMR requires good solubility and relatively low molecular mass to obtain high resolution spectra. Smaller molecules undergo rapid reorientation such that the interactions which would normally broaden the NMR spectrum are averaged out; for the larger molecules this orientational averaging is no longer efficient resulting in poorly defined spectra.

Solid state NMR offers a very different insight into the structure and dynamics of biological systems, and in particular is employed extensively in the study of membrane peptides and proteins in model membranes, an area not easily accessible to other techniques[1]. Spectra of solids are dominated by strong

Supramolecular Structure and Function 8, Edited by Pifat-Mrzljak
Kluwer Academic/Plenum Publishers, New York 2004

interactions which usually lead to the observation of broad, poorly resolved spectral lines. This is a result of the highly restricted motion in solids and is largely responsible for the fact that, in its infancy, solid state NMR was not widely used to interrogate bio-systems. The rapid development of technology and techniques over recent decades has altered this situation considerably: it is now possible to overcome many of the obstacles which previously hindered the solid state spectroscopist, and even to turn some of those obstacles into allies. As a result solid state NMR has evolved into a technique which allows us to probe information that is often difficult to obtain using any other method.

While this field of NMR cannot provide access to the entire 3D structure of biological molecules as readily as its liquid state partner, it is able to afford insight into structural and conformational changes that occur as a result of environmental interactions. This aspect makes solid state NMR particularly attractive in the study of the many membrane peptides, proteins and complexes which, on the NMR timescale (ms-ns depending on experiment type), can be considered as virtually immobile. Although the technique has been used for decades in the study of model lipid membranes, it is only recent advances in the technology of spectroscopy, peptide synthesis and protein labeling that have allowed its application to structural studies of membrane peptides, trans-membrane protein segments and membrane proteins[2,3]. In particular, solid state NMR allows the spectroscopist to investigate the peptide or protein interaction with lipid bilayers, a system far more representative of biological membranes than the detergent micelles or solvents commonly used in solution NMR.

The nucleus specific nature of NMR allows the investigator to specifically pinpoint an area of interest within the system. In phospholipid membranes, for example, it is possible to exclusively monitor the lipid head group environment by way of the phosphorous atoms, or similarly to selectively study the acyl chains[1]. The advent of isotopic labeling has enabled the spectroscopist to plant position specific spies into both the protein and the model membrane either by substitution, addition or enrichment. These infiltrators can report in a variety of ways on the conformation or orientation of proteins in a membrane environment: ^{15}N labeling is commonly used to investigate structure in aligned bilayer systems by simple measurement of the resonance frequency[4,5], ^{13}C labeling can be used with techniques such as rotational resonance (RR) to determine homonuclear distances[6], while multiple species labeling can reveal interaction between unlike nuclei by implementation of variations on the rotational echo double resonance (REDOR) technique[3]. ^{19}F labeling of the peptide or protein can be used to allow particular sites to be scrutinized and ^{2}H labeling can be applied to both the protein and the lipid to reduce spectral broadening due to the very strong magnetic interactions caused by the presence of protons[7].

There now exists a plethora of experimental NMR techniques which can be applied to biological systems; our aim here is to present some of the more commonly used aspects of solid state NMR, represent the type of results that can be attained and introduce some of the more fundamental concepts of the theory.

2. NMR INTERACTIONS

The NMR phenomenon depends on the intrinsic spin of the nucleus and its interaction with its surrounding environment[8]. In the presence of a magnetic field, a nucleus with non-zero spin can exist in one of a discrete number of available energy states. The energy splitting and therefore population difference between these two states is dependent on the type of nucleus and the strength of the static magnetic field, typically referred to as B_0. A great deal of insight can be gained into the atomic environment by perturbing the system from equilibrium with radio-frequency (rf) pulses and observing the way in which the nuclear spins relax back to their preferred state.

To a good approximation, the rapid tumbling of molecules in the liquid state has the effect of decoupling the nuclei from their surroundings such that spectra are recorded containing discrete peaks with narrow frequency distribution. The frequencies at which these peaks appear can, for example, reveal the nature of the bond or group to which the nucleus under inspection is attached and so are valuable in determination of the secondary structure. In the solid state, the spectra are complicated by the presence of strong couplings and anisotropic interactions. These couplings have the effect of smearing out the energy states that can be occupied by the nucleus and this effect translates to a broadening of the resonance line. There are 3 major contributors to this broadening: *dipolar coupling* which involves the magnetic interaction between neighbouring spins, *chemical shift anisotropy* (CSA) which results from variation in the local electronic environment and *quadrupole coupling*, an interaction specific to nuclei with spin $I > ½$ whose electric quadrupole moment interacts with the surrounding electric field gradient. These interactions manifest themselves as a broadening of the spectrum which can be as large as several hundred kHz, effectively swamping any useful structural information that might be available. There are also a number of weaker interactions such as the scalar coupling which, although less influential in the spectra, are no less important in some areas of spectroscopy.

3. COMMONLY USED TECHNIQUES

The domain of solid state NMR presents a bewildering array of experimental possibilities with pulse sequences becoming ever more intricate in order to explore many of the possible avenues available. The majority of these techniques, however, rely on variations of a few key aspects of NMR which manipulate the various couplings with great effect.

3.1 Magic Angle Spinning (MAS)

Somewhat conveniently, it turns out that the three key interactions mentioned previously all share a first order angular dependence on the second Legendre polynomial, ($3\cos^2\theta$-1), where θ is the polar angle made between the magnetic field and the axis system of the nuclear spin interaction. This is a very powerful relationship: it means that if the orientation can be averaged to the so called "magic angle" of 54.74° then the effects of interactions with this angular dependence can be averaged out[9]. In practice this averaging is achieved by fast rotation of the sample rotor about the magic angle (Fig. 1(a)).

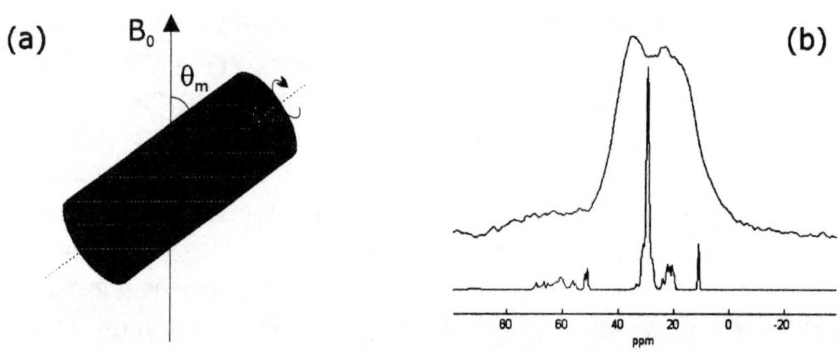

Figure 1. **a)** Schematic layout of the sample rotor oriented at the magic angle in the magnetic field. The rotor is spun about its long axis such that the position of the internuclear vector is averaged to the magic angle, and **b)** ^{13}C spectra for a phospholipid (DMPC) both static (top) and under MAS conditions at 6 kHz.

If the spinning speed is sufficiently fast with respect to the interaction frequency itself the couplings are averaged to 0 and the resonance line can be significantly narrowed. Under MAS the broad, featureless spectrum of Fig. 1(b) is transformed into a "solution like" spectrum in which the ^{13}C dipolar coupling is averaged out revealing the individual ^{13}C resonances according to their functional group association.

The most basic application of MAS with respect to biological systems is found in the simple recovery of the isotropic value of chemical shift, that is, the resonance frequency relative to that of a standard reference. These values allow resonances to be assigned to specific species and so are instrumental in determining the secondary structure of the protein or peptide. MAS can also be used in combination with sample alignment in "magic angle oriented sample spinning" which can be used to determine orientation and structure of α-helical species in membranes[10].

3.2 Cross Polarisation (CP)

While MAS can be sufficient to collapse the broadening in the ^{13}C spectrum of fluid phase lipid samples (Fig. 1(b)), it is often not possible to remove the heteronuclear dipolar coupling between the protons and the nuclei under inspection. In such cases it is necessary to "decouple" the ^{1}H spins. Decoupling can be achieved either by broadband irradiation at the proton resonance frequency or by applying selective decoupling pulse sequences; this causes the rapid reorientation of the proton spins and essentially does not allow them the time to interact with the nucleus of interest.

In biological solid state NMR it is often preferred to isotopically label the sample and inspect lower abundance species with wider chemical shift ranges and less intrusive couplings than their more abundant counterparts. ^{15}N ($I = ½$) is one such nucleus which is chosen over the more abundant ^{14}N ($I = 1$). Similarly ^{13}C is often targeted since ^{12}C has no net nuclear spin and therefore gives no resonance.

These dilute nuclei are often difficult to observe due to their weak magnetic moment, relatively low numbers and long relaxation times; these factors result in the need for very long acquisition times so as to obtain a reasonable signal, often in the order of many days. The CP technique (Fig. 2) enables magnetisation to be transferred from the abundant ^{1}H nuclei to the nearby weaker spins via satisfaction of the Hartmann Hahn match condition[8]. This not only increases the signal obtained from the target nuclei but also reduces the delay required between acquisitions since relaxation is now dependent on the relaxation rate of the protons, which tends to be much shorter. CP is commonly used in conjunction with MAS to further enhance spectral resolution.

3.3 Aligned Samples

In some instances it is not desirable to remove the effect of anisotropies in the system. For example, a spectrum may exhibit CSA broadening which represents spectral contributions from crystallites existing in all possible orientations – such a lineshape is known as a powder pattern. This pattern

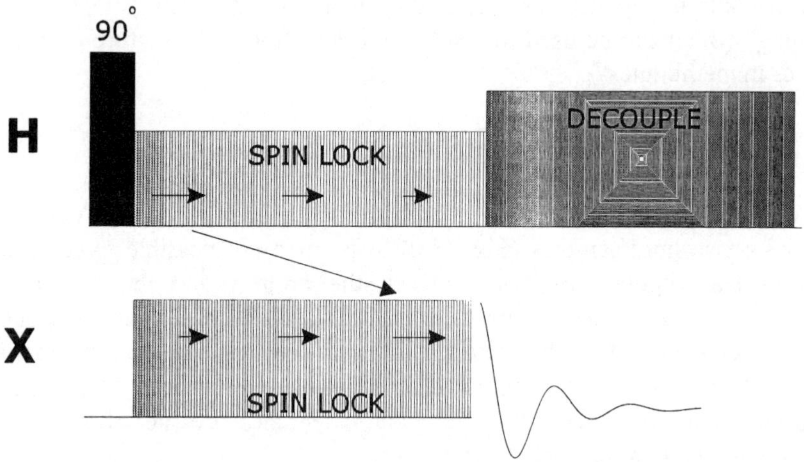

Figure 2. The CP pulse sequence. Under Hartmann Hahn match conditions whereby both nuclei have the same energy level splitting in the rotating frame, polarisation is transferred to the low abundance nuclei (X). Decoupling is commonly used during the acquisition of signal to remove the broadening effect of proton coupling.

can give information on aspects of the environment such as mobility, functional group or symmetry at the site.

In a ^{31}P CSA powder pattern such as that shown in Fig. 3(b), the frequency furthest to the left of the spectrum represents the lipid molecules with their long axis oriented parallel to the field, while the intensity at the right hand edge arises from lipids which are perpendicular to the field[4]. The ^{31}P CSA can give information about the lipid phase as well as changes in dynamics and order of the head group in model membrane systems with peptides or proteins incorporated[11].

Phospholipid bilayers can be effectively used as model biological membranes by aligning hydrated multilayers between glass plates. The long axis of the phospholipid molecule is oriented perpendicular to the glass surface and best results are achieved by orienting the bilayer surface perpendicular to the magnetic field. The spectra observed for the ^{31}P in the phospholipid head group (Fig. 4) can be used to demonstrate any alignment changes or disordering effects that incorporation of a peptide or protein may have on the lipid head group[12].

For α-helical peptides and proteins isotopically labeled with ^{15}N it is relatively straightforward to distinguish a trans-membrane orientation from a surface orientation in an aligned system by observing the chemical shift[5]. A trans-membrane segment has a chemical shift of around 220 parts per

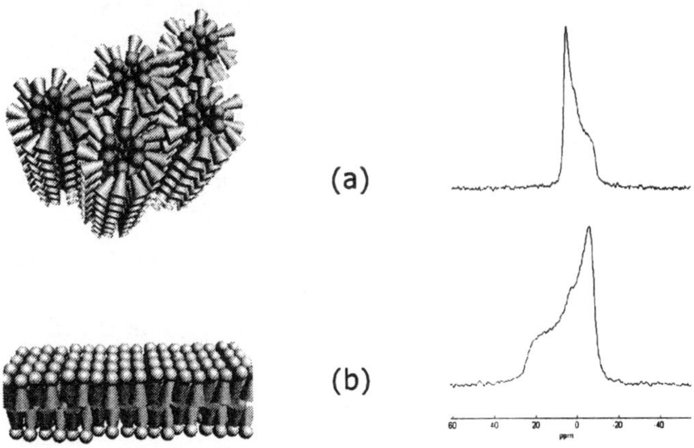

Figure 3. ^{31}P NMR spectra recovered from DMPC showing variation in CSA pattern dependent on the phase of the lipid. (a) Inverted hexagonal (H_{II}) phase, and (b) fluid bilayer (Lα) phase. For smaller vesicles, micelles and cubic phases an isotropic peak is observed.

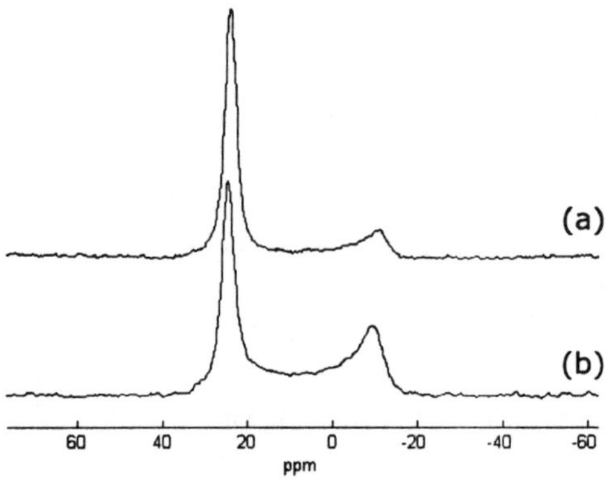

Figure 4. ^{31}P spectra acquired from aligned bilayers with phospholipid molecules parallel to the magnetic field, **(a)** DMPC, and **(b)** DMPC with maculatin peptide. The larger peak on the left is from aligned DMPC molecules while the smaller features are from non-aligned molecules or hydrated powder.

million (ppm) whereas the surface aligned segment has a shift of less than 100 ppm. Analogous studies have also been carried out on melittin and gramicidin A in aligned phospholipid bilayers using ^{13}C labeling to determine orientation and conformation[13,14].

Deuterium NMR is also commonly used in aligned systems to determine any effects of the protein-lipid interaction at the deuterated acyl chains which can give particular insight into the degree of penetration into the membrane system[1].

3.4 Rotational Resonance (RR)

MAS was described above as a powerful technique in determining structural information from biological systems, however, used with simple pulse sequences (such as single pulse – acquire experiments) the amount of information available is limited. In particular, MAS is used to average out the dipolar coupling interaction. Dipolar coupling is inversely proportional to the sixth power of internuclear distance[8] which makes it a highly sensitive probe for inspecting nuclear separation and is therefore a very desirable interaction to quantify. A number of multiple pulse NMR techniques exist which can be used to measure the dipolar coupling strength and thus internuclear distance under MAS conditions; one such technique is RR.

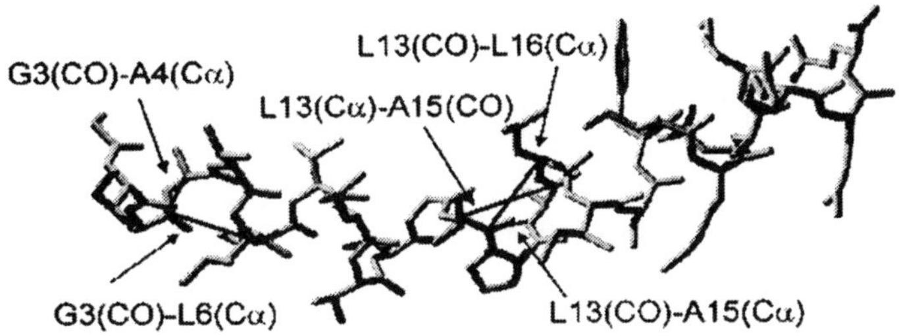

Figure 5. ^{13}C labeled sites in melittin used for RR determination of distances[6].

Rotational resonance measurements allow the determination of internuclear connectivities via the dipolar interaction between like spin species separated by distances of less than 10 Å. RR is preferentially used to measure the dipolar coupling between dilute spins such as ^{13}C (Fig. 5) which have fewer other connectivities to complicate their spectra; the condition which must be met for rotational resonance is simply that the spinning speed must be a multiple of the frequency difference between the two labeled spins. The resulting spectra reflect the exchange of magnetisation between the two spins as a function of mixing time and can be compared to simulation in order to determine an estimate of the dipolar coupling strength. This information can yield estimates of the distance between specifically labeled residues and thus can be highly instructive as to the nature of the structure.

3.5 Rotational Echo Double Resonance (REDOR)

Rotational echo double resonance is another MAS based technique used to determine internuclear distances; REDOR is more commonly applied to unlike nuclei and is often used to determine ^{13}C-^{15}N distances in bio-molecules[3]. In its simplest form, the REDOR experiment makes use of a simple spin-echo sequence on the observed spin wherein the magnetisation is transferred into the x-y plane by a 90° pulse and allowed to dephase for a time τ before application of an inverting 180° pulse which leads to the refocussing of the magnetisation a further time τ later. The spin echo refocusses any dephasing of spins which occurs due to the dipolar coupling (and other) interactions.

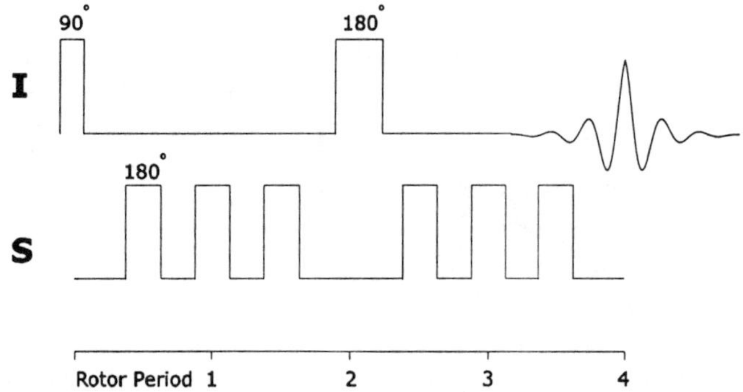

Figure 6. Basic $^{15}N\{^{13}C\}$ REDOR pulse sequence. Pulses on the "S" spin (^{13}C in this case) channel are rotor synchronised to cause maximum dephasing effect, signal is recorded for the ^{15}N (I) spins; decoupling may also be used during acquisition of the signal. This is one of many variations on the technique.

While the observe spin undergoes a spin echo in the REDOR experiment, pulses are selectively applied at the resonance frequency of the second spin species in order to hinder the refocussing of spins dephased by dipolar coupling between the two species[15]. As a result, the recovered echo intensity is reduced by an amount dependent on the dipolar coupling strength and data recovered from spectra as a function of evolution time can again be compared with simulation to determine dipolar coupling strength. There are many improvements and variations on the REDOR technique which make it a very versatile method in structure determination.

In solid state NMR there are a vast number of experimental possibilities: the techniques mentioned above represent a cross section of the basic methods used, in practice these and countless other techniques are used in combination to produce pulse sequences which can be uniquely tailored to a specific need. An excellent example of the successful combination of MAS and multidimensional NMR is given by the recent determination of the 3D structure of a relatively large protein in the solid state[7].

4. CONCLUSIONS

In the past two decades solid state NMR has exploded from its humble beginnings into a formidable methodology with enormous potential and flexibility. With giant leaps continuing to be made in both hardware and computational aspects of the technique as well as continual development of increasingly complex pulse sequences, solid state NMR will continue to provide a unique insight into the structure and behaviour of important biological molecules and their association with cell membranes.

ACKNOWLEDGEMENTS

The authors gratefully acknowledge financial support from the Australian Research Council.

REFERENCES

1. Gawrisch, K. and Koenig, B. W., 2002, Lipid-peptide interaction investigated by NMR. *Current Topics in Membranes* **52**: 163-190.
2. Drechsler, A. and Separovic, F., 2003, Solid-state NMR structure determination. IUBMB Life, **55**: 515-523.
3. Thompson, L. K., 2002, Solid-state NMR studies of the structure and mechanisms of proteins. *Current Opinions in Structural Biology* **12**: 661-669.
4. Bechinger, B., 1999, The structure, dynamics and orientation of antimicrobial peptides in membranes by multidimensional solid-state NMR spectroscopy. *Biochim. Biophys. Acta* **1642**: 157-183.
5. Marassi, F. M., 2002, NMR of peptides and proteins in oriented membranes. *Concepts in Magnetic Resonance* **14**: 212-224.
6. Lam, Y-H, Wassall, S. R., Morton, C. J., Smith, R. & Separovic, F., 2001, Solid-state NMR structure determination of melittin in a lipid environment. *Biophys. J.* **81**: 2752-2761.
7. Castellani, F., van Rossum, B., Diehl, A., Schubert, M., Rehbein, K. and Oschkinat, H., 2002, Structure of a protein determined by solid-state magic-angle-spinning NMR spectroscopy. *Nature* **420**: 98-102.
8. Ernst, R. R., Bodenhausen, G. and Wokaun, A., 1987, *Principle of Nuclear Magnetic Resonace in One and Two Dimensions,* Clarendon Press, Oxford.
9. Laws, D. D., Bitter, H.-M.L. and Jerschow, A., 2002, Solid-state NMR spectroscopic methods in chemistry. *Angew. Chem. Int. Ed.* **41**: 3096-3129.
10. Watts, A, 2002, Direct studies of ligand-receptor interactions and ion channel blocking. Molec. Membr. Biol.**19**: 267-275.
11. Bonev, B. B., Lam, Y-H, Anderluh, G., Watts, A., Norton, R. S. & Separovic, F., 2003, Effects of the eukaryotic pore-forming cytolysin equinatoxin II on lipid membranes and the role of sphingomyelin."*Biophys. J.* **84**: 2382-2392.
12. Marcotte, I., Wegener, K. L., Lam, Y-H, Chia, B. C. S., de Planque, M. R. R., Bowie, J. H., Auger, M. & Separovic, F., 2003, Interaction of antimicrobial peptides from Australian amphibians with lipid membranes. *Chem. Phys. Lipids,* **122**: 107-120.

13. Smith, R., Separovic, F., Milne, T. J., Whittaker, A., Bennett, F. M., Cornell, B. A. & Makriyannis, A., 1994, Structure and orientation of the pore-forming peptide, melittin, in lipid bilayers. *J. Mol. Biol.* **241**: 456-466.
14. Cornell, B. A., Separovic, F., Smith, R. and Baldassi, A. J., 1988, Conformation and orientation of gramicidin A in oriented phospholipid bilayers measured by solid state carbon-13 NMR. *Biophys. J.* **53**: 67-76.
15. Gullion, T. and Schaeffer, J., 1989, Rotational Echo Double Resonance NMR, *J. Magn. Reson.*, **81**, 196-200.

Multi-Frequency EPR Spectroscopy Studies of the Structure and Conformational Changes of Site-Directed Spin Labelled Membrane Proteins

HEINZ-JÜRGEN STEINHOFF
Department of Physics, University of Osnabrück, Barbarastrasse 7, 49069 Osnabrück, Germany

1. INTRODUCTION

Electron paramagnetic resonance (EPR) spectroscopy of site-directed spin labelled biomolecules (site-directed spin labelling, SDSL) has emerged as a powerful method for studying the structure and conformational dynamics of proteins and nucleic acids under conditions relevant to function[1-5]. In this technique, a spin label side chain is introduced at a selected site via cysteine substitution mutagenesis followed by modification of the unique sulfhydryl group with a specific paramagnetic nitroxide reagent (Fig. 1). The continuous wave (cw) EPR spectrum yields information about the nitroxide side chain mobility, the solvent accessibility, the polarity of its immediate environment and the distance between the nitroxide and another paramagnetic centre in the protein. EPR data analysis of a series of spin labelled variants of a given protein allows definition of elements of secondary structure, including their solvent exposure, to characterize protein topography and to determine orientations of individual segments of the protein. A complete analysis allows modelling of protein structures with a spatial resolution at the level of the backbone fold[2, 3, 6-9]. The method is applicable to any protein that retains its function after spin labelling. One of the most powerful properties of the method is its sensitivity to molecular dynamics: protein equilibrium fluctuations and conformational

Supramolecular Structure and Function 8, Edited by Pifat-Mrzljak
Kluwer Academic/Plenum Publishers, New York 2004

changes of functional relevance can be followed on a wide time scale ranging from picoseconds to seconds. The present report presents applications of this method to a selected set of membrane proteins: The light driven proton pump bacteriorhodopsin (BR) (cf. Fig. 2a) and the sensory rhodopsin – transducer complex, pSRII-pHtrII, responsible for phototaxis may be regarded as model systems for ion pumps and signal receptors. The Na^+/proline transporter PutP is a model for the Na^+/substrate symporter family (SSSF) which currently comprises more than 200 similar proteins of pro- and eukaryotic origin[10]. Investigation of the structure and conformational dynamics of these classes of membrane proteins represents one of the current major challenges.

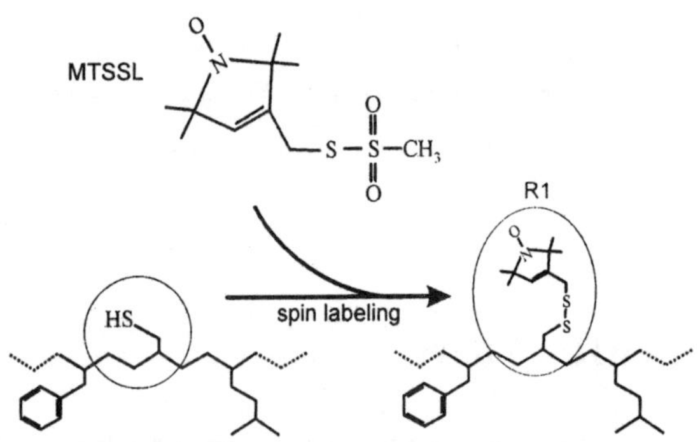

Figure 1. The reaction of the methanethio-sulfonate spin label (MTSSL)[11] with a sulfhydryl group generates the spin label side chain R1.

2. SPIN LABEL DYNAMICS AND SOLVENT ACCESSIBILITY

The relationship between the nitroxide side chain mobility and the protein secondary and tertiary structure has been extensively reviewed[12-16]. The term "mobility" is used here in a general sense and includes effects due to the motional rate, amplitude and anisotropy of the nitroxide reorientation. Weak interaction between the nitroxide and neighbouring side chain or backbone atoms as found for helix surface sites or loop regions results in a high degree of mobility. In this case the apparent hyperfine splitting and the line width are small. In turn, strong interaction of the nitroxide group with adjacent side chain or backbone atoms restricts its reorientational motion. Hence, tertiary contact or buried sites are characterized by an increased apparent hyperfine splitting and line width. Figure 2b shows an example of three spin labelled sites of bacteriorhodopsin. Two sites, positions 163 and 164, are located in the loop connecting helices E and F. The spin label side chain at position 164 is oriented to the aqueous phase. Nearly unhindered reorientational motion of the nitroxide is possible within a large motion cone as revealed by molecular dynamics simulations[17]. This motion leads to considerable averaging of the anisotropic contribution of the hyperfine interaction. Hence, the apparent hyperfine splitting is expected to be small in agreement with the experimental result. The side chain at position 163 is oriented away from the aqueous phase into the direction of helices B and C. The reorientational motion of the nitroxide is more restricted. Consequently, averaging of the anisotropic hyperfine interaction is less complete. The nitroxide at position 170 is buried within the protein. The nitroxide motion is completely restricted and the EPR spectrum resembles that of a nitroxide powder spectrum. Depending on the length and flexibility of the linker between the nitroxide and the protein backbone and possible hydrogen bonding of the nitroxide to the protein the flexibility of the protein backbone itself contributes to the overall nitroxide mobility. Numerous examples have shown that the analysis of the nitroxide dynamics of a series of spin labelled protein variants uncovers the secondary structure and provides important information about tertiary interaction[2, 16, 18, 19].

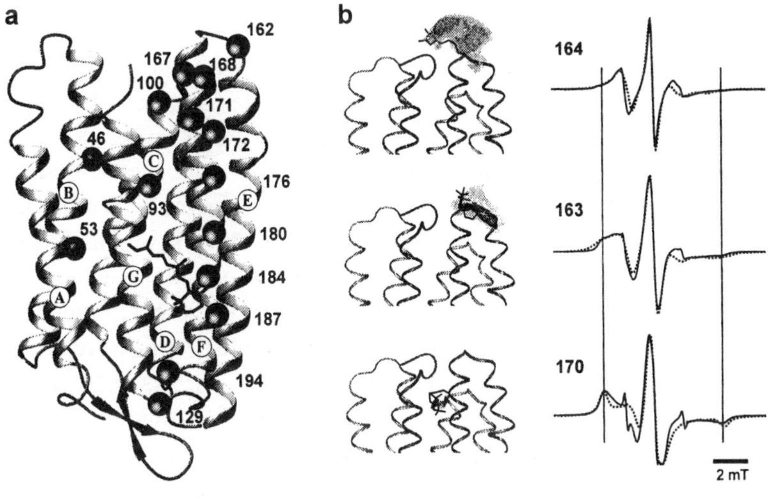

Figure 2. (a) Structure of bacteriorhodopsin. A selection of the spin labelled positions discussed in this text is highlighted. *(b)* Motional freedom of the spin label side chains at different positions in the E-F loop and in helix F of bacteriorhodopsin. The dotted area represents the accessible space of the nitroxide covered within a high temperature (600K) molecular dynamics simulation of 6 ns length[17]. The different degree of mobility of the nitroxides is reflected in the variation of the apparent hyperfine splitting of the corresponding experimental spectra (T = 293K) (continuous line) and the spectra calculated on the basis of the molecular dynamics trajectories (dotted).

The collision frequency of the nitroxide side chains with freely diffusing paramagnetic probe molecules provides additional structural information. The collision frequency of such a probe depends on the product of its translational diffusion coefficient and its local concentration. Molecular oxygen and water soluble paramagnetic Ni(II) complexes or chromium oxalate have been frequently used and are ideally suited because of their sizes and solubility properties[20-22]. In a water/membrane system these molecules are partitioned between the water and the hydrophobic phase according to their polarity. Polar metal complexes preferentially partition into the aqueous phase, whereas apolar oxygen exhibits a maximum value of the product of concentration and diffusion coefficient in the centre of the membrane bilayer[23]. The determination of the collision frequency of nitroxide side chains with these paramagnetic reagents in solution allows identification of the side chain orientations with respect to the protein-water or protein-lipid interface. Continuous wave power saturation has been shown to provide an easy and reliable means for the quantification of the collision frequencies[23].

3. POLARITY OF THE SPIN LABEL MICRO-ENVIRONMENT

The application of high-field EPR techniques with Lamor frequencies exceeding 90 GHz has considerably enhanced the Zeeman resolution of rigid-limit spectra of disordered spin labelled samples. The principal g-tensor components and their modulation due to solute-solvent interactions can be determined with high accuracy[24-26]. The structural information is contained in the variation of the polarity in the spin label micro-environment. A polar environment shifts the tensor component g_{xx} of a nitroxide to smaller values whereas the hyperfine tensor component A_{zz} is increased[27, 28] (cf. Fig.3). Hence, both tensor components can be regarded as polarity indexes. In addition, a plot of g_{xx} *vs.* A_{zz} allows discrimination between protic and aprotic environment due to the different sensitivities of these tensor components towards the influence of hydrogen bonding to the NO group[28, 29]. In a sequence of a regular secondary structure with anisotropic solvation, the local water density in the vicinity of the spin labelled site and hence the polarity index values are a periodic function of sequence number, similar to the behaviour of the nitroxide mobility or its accessibility for water soluble paramagnetic ions.

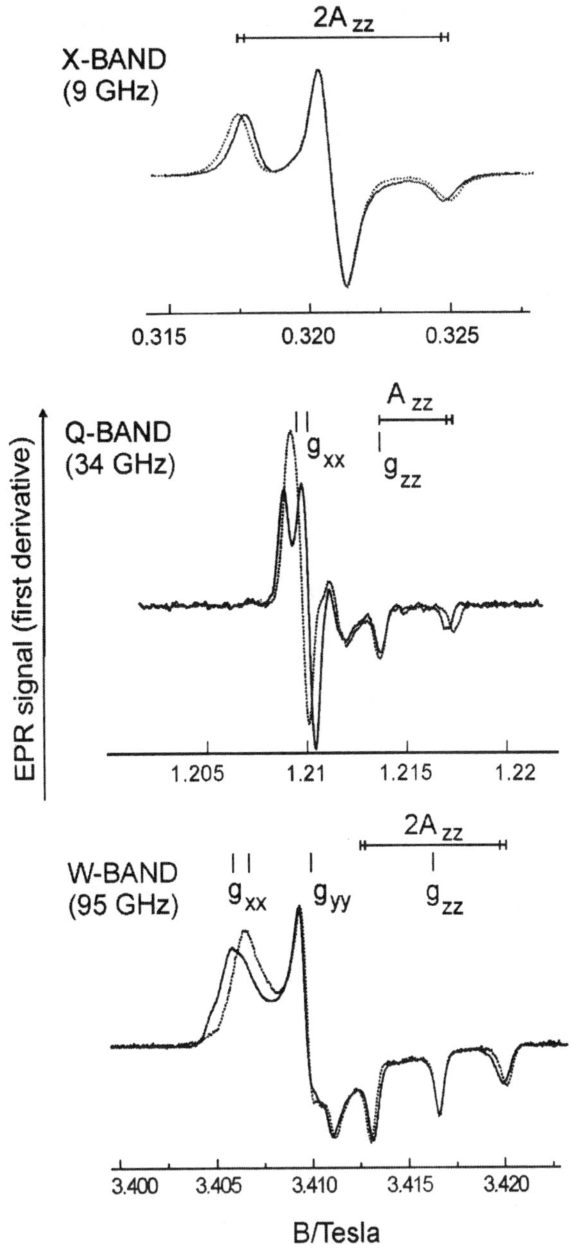

Figure 3. EPR spectra of MTS spin label in micro-environments of low (continuous line) and high (dotted line) polarity: *(top),* X-band spectra of bacteriorhodopsin variants L93R1 (protein interior, continuous line) and K129R1 (extracellular surface, dotted line)[30]; *(middle),* Q-band spectra of the spin label in frozen polystyrene/toluene (continuous line) and ethanol/water mixtures (dotted line); *(bottom),* W-band spectra of bacteriorhodopin variants T46R1 (protein interior, continuous line) and M163R1 (cytoplasmic surface, dotted line)[29].

3.1 Polarity profile along the bacteriorhodopsin proton channel

Sample spectra of spin label side chains attached along the bacteriorhodopsin proton channel are shown in figure 4. The variation of g_{xx} with the nitroxide binding site is revealed by the shift of the position of the low-field maximum. This shift is observable already in the Q-band (34 GHz) spectra (Reyher and Steinhoff, unpublished) and is clearly resolved in W-band (95 GHz) spectra[29]. Fittings of simulated powder spectra to the experimental data yielded the components of the g-tensor and of the hyperfine tensor with high accuracy (see Figs. 5 and 6). The plot of g_{xx} *vs.* nitroxide position along the proton channel reveals distinct variations in the polarity of the nitroxide micro-environment (Fig. 5a). The high polarity in the environment of residues at the cytoplasmic and extracellular surfaces, positions 162, 163, 166, 168 and 129, is clear evidence that these nitroxides are accessible to water, which is in agreement with the structure. The environmental polarity reaches its minimum close to position 46 between the proton donor D96 and the retinal, again in agreement with the protein structure. The g_{xx} behaviour provides quantitative information about the hydrophobic barrier the proton has to overcome on its way through the protein.

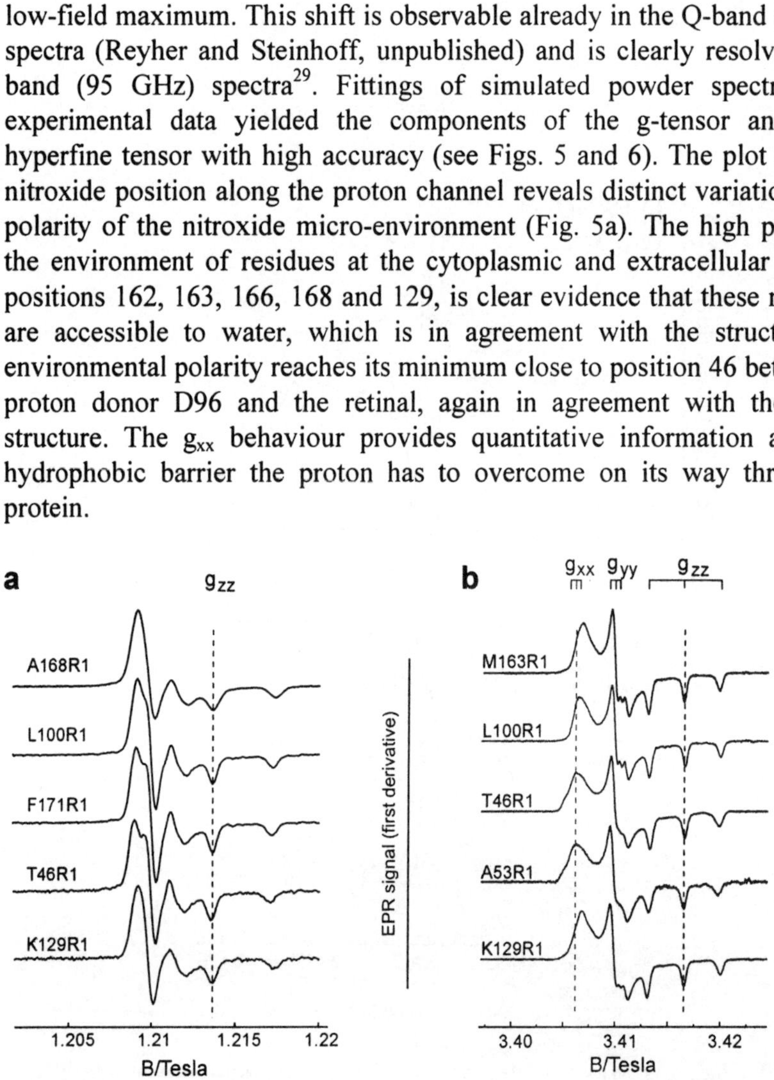

Figure 4. (a) Q-band (34 GHz) and *(b)* W-band (95GHz) EPR spectra[29] of spin labels introduced along the proton channel of bacteriorhodopsin (cf. Fig. 2a). The variation of the local polarity is revealed by the different shapes of the Q-band spectra at their low field resonance position (1.208 T) and by a clear shift of the respective resonance peak at the g_{xx}-position in the W-band spectra (3.46 T).

3.2 Polarity profile along the protein-lipid interface

The polarity variation along the lipid-exposed surface of transmembrane protein elements can be used to locate the spin labelled sites with respect to the lipid-water interface. These data provide information on the topology of membrane protein elements, their orientation with respect to the lipid phase[31] and possible water penetration along the protein-lipid interface. Experimental results for helix F of BR are shown in figure 5b. The component A_{zz} of the hyperfine tensor reveals polar micro-environments for sites 162 and 168 at the cytoplasmic surface of BR and at position 129 close to its extracellular surface. As discussed above, the nitroxides attached to these sites are accessible for water molecules in agreement with the protein structure. The minimum values for A_{zz} are found for the nitroxides at positions 180, 184 and 187. According to the BR structure these sites are oriented to the lipid phase and located close to the centre of the membrane (cf. Fig. 2a). The values of A_{zz} found for positions 172 and 176 are between these polar and non polar limits. The behaviour of this polarity profile along the transmembrane helix F of BR is in line with the polarity profile found in frozen protein free hydrated lipid bilayers[32]. Restricted water penetration along the protein lipid interface might be the explanation for the relatively high polarity values in the vicinity of positions 172 and 176.

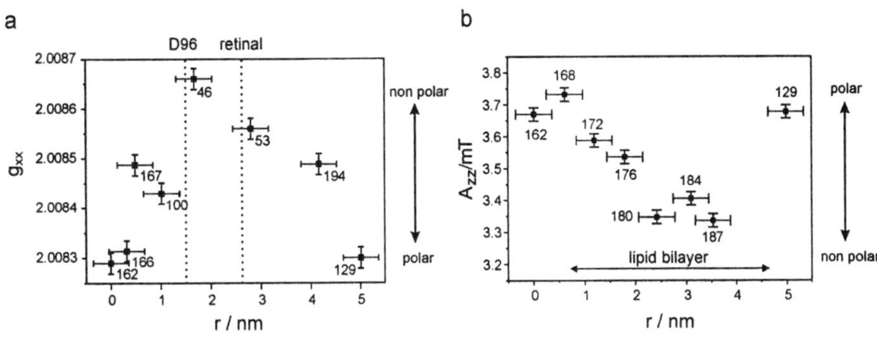

Figure 5. (a) The tensor element g_{xx} of spin labels oriented towards the aqueous phase or into the proton channel of BR as a function of the nitroxide location with respect to position 164[29]. The dotted lines indicate the positions of D96 and the Schiff-base of retinal, respectively. *(b)* The hyperfine tensor component A_{zz} for spin label side chains bound to the surface of helix F of BR as a function of the nitroxide location with respect to position 164[33] (Wegener, Pfeiffer, Steinhoff, unpublished). Here, the nitroxides are oriented either to the aqueous or to the lipid phase. The purple membrane spans the range from approximately 0.6 to 4.6 nm. The uncertainty of the nitroxide positions (horizontal error bars) was estimated from molecular modelling of different spin label side chain orientations with respect to the helix. Measurements were performed with frozen suspensions of purple membrane (170 K).

The analysis of g_{xx} as a function of the hyperfine tensor component A_{zz}, provides additional information on the proticity of the micro-environment of the spin label side chain. This dependence is plotted in Fig. 6 for the various spin label positions in BR. As pointed out earlier[29], the plot is suggestive of straight-line correlations. Theoretically, both g_{xx} and A_{zz} are expected to be linearly dependent on the π spin density ρ_O at the oxygen atom of the nitroxide group[28]. For A_{zz} this is evident from the relation $A_{zz} \propto \rho_N$ and the condition $\rho_N + \rho_O \approx 1$. For g_{xx}, however, apart from a direct proportionality to ρ_O, there is an additional dependence on specific electronic properties of the oxygen lone-pair orbitals, such as their degree of s, p_x, p_y –hybridization and their orbital energy. The lone-pair orbital energy E_n affects g_{xx} via the excitation energy $\Delta E_{n\to\pi^*} = E_{\pi^*} - E_n$ and is known to be sensitive to a polar environment, e.g. water, and to be particularly sensitive to direct H-bonding of the lone pairs to water or to polar amino acid residues[28]. Deviations from the straight line connecting the data of the polar (ethanol/water) and apolar (polystyrene/toluene) limits in the direction of higher/lower g_{xx}-values may therefore indicate a more aprotic/protic micro-environment of the nitroxide[28].

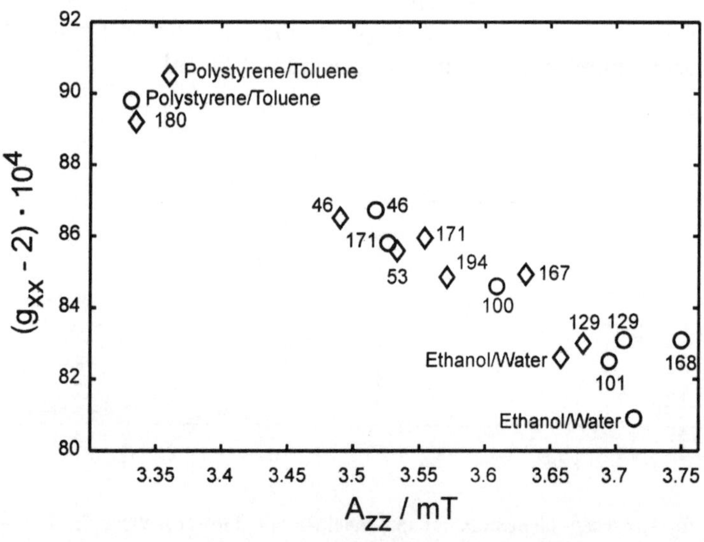

Figure 6. Plot of g_{xx} *vs.* A_{zz} of the nitroxide for various spin label positions in BR and for MTSSL in polystyrene/toluene or ethanol/water mixtures. Experimental errors for Q-band data (circles, Reyher and Steinhoff, unpublished) were estimated to be less than $\pm 5 \cdot 10^{-5}$ for g_{xx} and ±0.05 mT for A_{zz}. For W-band (diamonds) (data taken from Steinhoff et al.[29]) experimental errors are less than $\pm 2 \cdot 10^{-5}$ for g_{xx} and ±0.02mT for A_{zz}. Q-band and W-band data coincide within experimental error. The plot reveals that the polarity and proticity properties at the helix-lipid interface close to the centre of the purple membrane (position 180 in helix F of BR) resemble those of a polystyrene/toluene mixture.

4. INTER-SPIN DISTANCE MEASUREMENTS

The simultaneous exchange of two native amino acids by cysteines followed by modification with spin labels allows determination of inter-residual distances[34, 35] and thus provides a strategy for deducing proximity of selected secondary structural elements.

The spin-spin interaction between two spin labels attached to a protein is composed of static dipolar interaction, modulation of the dipolar interaction by the residual motion of the spin label side chains and exchange interaction. The static dipolar interaction leads to considerable broadening of the cw EPR spectrum if the inter-spin distance is less than 2 nm. For unique orientations of the nitroxides relative to each other as found for spin labels introduced at buried sites in the rigid limit a rigorous solution of the spin Hamiltonian of the system can be obtained. Spectra simulations yield the distance between the nitroxides and the Euler angles describing their relative orientation and that of the inter-spin vector relative to the magnetic field[34, 36]. For surface sites, the nitroxides may adopt a statistical distribution of distances and relative orientations. Values of the inter-spin distances can be determined from a detailed line shape analysis of spectra measured below 200 K[37-39] or in solutions of high viscosity[40] using spectra convolution or deconvolution techniques. Here, the superposition of the powder spectra of the two nitroxides is regarded as to be convoluted with the Pake pattern resulting from the dipolar interaction within the nitroxide pair. The nitroxides and the inter-spin vector are assumed to be randomly oriented in space. Since the spin labelling efficiency is less than 100% a variable fraction of singly spin labelled protein has to be accounted for. The lower limit for reliable distance determination using the above methods is given by the increasing influence of exchange interaction for inter-spin distances less than 0.8 nm due to partial overlap of the nitrogen pi-orbitals of the two interacting nitroxides.

The method of inter-spin distance determination has been successfully applied to a number of proteins, including rhodopsin[41, 42], lac permease[43], the KcsA potassium channel[8, 44] and alpha-crystallin[7]. In addition, changes in dipolar interaction can result in large spectral changes, making it straightforward to monitor conformational changes[44-46] (see below).

For inter-spin distances exceeding 2 nm the line broadening due to dipolar interaction is much less than the influence of other homogeneous and inhomogeneous contributions. Here pulsed EPR techniques are much more powerful. These techniques include pulsed ELDOR techniques such as four-pulse double electron-electron resonance (DEER)[47], the 2 + 1 pulse sequence[48], multiple-quantum EPR[49], and single-frequency techniques for refocusing (SIFTER) electron-electron couplings[50, 51].

Metal ion - nitroxide interactions in metallo-proteins or engineered copper-ion-binding sites allow estimation of intramolecular distances also at room temperature[52, 53]. A single metal ion provides a reference site for the estimation of distances to multiple nitroxide sites[54].

One attractive feature of site directed spin labelling is the ability to time resolve changes in any of the parameters discussed above. Changes in the protein secondary structure, protein tertiary fold or domain movements can be followed with up to 0.1 ms resolution with conventional EPR instrumentation and detection schemes (field modulation). Important examples found in the literature include the detection of rigid body helix motion in both rhodopsin and bacteriorhodopsin[45, 55-58], domain movements in T4 lysozyme[1], structural reorganization in colicin E1[59] upon membrane binding and conformational changes during signal transfer from sensory rhodopsin pSRII to the transducer pHtrII [9].

4.1 Inter-spin distances and distance changes in spin labelled bacteriorhodopsin

As an example for the elucidation of protein structure and conformational dynamics using dipole-dipole interaction within doubly spin labelled protein variants, an application of the method to BR is described. After light excitation BR undergoes a catalytic cycle, the so called photocycle, during which a proton is pumped from the cytoplasmic to the extracellular side of the membrane. The proton transport is accompanied by conformational changes as revealed by several different biophysical techniques (for a summary see, e.g., Subramaniam et al.[60, 61]). EPR spectroscopy of spin labelled BR mutants uncovered the time course of this conformational change[56-58, 62, 63]. Transient alterations of the spin label mobility during the BR photocycle were found in the loop regions A-B, C-D and E-F on the cytoplasmic surface of the protein. The results indicated a transient outward movement of the cytoplasmic part of helix F during the so called M intermediate of the photocycle which is characterized by a deprotonated Schiff base.

For a quantitative analysis of the structural changes pairs of spin labels were introduced into the C-D loop and the E-F loop. Five of these spin labelled positions are indicated in figure 7d. Corresponding EPR spectra of doubly spin labelled BR mutants are depicted in figure 7a and 7b. X-band[63] and Q-band spectra (A. Holt, Bachelor thesis, unpublished) of doubly spin labelled BR mutants are compared to the superposition of the spectra of the corresponding singly spin labelled samples. The doubly labelled samples show broader lines and consequently smaller amplitude due to spin-spin interaction. This increase of the line width is especially pronounced in the rising edge of the low field region of the Q-band spectra. Fitting of simulated powder spectra to the experimental data yielded inter-spin distance values and the fraction of singly labelled protein in the samples[63]. The values determined from X- and Q- band spectra for each double mutant agree within experimental error (A. Holt, Bachelor thesis, unpublished). A comparison of these values with inter-nitroxide distances determined from molecular dynamics simulations of the spin labelled protein reveals also nice agreement. Hence, the structure of the cytoplasmic moieties of helices C, E and F in the native purple membrane environment resembles that of the crystal structure. The fitting uncovered an amount of between 40% and 50% of non-interacting nitroxides. This percentage is in agreement with the experimentally determined spin labelling efficiency of between 70% and 80%.

Inter-spin distance changes during the photocycle were detected by freezing BR in different intermediates. The intermediate showing the largest conformational change was stabilized by illuminating the sample at 230 K with $\lambda > 580$ nm and fast cooling to 170 K in the dark. FT-Raman spectroscopy showed that this protocol lead to accumulation of the M intermediate to approximately 75%[63]. Compared to the ground state spectrum the line width of the M intermediate state spectrum of V101R1&A168R1 is considerably reduced. The inter-spin distance was calculated from the experimental data to increase by at least 0.1 nm. Concurrently, the dipolar line broadening visible in the EPR spectra of L100R1&S226R1 increased (data not shown). The corresponding inter-spin distance decrease amounts to 0.2 nm. No changes of the inter-spin distances were detected for V101R1&A160R1. These results reveal an outward movement of the cytoplasmic end of helix F and an inward tilt of helix G into the direction of the proton channel as shown in figure 7d. There is no distance change detectable between the positions 101 and 160, hence helices C and E are most probably not affected in the observed conformational change.

The time course of the helix movements in BR was resolved at room temperature by detection of the EPR signal amplitude changes at fixed B-field positions. This is again exemplified for the double mutant V101R1&A168R1 which monitors the outward movement of helix F (figure 7c).

After light activation a transient increase of the amplitude of the centre line due to a transient decrease of the dipolar interaction was observed. Time constants of the rise and decay of the transient EPR signal were determined from fitting of a superposition of exponentials. A comparison with the respective time constants for the photocycle intermediates of BR determined from optical and FTIR spectroscopy revealed that the observed

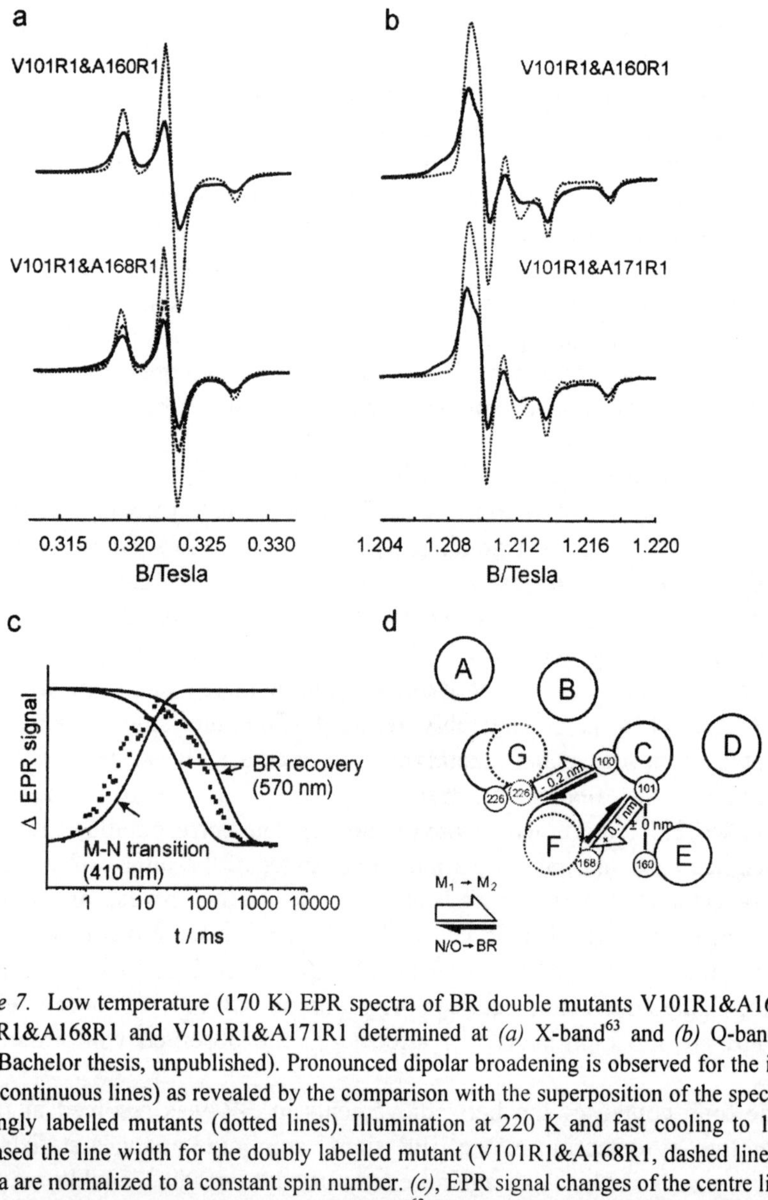

Figure 7. Low temperature (170 K) EPR spectra of BR double mutants V101R1&A160R1, V101R1&A168R1 and V101R1&A171R1 determined at *(a)* X-band[63] and *(b)* Q-band (A. Holt, Bachelor thesis, unpublished). Pronounced dipolar broadening is observed for the initial state (continuous lines) as revealed by the comparison with the superposition of the spectra of the singly labelled mutants (dotted lines). Illumination at 220 K and fast cooling to 170 K decreased the line width for the doubly labelled mutant (V101R1&A168R1, dashed line). All spectra are normalized to a constant spin number. *(c)*, EPR signal changes of the centre line of sample V101R1&A168R1 after photo-excitation[63] (squares, X-band, T=293 K). The solid lines show the time course of the M to N transition and two dominating rates of the recovery of the BR initial state determined by a multi-exponential fit of transients determined at 410 and 570 nm, respectively. The inter-spin distance between positions 101 and 168 increases prior to the M decay and returns into its initial value with a time course that can be fitted with the two dominating time constants of the BR initial state recovery. *(d)*, Top view of the cytoplasmic surface of BR with indicated inter-spin distances for the BR initial state and distance changes occurring during the photocycle. According to the EPR inter-spin distance measurements the outward movement of the cytoplasmic terminus of helix F by 0.1 nm is accompanied by an inward shift of the cytoplasmic terminus of helix G by 0.2 nm[63].

conformational change occurs in phase or slightly prior to the decay of the M intermediate[63]. Hence, the trigger for this conformational change must be independent from the reprotonation of the Schiff base. The recovery of the ground state conformation of helix F can be satisfactorily fitted with the two dominating relaxation time constants of the optical transient determined at 570 nm which indicates the back reaction of the retinal to its initial state.

4.2 Protein-protein interaction: the sensory rhodopsin – transducer complex

The archeabacterial photoreceptors sensory rhodopsin I (SRI) and II (SRII) provide the initial signal which enables the cells to seek favourable light conditions (for reviews, see, e.g., Spudich et al.[64, 65]). The two photoreceptors are tightly complexed to receptor specific transducers (HtrI and HtrII). A light induced signal is transferred from the receptor to the cytoplasmic domain of the transducer which activates a two-component signalling cascade well known from bacterial chemotaxis. Both transducers contain two membrane spanning helices which are thought to be involved in the signal transfer.

A structural model of the pSRII - pHtrII complex from *Natronobacterium pharaonis* was determined from an EPR analysis of a set of singly and doubly spin labelled protein variants reconstituted into purple membrane lipid bilayers (Fig. 8)[9]. Inter-spin distances were obtained for 26 pairs of spin labels introduced into the transmembrane helices TM1 or TM2 of the transducer pHtrII and helices F or G of the receptor pSRII. Considerable spin-spin interactions observed for the singly labelled transducer with the nitroxide side chain located at positions 78 or 82 confirmed a dimeric arrangement of TM2 and TM2'. The strongest interaction corresponding to the closest distances (less than 1.2 nm) between pSRII and pHtrII were observed between positions 25 and 210, 81 and 210, and between 81 and 211 indicating that both TM1 and TM2 are close to helix G. According to these inter-spin distance values and the distances between the other pairs TM2 had to be located between the C-terminal helices F and G as depicted in figure 8[9]. The results revealed a quaternary complex between two copies of pHtrII and pSRII each, forming a structure with an apparent two-fold symmetry. The crystal structure of the complex confirmed the EPR derived model[66].

On light excitation the pSRII receptor reverts to the long lived M-intermediate which was proposed to represent the signalling state[67]. During this reaction helix F of the receptor moves outwardly similar to the conformational change found in BR[68] (see also Fig. 7d). This helix movement of the receptor is transferred to the transducer as revealed by comparing the inter-residual distance changes observed between transducer /

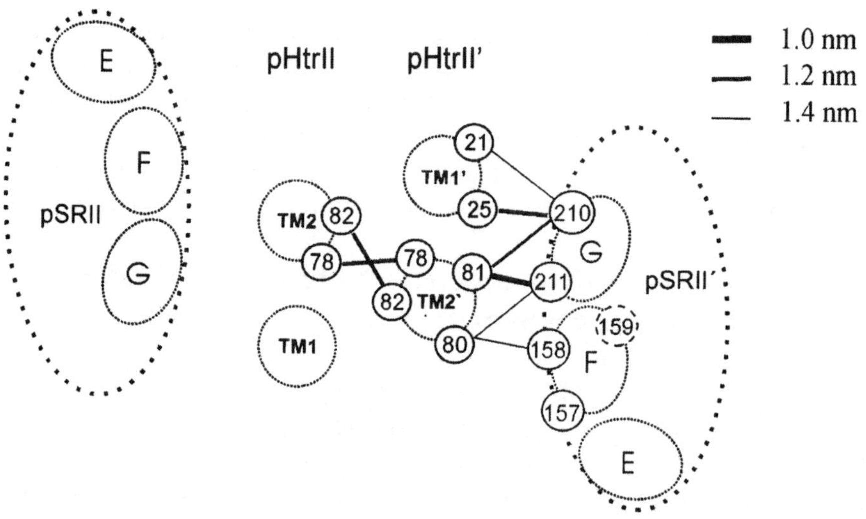

Figure 8. Schematic model of the transmembrane region of the 2:2-complex of pSRII with its transducer pHtrII (view from the cytoplasm). The transmembrane helices of the transducer, TM1 and TM2, and helices E, F and G of the receptor were oriented with respect to each other based on dipolar spin-spin interactions and derived distance constraints[9]. Strong dipolar interactions due to inter-spin distances less than 1.4 nm between spin labelled positions of pSRII and pHtrII or within the pHtrII dimer are coded by the thickness of connecting lines.

transducer as well as receptor / transducer[9]. Low temperature experiments (T = 170 K) with the receptor trapped in the M intermediate showed decreases and increases of the interaction strengths between the TM2 and TM2' and between TM2 and helix F of pSRII. Residues S158R1 and K157R1 of helix F approached TM2-residues A80R1 and T81R1 in M. Concurrently, the strong dipolar interaction between the two V78R1 in neighbouring TM2 helices was significantly reduced revealing a rearrangement of TM2.

In figure 9 the time course of the EPR signal change of V78R1 at fixed B-field position is compared to that of L159R1. The nitroxides at positions V78R1 of TM2 and V78R1' of TM2' face each other leading to considerable dipolar broadening of the spectra. L159R1 is located in helix F and oriented towards the inside of the receptor (cf. figure 8). An optical trace monitoring the depletion and reformation of the pSRII ground state (measured at 500 nm) is also depicted in figure 9. The kinetic EPR difference spectra recorded at room temperature represented features of a transient increase of the inter-spin distance for sample V78R1 and a transient mobilization for L159R1[9]. Thus, the three signals shown in figure 9 record events occurring at the level of the retinal chromophore (500 nm optical trace), of helix F (L159R1, EPR trace), and of TM2 (V78R1, EPR trace) which allows to follow the signal transfer in the sequence retinal - helix F - TM2 and vice versa. After absorption of a photon by the retinal the activated state remains unchanged for about 100 ms. With the reformation of the ground state, the reaction of the receptor seems to be decoupled from that of the transducer. The movement of helix F into the original position seems to precede the recovery of TM2 position. This decoupling allows the system to modulate the activation/deactivation of the transducer, thereby enabling the bacteria to respond to external stimuli adequately[9].

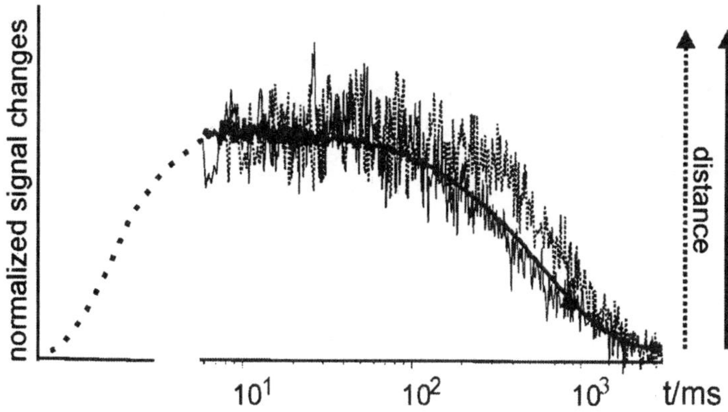

Figure 9. EPR transients (noisy lines) and the corresponding optical traces recorded at 500 nm (continuous line) of the receptor – transducer complex pSRII - pHtrV78R1 (dotted line), and of L159R1-pHtrII (continuous line) after light activation (T=293K)[9]. The EPR signal changes in the complex uncover a transient increase of the distance between positions 78 and 78' of the two transmembrane helices of the transducer dimer, TM2 and TM2' (see Fig. **8**). The transient increase of the mobility of the spin label side chain L159R1 indicates an outward movement of helix F of pSRII[9] similar to that found in BR[63]. The rise of the EPR signals could not be resolved in this experiment.

4.3 Inter-spin distance determination by four-pulse DEER in PutP

Double electron-electron resonance (DEER)[47], in combination with site-directed spin labelling, has been applied to determine details of the so far unknown structure of the Na^+/proline transporter PutP of Escherichia coli[69]. PutP is a member of the Na^+/substrate symporter family (SSSF) which currently comprises more than 200 similar proteins of pro- and eukaryotic origin[10]. These integral membrane proteins utilize the Na^+ electrochemical gradient to drive the transport of a variety of substrates like, e.g., sugars, amino acids and vitamins. Biochemical and biophysical studies suggest a secondary structure model according to which PutP contains 13 transmembrane helices (TMs) with the N terminus located on the periplasmic side of the membrane and the C terminus facing the cytoplasm[70, 71] (see also figure 10).

EPR spectroscopy data of singly spin labeled PutP variants[71] in the presence and absence of Na^+ or proline lead to the conclusion that binding of ligands induces conformational alterations that involve at least part of TM II and the preceding cytoplasmic loop (loop L2). Further investigations of the arrangement of the cytoplasmic loops were performed with the double cysteine PutP derivatives L37C&A107C and L37C&D187C. These variants were used to determine distances between the cytoplasmic loops L2 and L4 and between the loops L2 and L6. In addition the doubly spin labeled PutP derivative A107C&S223C provide nitroxides positions on different sides of the lipid membrane (loops L4 and L7).

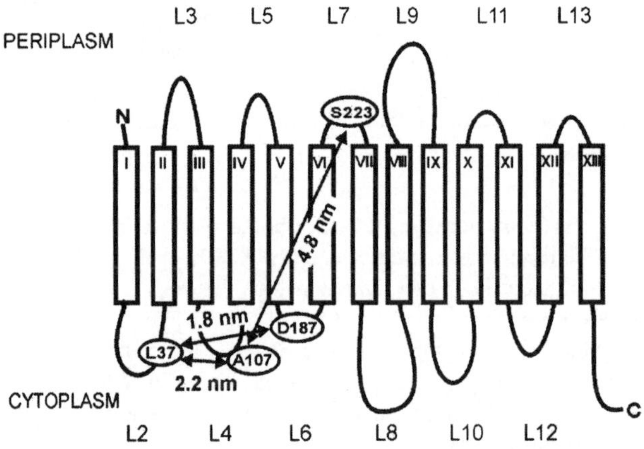

Figure 10. Secondary structure model of PutP with the spin labelled sites highlighted. Given are the mean distances between these sites as determined by four-pulse DEER[69].

Cw EPR experiments showed that the inter-spin distances within these three samples exceed 1.8 nm. Hence, DEER experiments were performed which were shown to provide inter-spin distance resolution in the range from 2 to 8 nm[72]. By using a model-free direct transformation of the DEER time domain data with the crosstalk-corrected approximate Pake transformation (APT)[73] a broad distance distribution was obtained for the doubly spin labeled mutant L37R1&D187R1 (Fig. 11). The maximum of the distribution was found to be close to the lower distance limit of the pulse DEER method (~1.8 nm) and the asymmetric width of the distribution was estimated to approach 0.8 nm. The results suggest that the loops L2 and L6 are in close proximity and that these regions are very poorly ordered. Although L37 itself is not important for PutP function, R40 in its vicinity plays an important role in coupling of Na^+ and proline transport[74]. Similarly, D187 in loop L6 is crucial for an efficient coupling of the transport of both ligands[75]. Both residues are highly conserved within the SSSF. The DEER measurements provided evidence that R40 and D187 may interact with each other during the transport cycle. This is even more pronounced by the observation that the distance between positions 37 and 187 increased upon Na^+ binding (cf. Fig. 11). In contrast to that, Na^+ and proline did not significantly influence the DEER signal of L37R1&A107R1. This indicates that the relative arrangement of loops L2 and L4 are only minor modulated by the ligands. The large value of 4.8 nm for the inter-spin distance determined between position 107 in loop 4 and position 223 in loop 7 is strong evidence that these positions are located on opposite sides of the membrane. The above results demonstrate that four-pulse DEER spectroscopy is a powerful means to investigate the structure and conformational changes of integral membrane proteins reconstituted in liposomes.

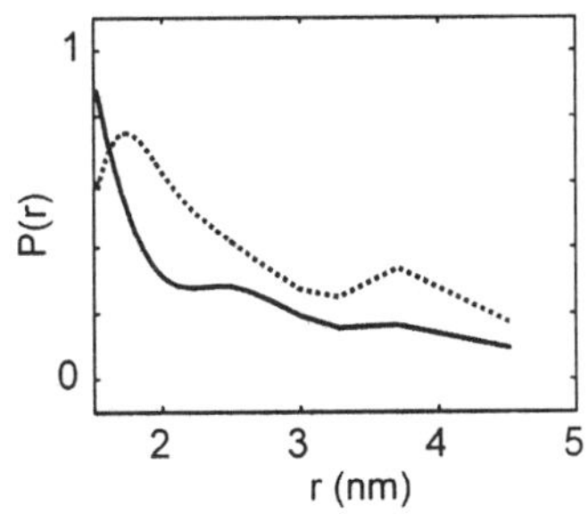

Figure 11. Inter-spin distance distribution, P(r), for spin label side chains L37R1 and D187R1 in the absence (continuous line) and presence of Na^+ (dotted line) as determined by pulse DEER (modified[69]). The shift of the maximum of P(r) reveals that loops L2 and/or L4 of PutP undergo a Na^+ induced conformational change.

ACKNOWLEDGEMENTS

We gratefully acknowledge the support of the Deutsche Forschungsgemeinschaft (SFB431-P18).

REFERENCES

1. Hubbell, W.L., H.S. McHaourab, C. Altenbach, and M.A. Lietzow, 1996, Watching proteins move using site-directed spin labeling. *Structure* **4**: 779-783.
2. Hubbell, W.L., A. Gross, R. Langen, and M.A. Lietzow, 1998, Recent advances in site-directed spin labeling of proteins. *Curr. Opin. Struct. Biol.* **8**: 649-656.

3. Hubbell, W.L., D.S. Cafiso, and C. Altenbach, 2000, Identifying conformational changes with site-directed spin labeling. *Nat. Struct. Biol.* **7**: 735-739.
4. Feix, J.B. and C.S. Klug, Site-directed spin labeling of membrane proteins and peptide-membrane interactions, in *Spin Labeling: The Next Millenium*, L.J. Berliner, Editor. 1998, Plenum Press: New York. p. 251-281.
5. Steinhoff, H.-J., 2002, Methods for study of protein dynamics and protein-protein interaction in protein-ubiquitination by electron paramagnetic resonance spectroscopy. *Frontiers in Bioscience* **7**: c97-110.
6. Mchaourab, H.S. and E. Perozo, Determination of protein folds and conformational dynamics using spin-labeling EPR spectroscopy., in *Distance Measurements in Biological Systems by EPR*, L.J. Berliner, Eaton, S.S., Eaton, G.R., Editor. 2000, Kluwer: New York.
7. Koteiche, H.A. and H.S. Mchaourab, 1999, Folding pattern of the a-crystallin domain in aA-crystallin determined by site-directed spin labeling. *J. Mol. Biol.* **294**: 561-77.
8. Perozo, E., D.M. Cortes, and L.G. Cuello, 1998, Three-dimensional architecture of a K+ channel: implications for the mechanism of ion channel gating. *Nature of Structural Biology* **5**: 459-469.
9. Wegener, A.A., J.P. Klare, M. Engelhard, and H.-J. Steinhoff, 2001, Structural insights into the early steps of receptor-transducer signal transfer in archaeal phototaxis. *EMBO J.* **20**: 5312-5319.
10. Jung, H., 2001, Towards the molecular mechanism of Na+/solute symport in prokaryotes. *Biochim. Biophys. Acta* **1505**: 131-143.
11. Berliner, L.J., J. Grunwald, H.O. Hankovszky, and K. Hideg, 1982, A novel reversible thiol-specific spin label: papain active site labeling and inhibition. *Anal. Biochem.* **119**: 450-455.
12. Berliner, L.J., *Spin Labeling : Theory and Applications*. 1976, New York: Academic Press.
13. Berliner, L.J., *Spin Labeling II: Theory and Applications*. 1979, New York: Academic Press.
14. Berliner, L.J. and J. Reuben, *Biological Magnetic Resonance. Vol. VIII: Spin Labeling Theory and Applications*, ed. L.J. Berliner and J. Reuben. 1989, New York: Plenum Press.
15. Berliner, L.J., *Spin Labeling: The Next Millenium*. 1998, New York: Academic Press.
16. Mchaourab, H.S., M.A. Lietzow, K. Hideg, and W.L. Hubbell, 1996, Motion of spin-labeled side chains in T4 lysozyme. Correlation with protein structure and dynamics. *Biochemistry* **35**: 7692-704.
17. Steinhoff, H.J., M. Müller, C. Beier, and M. Pfeiffer, 2000, Molecular dynamics simulation and EPR spectroscopy of nitroxide side chains in bacteriorhodopsin. *J. Molecular Liquids* **84**: 17-27.
18. Pfeiffer, M., T. Rink, K. Gerwert, D. Oesterhelt, and H.J. Steinhoff, 1999, Site-directed spin-labeling reveals the orientation of the amino acid side-chains in the E-F loop of bacteriorhodopsin. *J. Mol. Biol.* **287**: 163-171.
19. Mchaourab, H.S., T. Kalai, K. Hideg, and W.L. Hubbell, 1999, Motion of spin-labeled side chains in T4 lysozyme: Effect of side chain structure. *Biochemistry* **38**: 2947-2955.
20. Altenbach, C., T. Marti, H.G. Khorana, and W.L. Hubbell, 1990, Transmembrane protein structure: spin labeling of bacteriorhodopsin mutants. *Science* **248**: 1088-1092.
21. Farahbakhsh, Z., C. Altenbach, and W.L. Hubbell, 1992, Spin labeled cysteines as sensors for protein-lipid interaction and conformation in rhodopsin. *Photochem. Photobiol.* **56**: 1019-1033.
22. Hubbell, W.L. and C. Altenbach, 1994, Investigation of structure and dynamics in membrane proteins using site-directed spin labeling. *Curr. Opin. Struct. Biol.* **4**: 566-573.
23. Altenbach, C., D.A. Greenhalgh, H.G. Khorana, and W.L. Hubbell, 1994, A collision gradient method to determine the immersion depth of nitroxides in lipid bilayers: application to spin-labeled mutants of bacteriorhodopsin. *Proc. Natl. Acad. Sci. U S A* **91**: 1667-1671.

24. Burghaus, O., M. Rohrer, T. Gotzinger, M. Plato, and K. Möbius, 1992, A novel high-field/high-frequency EPR and ENDOR spectrometer operating at 3 mm wavelength. *Measurement Science & Technology* **3**: 765-774.
25. Prisner, T.F., A. Vanderest, R. Bittl, W. Lubitz, D. Stehlik, and K. Möbius, 1995, Time-Resolved W-Band (95 Ghz) Epr Spectroscopy of Zn-Substituted Reaction Centers of Rhodobacter Sphaeroides R-26. *Chem. Phys.* **194**: 361-370.
26. Huber, M. and J.T. Törring, 1995, High-field EPR on the primary electron donor cation radical in single crystals of heterodimer mutant reaction centers of photosynthetic bacteria - first characterization of the G-tensor. *Chem. Phys.* **194**: 379-385.
27. Stone, A.J., 1963, *Proc. Roy. Soc. London* **A271**: 424-434.
28. Plato, M., H.-J. Steinhoff, C. Wegener, J.T. Törring, A. Savitsky, and K. Möbius, 2002, Molecular orbital study of polarity and hydrogen bonding effects on the g and hyperfine tensors of site directed NO spin labeled bacteriorhodopsin. *Molecular Physics* **100**: 3711-3721.
29. Steinhoff, H.-J., A. Savitsky, C. Wegener, M. Pfeiffer, M. Plato, and K. Möbius, 2000, High-field EPR studies of the structure and conformational changes of site-directed spin labeled bacteriorhodopsin. *Biochim. Biophys. Acta* **1457**: 253-262.
30. Steinhoff, H.J., M. Pfeiffer, T. Rink, O. Burlon, M. Kurz, J. Riesle, E. Heuberger, K. Gerwert, and D. Oesterhelt, 1999, Azide reduces the hydrophobic barrier of the bacteriorhodopsin proton channel. *Biophys. J.* **76**: 2702-2710.
31. Kveder, M., A. Kriško, G. Pifat, and H.-J. Steinhoff, 2003, The study of structural accessibility of free thiol groups in human low-density lipoproteins. *Biochim. Biophys. Acta* **1631**: 239-245.
32. Griffith, O.H., P.J. Dehlinger, and S.P. Van, 1974, Shape of the hydrophobic barrier of phospholipid bilayers (Evidence for water penetration in biological membranes). *J Membrane Biol* **15**: 159-192.
33. Wegener, C., *PhD thesis*. 2002, Bochum: Ruhr-Universitaet Bochum.
34. Hustedt, E.J. and A.H. Beth, 1999, Nitroxide spin-spin interactions: applications to protein structure and dynamics. *Annu. Rev. Biophys. Biomol. Struct.* **28**: 129-153.
35. Eaton, G.R., S.S. Eaton, and L.J. Berliner, *Distance Measurements in Biological Systems*. 2000, New York: Kluwer.
36. Hustedt, E.J., A.I. Smirnov, C.F. Laub, C.E. Cobb, and A.H. Beth, 1997, Molecular distances from dipolar coupled spin-labels - the global analysis of multifrequency continuous wave electron paramagnetic resonance data. *Biophys. J.* **72**: 1861-1877.
37. Steinhoff, H.-J., O. Dombrowsky, C. Karim, and C. Schneiderhahn, 1991, Two dimensional diffusion of small molecules on protein surfaces: an EPR study of the restricted translational diffusion of protein-bound spin labels. *Eur. Biophys. J.* **20**: 293-303.
38. Rabenstein, M.D. and Y.K. Shin, 1995, Determination of the distance between two spin labels attached to a macromolecule. *Proc. Natl. Acad. Sci. USA* **92**: 8239-8243.
39. Steinhoff, H.J., N. Radzwill, W. Thevis, V. Lenz, D. Brandenburg, A. Antson, G. Dodson, and A. Wollmer, 1997, Determination of interspin distances between spin labels attached to insulin: comparison of electron paramagnetic resonance data with the X- ray structure. *Biophys. J.* **73**: 3287-3298.
40. Altenbach, C., K.J. Oh, R.J. Trabanino, K. Hideg, and W.L. Hubbell, 2001, Estimation of inter-residue distances in spin labeled proteins at physiological temperatures: experimental strategies and practical limitations. *Biochemistry* **40**: 15471-82.
41. Cai, K.W., R. Langen, W.L. Hubbell, and H.G. Khorana, 1997, Structure and Function in Rhodopsin - Topology of the C-Terminal Polypeptide Chain in Relation to the Cytoplasmic Loops. *Proc. Natl. Acad. Sci. USA* **94**: 14267-14272.

42. Altenbach, C., K. Yang, D.L. Farrens, Z.T. Farahbakhsh, H.G. Khorana, and W.L. Hubbell, 1996, Structural features and light-dependent changes in the cytoplasmic interhelical E-F loop region of rhodopsin - a site-directed spin-labeling study. *Biochemistry* **35**: 12470-12478.
43. Voss, J., W.L. Hubbell, and H.R. Kaback, 1998, Helix Packing in the Lactose Permease Determined By Metal-Nitroxide Interaction. *Biochemistry* **37**: 211-216.
44. Perozo, E., D.M. Cortes, and L.G. Cuello, 1999, Structural rearrangement underlying K^+-channel activation gating. *Science* **285**: 73-78.
45. Farrens, D.L., C. Altenbach, K. Yang, W.L. Hubbell, and H.G. Khorana, 1996, Requirement of rigid-body motion of transmembrane helices for light activation of rhodopsin. *Science* **274**: 768-770.
46. Tiebel, B., N. Radzwill, L.M. Aung-Hilbrich, V. Helbl, H.J. Steinhoff, and W. Hillen, 1999, Domain motions accompanying Tet repressor induction defined by changes of interspin distances at selectively labeled sites. *J. Mol. Biol.* **290**: 229-240.
47. Pannier, M., S. Veit, A. Godt, G. Jeschke, and H.W. Spiess, 2000, Dead-time free measurement of dipole-dipole interactions between electron spins. *J. Magn. Reson.* **142**: 331-340.
48. Kurshev, V.V., A.M. Raitsimring, and Y.D. Tsvetkov, 1989, Selection of dipolar interaction by the 2+1 pulse train ESE. *J. Magn. Reson.* **81**: 441-454.
49. Borbat, P.P. and J.H. Freed, 1999, Multi-quantum ESR and distance measurements. *Chem. Phys. Lett.* **313**: 145-154.
50. Jeschke, G., M. Pannier, A. Godt, and H.W. Spiess, 2000, Dipolar spectroscopy and spin alignment in electron paramagnetic resonance. *Chem. Phys. Lett.* **331**: 234-252.
51. Jung, K., J. Voss, M. He, W.L. Hubbell, and H.R. Kaback, 1995, Engineering a Metal Binding Site Within a Polytopic Membrane Protein, the Lactose Permease of Escherichia Coli. *Biochemistry* **34**: 6272-6277.
52. Leigh, J.S., 1970, ESR rigid lattice line shape in a system of two interacting spins. *J. Chem. Phys.* **52**: 2608-2612.
53. Voss, J., L. Salwinski, H.R. Kaback, and W.L. Hubbell, 1995, A method for distance determination in proteins using a designed metal ion binding site and site-directed spin labeling - evaluation with T4 lysozyme. *Proc. Natl. Acad. Sci. USA* **92**: 12295-12299.
54. Voss, J., W.L. Hubbell, and H.R. Kaback, 1995, Distance determination in proteins using designed metal ion binding sites and site-directed spin labeling - application to the lactose permease of escherichia coli. *Proc. Natl. Acad. Sci. USA* **92**: 12300-12303.
55. Farahbakhsh, Z., K. Hideg, and W.L. Hubbell, 1993, Photoactivated conformational changes in rhodopsin: a time resolved spin label study. *Science* **262**: 1416-1420.
56. Steinhoff, H.J., R. Mollaaghababa, C. Altenbach, K. Hideg, M. Krebs, H.G. Khorana, and W.L. Hubbell, 1994, Time-resolved detection of structural changes during the photocycle of spin-labeled bacteriorhodopsin. *Science* **266**: 105-107.
57. Thorgeirsson, T.E., W.Z. Xiao, L.S. Brown, R. Needleman, J.K. Lanyi, and Y.K. Shin, 1997, Transient channel-opening in bacteriorhodopsin - an EPR study. *J. Mol. Biol.* **273**: 951-957.
58. Xiao, W., L.S. Brown, R. Needleman, J.K. Lanyi, and Y.-K. Shin, 2000, Light-induced rotation of a transmembrane alpha-helix in bacteriorhodopsin. *J. Mol. Biol.* **304**: 715-721.
59. Shin, Y., C. Levinthal, F. Levinthal, and W.L. Hubbell, 1993, Colicin E1 binding to membranes: time resolved studies of spin-labeled mutants. *Science* **259**: 960-963.
60. Subramaniam, S., I. Lindahl, P. Bullough, A.R. Faruqi, J. Tittor, D. Oesterhelt, L. Brown, J. Lanyi, and R. Henderson, 1999, Protein conformational changes in the bacteriorhodopsin photocycle. *J. Mol. Biol.* **287**: 145-161.
61. Subramaniam, S. and R. Henderson, 2000, Crystallographic analysis of protein conformational changes in the bacteriorhodopsin photocycle. *Biochim. Biophys. Acta* **1460**: 157-165.

62. Rink, T., M. Pfeiffer, D. Oesterhelt, K. Gerwert, and H.J. Steinhoff, 2000, Unraveling photoexcited conformational changes of bacteriorhodopsin by time resolved electron paramagnetic resonance spectroscopy. *Biophys. J.* **78**: 1519-1530.
63. Radzwill, N., K. Gerwert, and H.-J. Steinhoff, 2001, Time-resolved detection of transient movement of helices F and G in doubly spin-labeled bacteriorhodopsin. *Biophys. J.* **80**: 2856-2866.
64. Spudich, J.L., 1998, Variations on a molecular switch - transport and sensory signalling by archaeal rhodopsins. *Mol. Microbiol.* **28**: 1051-1058.
65. Spudich, J.L., C.S. Yang, K.H. Jung, and E.N. Spudich, 2000, Retinylidene proteins: Structures and functions from archaea to humans. . *Annu. Rev. Cell Dev. Biol.* **16**: 365-392.
66. Gordeliy, V.I., J. Labahn, R. Moukhametzianov, R. Efremov, J. Granzin, R. Schlesinger, G. Büldt, T. Savopol, A.J. Scheidig, J.P. Klare, and M. Engelhard, 2002, Molecular basis of transmembrane signalling by sensory rhodopsin II-transducer complex. *Nature* **419**: 484-487.
67. Yan, B., T. Takahashi, R. Johnson, and J.L. Spudich, 1991, Identification of signaling states of a sensory receptor by modulation of lifetimes of stimulus-induced conformations: the case of sensory rhodopsin II. *Biochemistry* **30**: 10686-92.
68. Wegener, A.A., I. Chizhov, M. Engelhard, and H.J. Steinhoff, 2000, Time-resolved Detection of Transient Movement of Helix F in Spin-labelled Pharaonis Sensory Rhodopsin II. *J. Mol. Biol.* **301**: 881-891.
69. Jeschke, G., C. Wegener, M. Nietschke, H. Jung, and H.-J. Steinhoff, 2004, Inter-residual distance determination by four-pulse DEER in an integral membrane protein: the Na+/proline transporter PutP of Escherichia coli. *Biophys. J.,* in press.
70. Jung, H., R. Rübenhagen, S. Tebbe, K. Leifker, N. Tholema, M. Quick, and R. Schmid, 1998, Topology of the Na+/Proline transporter of Escherichia coli. *J. Biol. Chem.* **273**: 26400-26407.
71. Wegener, C., S. Tebbe, H.J. Steinhoff, and H.R. Jung, 2000, Spin labeling analysis of structure and dynamics of the Na+/proline transporter of Escherichia coli. *Biochemistry* **39**: 4831-4837.
72. Jeschke, G., 2002, Distance measurements in the nanometer range by pulse EPR. *Chem. Phys. Chem.* **3**: 927-932.
73. Jeschke, G., A. Koch, U. Jonas, and A. Godt, 2002, Direct conversion of EPR dipolar time evolution data to distance distributions. *J. Magn. Reson.* **155**: 72-82.
74. Quick, M., S. Stölting, and H. Jung, 1999, Role of conserved Arg40 and Arg117 in the Na+/proline transporter of Escherichia coli. *Biochemistry* **38**: 13523-13529.
75. Quick, M. and H. Jung, 1998, A conserved aspartate residue, Asp187, is important for Na+-dependent proline binding and transport by the Na+/Proline transporter of Escherichia coli. *Biochemistry* **37**: 13800-13806.

Identification of Protein Structure and its Modifications by Electrospray Mass Spectrometry in Proteomics

JASNA PETER-KATALINIĆ
Institut für Medizinische Physik und Biophysik, Universität Münster, Germany

1. INTRODUCTION

In the post genome era a great deal of efforts is invested in the characterization of proteins and their compositional analysis, essential for understanding of their biological function and mechanisms of interaction[1,2]. Research in proteomics in the next logical step after genomics in understanding life processes at the molecular level. Advanced genome sequence information provides only an initial information on protein structure and an incomplete prediction of protein function[3]. Proteomics encompasses knowledge of the structure, function and expression of all proteins in the biochemical or biological context of all organisms. Two-dimensional polyacrylamide gel electrophoresis (2D PAGE), described in its modern form in 1975[4,5] became the dominant tool for the study of complex protein mixtures. Thousands of proteins can be resolved on a single gel, and qualitative and quantitative differences between samples be detected. Identification of protein spots from 2D PAGE has been originally addressed by automatized Edman degradation analysis[6] requiring orders of magnitude of protein material higher than that being purified on one 2D gel. The identification procedure was greatly advanced by developments in mass spectrometry (MS), electrospray ionization (ESI)[7] and matrix-assisted laser desorption ionization (MALDI)[8]. These techniques were shown to be significantly more sensitive and able to cope with the structural

Supramolecular Structure and Function 8, Edited by Pifat-Mrzljak
Kluwer Academic/Plenum Publishers, New York 2004

identification of complex peptide mixtures, obtained directly from separated protein spots after enzymatic proteolysis and extraction from the gel[9-12] The expressions "peptide map" and "sequence tag" arose from detection of peptide molecular ions in complex mixtures, primarily by MALDI-TOF and by obtaining partial or full sequences of them by tandem MS, primarily by ESI analysers[13-16]. Modern technology features robotic cutting of protein spots from a 2D PAGE followed by automatic enzymatic digestion and transfer for analysis by MALDI and/or ESI-MS and MS/MS.

Identification of proteins provides a tool to verify sequences obtained by genome sequencing. Besides, the information, which is not available from the gene sequences, namely on alternative splicing, post-translational processing and protein modifications, which determine the protein function, is deduced from the sequence data obtained on the protein level. The mandate of proteomics according to the National Research Council Steering Committee (USA) conclusions from 2002[17] is defined following the issues of:

a) protein separation and identification
b) determination of protein structure and function
c) knowledge on metabolic pathways and post-translational modifications
d) development of new platforms and technologies
e) integration of bioinformatics and finally
f) the applications to clinical proteomics.

Combining existing highly developed technologies, like in mass spectrometry and hyphenated techniques to MS, with potential of data base search and bioinformatic, protein identification became the primary target of interest to be introduced on a broad scale in biochemical laboratories. In determination of protein structure, mass spectrometry can provide the essential service of *de-novo* sequencing to establish the structural base for functional studies.

The global proteome characterization is based on analysis of peptide mixtures released from protein digests and their amino acid sequence. Therefore, the separation and identification of peptides and their composition in amino acids represent a crucial step in elucidation of protein structure.

2. MASS SPECTROMETRY IN PROTEOMICS

2.1 Identification of proteins by mass mapping after gel electrophoresis and in-gel digestion

A significant break-through in proteomics was the introduction of peptide mass fingerprinting using MALDI-TOF MS[18]. The possibility to identify a protein from a complex mixture by determining the mass of its tryptic peptides at the low fmol level represents a clear advantage of two orders of magnitude over the Edman degradation concerning the sample consumption. Additional arguments in favor of the MS procedure, the speed and significantly lower running cost, and the availability of genome sequences contributed to reduce this approach during the last few years to the mass values alignment in comparison to the sequences deposited in databases.

The general procedure includes the reduction and alkylation of proteins in the mixture prior to gel separation. Gel spots can be stained by Coomassie blue, silver or Sypro Ruby, cut out and destained[19]. After proteolysis using an appropriate protease the generated mixture of peptides is extracted from the gel, desalted on a nanocolumn and admitted to the MALDI target to cocrystallize with the organic matrix for the mass analysis. The mass spectrum provides in an optimal case all peptide fragments, appearing as molecular $[M+H]^+$ ions in the positive ion analysis. Accurate mass determination under the use of internal calibrants should provide in standard TOF analyzers an accuracy better then 0.1 Da. By submitting the data to the search engines for protein identification by peptide alignment a hit list of potential candidates is created, from which the clues about the protein identity can be drawn.

Practical drawback of this procedure is a distinct ion formation capability of single peptide in the proteolytic mixture, depending on its constitution, influencing their abundance in the spectrum and a type and number of their co- and posttranslational modifications. Another obstacle is a difficult detection of ionic components of minor abundance in complex mixtures. According to the presence of acidic functional groups, $[M+Na]^+$ and/or $[M+K]^+$ can be formed instead, particularly in phosphorylated and glycosylated peptides. Misscuts by proteases leave a modified peptide structure behind (e.g. at the two adjacent Arg or Lys moieties) which can be only partially tolerated by search engines. Common modifications as cysteine derivatization, phosphorylation and oxidation of methionines can be included in the search program and handled properly.

Four common search engines including their URL are ProFound (http://www.proteometrics.com), Mascot (http://www.matrixscience.com), PeptideSearch (http://www.mann.embl-heidelberg.de) and MSFit (http://prospector.ucsf.edu).

The sensitivity of MALDI-TOF MS for identification of a single spot protein from a silver-stained 2D gel[20] according to the peptide map is not always sufficient. In such cases several parallel gel can be run and analog spots pooled.

2.2 MS sequencing of peptides by electrospray low energy collision induced dissociation (ESI-CID)

In electrospray MS analysis deliver a different picture than that of the MALDI is obtained from complex peptide mixtures, according to their inherent property of multicharged ions formation. For proteomics-type experiments nanospray (nanoESI)[21] is favored for sample admission allowing a nanoflow infusion and therefore a much lower consumption of material, than in electrospray with continuous infusion and higher flow values. In general, 5-10 µl of the sample containing less than 1 pmol in average, are applied to the capillary of 100 nm diameter to be inserted into the ion source. To provide electric contact the capillaries are either metal-coated or a contact is formed via a wire inserted in the solution[22].

Sequencing of peptides include three steps: (a) identification of a precursor ion (b) selection of a precursor ion inclusive the CID experiment and (c) identification of the peptide sequence/ interpretation of fragment ions. For precursor ion screening MS1 is performed over a broad range to encompass all ions present and to inspect the ion density which may be present due to overlapping of peptide molecular ions with similar m/z value due to formation of multicharge states or partial overlapping of isotopic patterns in ions with similar m/z values. The higher the resolution of the analyzer, the easier is the isolation of the precursor ion for fragmentation analysis. The subsequent MS/MS sequencing is regulated by adjusting the collision energy and the pressure of the gas in the collision cell. The adjustment of these parameters is of high relevance for sequencing of modified peptides, in particular by glycosylation, where the glycosidic bond at the sugar attachment site can break prior to the cleavage of the peptide bond, and the information on the glycosylation site is therefore lost[23]. In case of the identification of a protein, whithout getting into the details of structure modifications, and if the protein of interest is already present in a protein database, usually a short amino acid sequence is sufficient to identify it ("sequence tag")[24]. The process of peptide fragmentation occur basically at three sites within an amino acid unit delivering 6 fragment ions, three from the N-terminus (a, b, c) and three from the C-terminus (x, y, z) (Fig. 1). In most cases, however, under the low energy CID fragmentation the b- and y-series of ions were shown to be better stabilized than the others, making the interpretation of fragment ions somewhat easier.

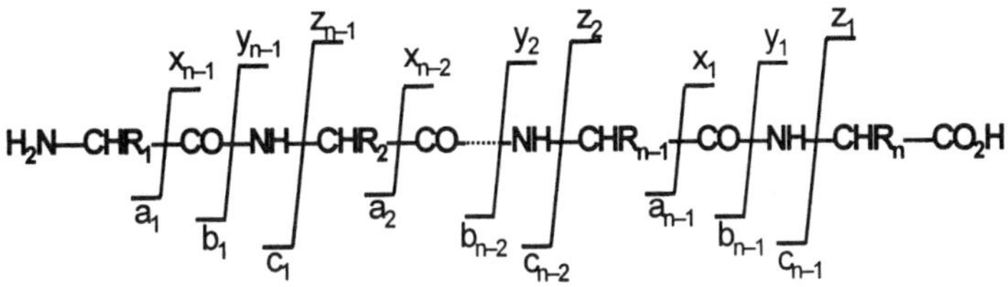

Figure 1. General rule for peptide fragmentation patterns and the nomenclature of their fragment ions[25].

For determination of O-glycosylation sites, a novel strategy using nucleophilic substition by amines was shown to provide more stable mark by introducing the amino moieties into the peptide chain at the sites, which were modified by O-glycans, more convenient for sequencing (Fig. 2)[26].

Most of the tryptic peptides to be successfully sequenced by low energy CID show an average size of 1000-2000 Da, allowing a short peptide chain segment of the sequence for identification in known genoms. For identification of the C-terminus of the crustacean peptide hormone from *O. limosus* the full series of Y-type ions, y_{13}-y_{21}, was obtained[27] (Fig. 3).

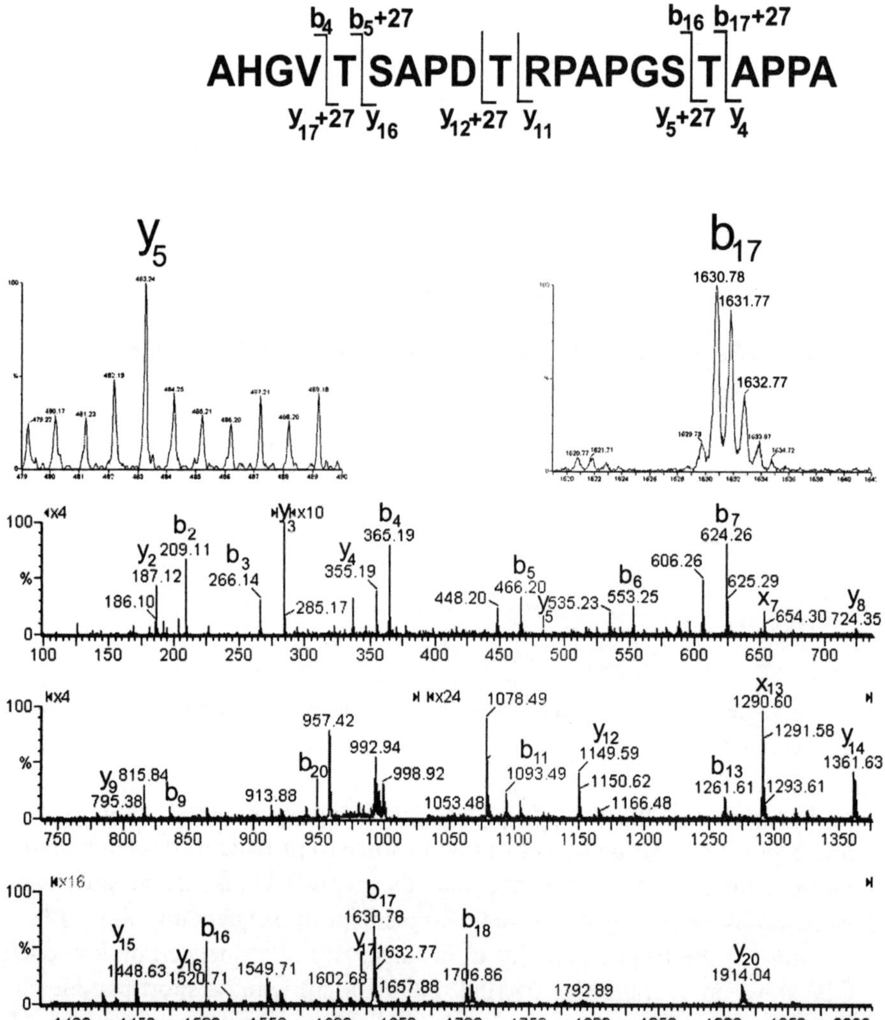

Figure 2. Fragmentation scheme of MUC1 peptides after de-O-glycosylation using methylamine (upper line). ESI low energy CID (MS/MS) of the peptide H2: b_{17} and y_5 identified Thr17 as a site of O-glycosylation. Adapted from[26].

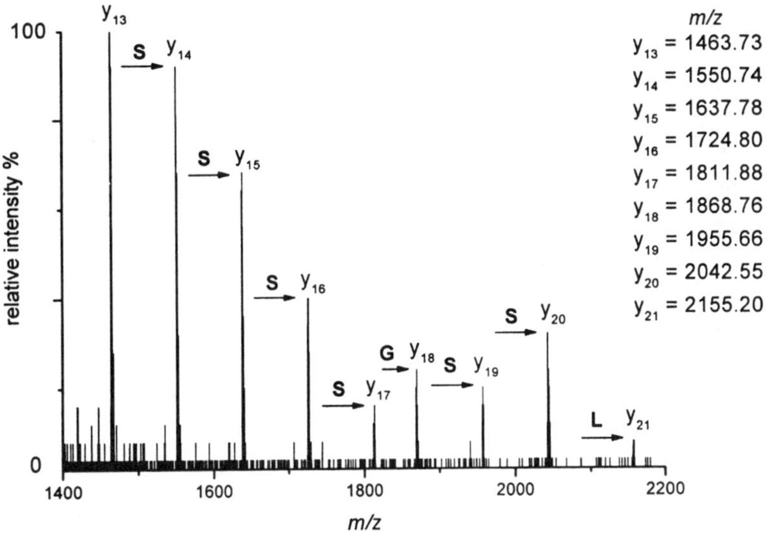

Orl-CPRPA*(1-33): **RSVE**GSSRMERL**LSSGSSSSEPLSFLSQDQSVN**

Figure 3: Part of the low-energy QTOF CID spectrum and amino acid sequence of the quadruply charged molecular ion of CPRPA* at *m/z* 886.63 (LC fraction 3) from *O. limosus*. Identified amino acids are in bold face. Amino acid sequence coverage of 76% was achieved for both known CPRPs. The m/z values of the y-ions obtained are listed in the inset[27].

3. HYPHENATED TECHNIQUES FOR MASS SPECTROMETRY IN PROTEOMICS

3.1 Characterization of peptides by capillary zone electrophoresis and ESI-MS

A number of proteins, located in extracellular matrix, is covalently linked to the glycosaminoglycans forming proteoglycans, general constituents of the connective tissues. Hyphenated techniques were developed to enable high sequence coverage of peptide mixtures obtained after enzymatic digestion of a protein from a 2D-gel spot. The identification by MALDI- or ESI-MS according to their molecular ion values, submitted to a database searching for peptide identification according to their mass, is largely enhanced after separation. For separation of peptides high performance liquid chromatography (HPLC) in combination with mass spectrometry is advantageous in particular for low abundant molecules, present in minute sample quantity.

High-efficiency capillary isoelectric focusing (CIEF) separation of peptides, by which a simultaneous sample concentration and separation can be achieved, was recently proposed by *Shen et al*.[28] at sensitivity higher than 1ng/nl in CIEF/UV detection. Due to their efficient performance, hyphenated CE/MS methods have been applied for characterization of peptides[29-33] and glycoconjugates[34-37]. Since CE has been currently introduced to proteome analysis, different type of buffers for CE/ESI-MS were implemented and optimized according to their compatibility with the desolvation requirements of ESI desorption for MS analysis. Therefore, the efforts were focused to the search for volatile electrolytes showing on the one hand no significant background in mass spectra, and providing sufficient separation and resolution on the other. Ammonium acetate and acetic acid solutions are currently the most suitable and widely used electrolytes in the CE/MS of peptides and proteins. However, further efforts were required for analytical studies in this field toward optimization of novel options for CE/ESI-MS compatible buffer systems and implementation of improved strategies in the proteome analysis, in particular volatile electrolytes[38]. Additionally, another CE/MS practical approach as a prerequisite for structural analysis of components in complex peptide mixtures and their amino acid sequence was developed by a "shot-gun proteolysis" possibly useable as a general tool in proteomics. The CE/MS analysis of the mixture generated by acid-catalyzed peptide hydrolysis has been designed to allow both options, for determination of the peptide amino acid composition and for the ladder sequencing as well[39]. Using the advantage of amino acid and peptide cationization at acidic pH, the positive ion mode ESI-MS mass spectra is favorable. The nanoESI parameters as capillary and sampling cone potential, source block temperature, desolvation and nebulizer gas pressure are to be optimized for a stable spray, minimization of the in-source fragmentation of peptide components and for enhancement of the decomposition of the analyte-buffer clusters. Lower signal to noise ratio, caused by reduced concentration of the analyte in the CE collected fractions, can be compensated by longer recording.

For identification of peptides by MS/MS using low energy CID with Ar as a collisional gas, the collision energy and the gas pressure should be adjusted for an optimal degree of fragmentation of the precursor ion. For screening under mild ESI ion source desorption conditions such as a low ESI capillary potential within 800-1000V range and a sampling cone potential of

20-50V favorable for ionization yield of peptide components collected from the CE buffer can be achieved. The desolvation temperature set to 60-80°C, as well as the use of nebulizer gas, enhances the decomposition of the analyte-buffer clusters and the detection of free peptides as both singly and multiply charged molecular ions.

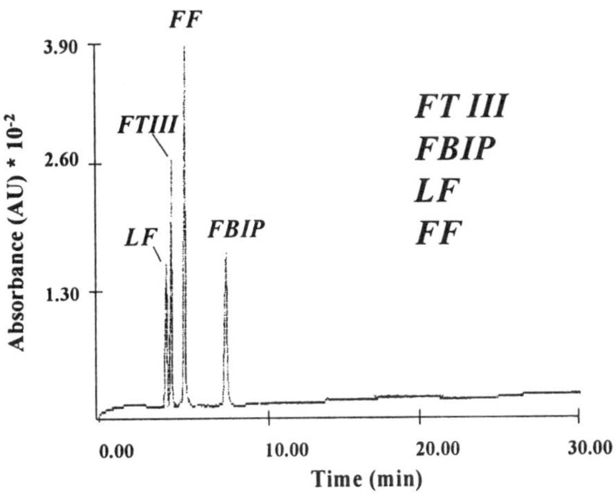

Figure 4. CE/UV profile of the equimolar mixture of *FTIII*, *FBIP*, *LF* and *FF*. Fused silica CE capillary [50μm i.d.x 375 μm o.d, 100 cm length]; CE voltage 30kV; BGE 0.5M H_2O/MeOH ammonium formiate; pH 2.2, substrate concentration was 2 mg/ml buffer (in 10μl); 3s injection by pressure; detection UV (200nm). Reprinted from[39] with permission.

Separation and identification of a complex peptide mixture containing four peptides from extracellular matrix has been investigated. In the UV profile, depicted in Fig. 4, a relatively short migration time, good sensitivity and resolving power are documented. Using the nanoESI-QTOF-MS, *FTIII* and *LF* were identified in the first and *FBIP* and *FF* in the second fraction (Fig. 5).

The crucial step in characterization of peptides is the analysis of their amino acid composition and sequence. Besides the optimization of the CE separation and off-line CE/ESI-MS method for detection, identification of peptides is usually achieved in the tandem MS mode to obtain characteristic fragment ions assignable to the sequence. In this context, we developed an alternative approach for compositional analysis of peptides, in order to complete the structural information obtained by tandem MS experiments, the "shotgun proteolysis". Building-block amino acid composition obtained by acid hydrolysis prior to the direct MS of the hydrolyzate (Fig. 6).

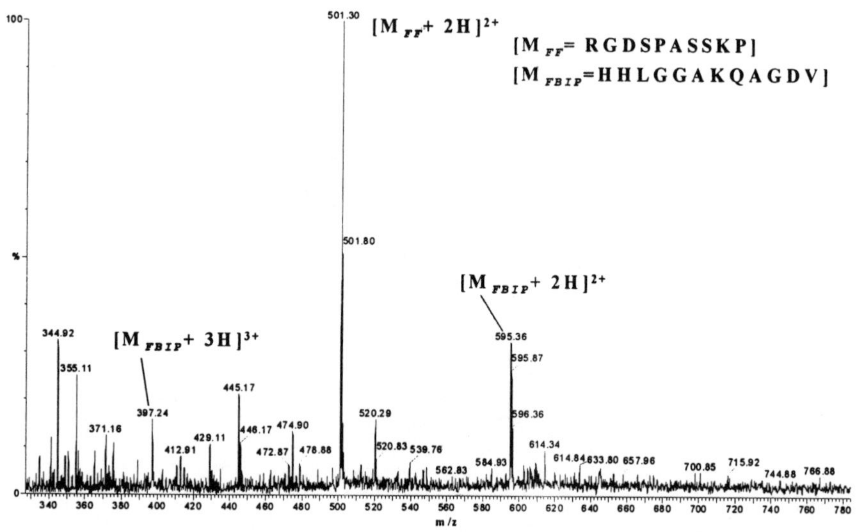

Figure 5. (+) nanoESI-QTOF-MS of the second CE collected fraction at min 11 from the *FTIII, FBIP, LF* and *FF* mixture separation. Reprinted with permission from[39].

New developments in CE and CE/ESI-MS analysis offer high resolution, separation efficiency, sensitivity and reproducibility parameters for detecting and sequencing of CE-separated peptides in complex peptide and amino acid mixtures in proteome analysis. Peptides are identified by nanoESI-QTOF-MS by mapping of their molecular ions in the MS mode and by their fragmentation using low energy CID in the MS/MS mode. Additional advantage of this protocol is the identification of the amino acid composition in peptides adding another dimension of the CE and CE/MS.

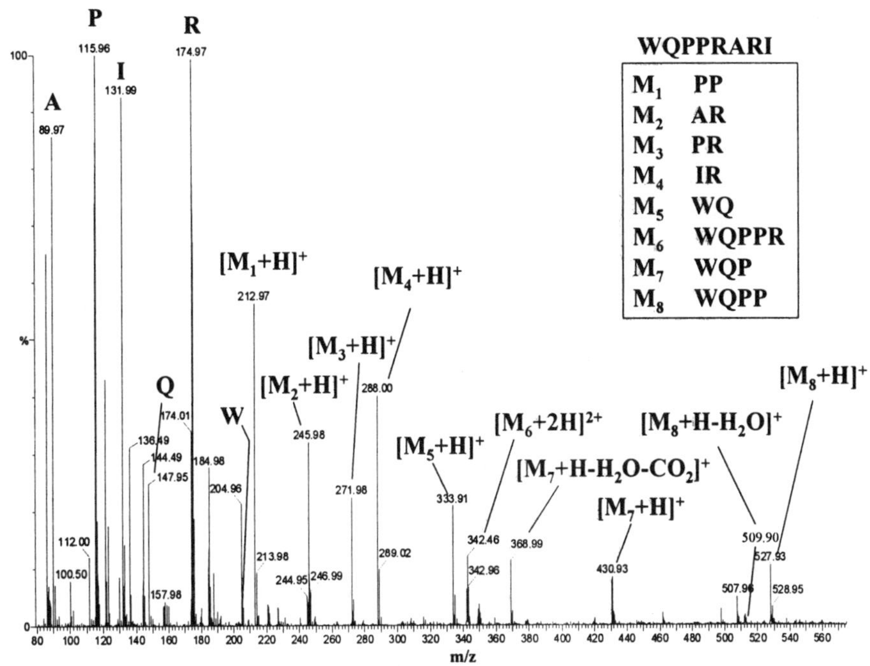

Figure 6. (+) nanoESI-QTOF-MS of the total mixture obtained by acid hydrolysis of *WQPPRARI*. Reprinted from[39] with permission.

3.2 Identification of peptides by on-line liquid chromatography tandem mass spectrometry

Another widespread method for separation and identification of peptides is high performance liquid chromatography (HPLC) in combination with MS, in particular for high throughput (HPT) screening[40-44]. Peptides lower in mass than 10 kDa are not suitable for the 2D gel electrophoresis, therefore hyphenated LC-MS and MS/MS techniques represent a powerfull tool for their screening and identification. Using reversed phase (RP) for separation and columns from analytical to capillary a high gain of four order of magnitude in sensitivity was observed. In some cases multidimensional chromatography coupled to MS was designed to overcome the complexity of the peptide samples, like those from urine[45-47].

This sensitive analytical strategy was used for direct characterization of neuropeptides from the gland neurosecretory system of Crustacea is a nanoflow liquid chromatography system coupled to the quadrupole time-of-flight tandem mass spectrometer (nanoLC-QTOF MS/MS)[27]. It was possible to reveal the existence and structural identity of four hormone precursor related peptide-variants and two new genetic variants of the pigment-dispersing hormone, not detected before by conventional chromatographic systems, molecular cloning or immunochemical methods. The limitations of these established methods to identify and characterize new peptides are the relatively low sensitivity in the case of conventional chromatography, the necessity of sufficient sequence homology in the case of molecular cloning and the availability of specific antibodies in the case of immunochemical methods. In contrast to this, the combination of nanoscale liquid chromatography and mass spectrometry capable of tandem MS analysis (nanoLC-MS/MS) was designed for detection and sequencing of known and unknown peptides independent of their sequence homology, and represents a powerful tool to discover new peptide hormones in biological systems, due tc its sensitivity, accuracy and speed.

Nanoscale liquid chromatography mass spectrometric (nanoLC-MS) analysis of SG extract aliquots corresponding to 0.5 equivalents of a sinus gland was performed using the UltiMate capillary LC system equipped with a FAMOS autosampler (LC Packings, Amsterdam, Netherlands) coupled to a hybrid quadrupole orthogonal acceleration time-of-flight tandem mass spectrometer (Q-TOF, Micromass, Manchester, UK). The LC-MS device was adjusted with a PicoTip™ (New Objective, Woburn, MA) fitted on a Z-spray (Micromass) interface. Chromatographic separations were performed on a reversed-phase (RP) capillary column (Pepmap C18, 75 μm i.d., 15 cm length, LC Packings) with a flow rate of 200 nL/min. The chromatography was carried out using a linear gradient from 5 – 50% solvent B in 1 h and from 5 – 80% solvent B in 1,5 h (solvent A: H_2O/formic acid, 99.9/0.1, v/v; solvent B: H_2O/acetonitril/FA, 19.92/80/0.08, v/v/v) (Fig. 7). Eluting peptides were detected at a wavelength of 210 nm prior to analysis by electrospray ionisation (ESI)-MS/MS in a quadrupole time-of-flight (Q-TOF) mass spectrometer using a nanoelectrospray ion source. Data acquisition was controlled by Masslynx™ software (Micromass) using a manual acquisition mode for MS and MS/MS experiments. Typically, the capillary voltage was set to 1800 V and the counter electrode was set to 40 V. Low-energy CID was performed using argon as a collision gas (pressure in the collision cell was set to 3-4 $\times$ 10^{-5} mbar) and the collision energy was optimised manually for all precursor ions (in the range of 25 to 35 eV).

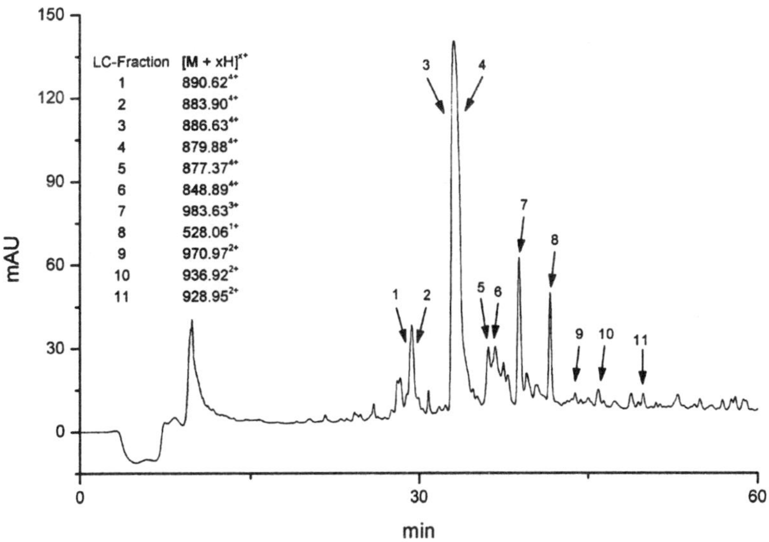

Figure 7. NanoLC-separation UV-profile of the peptide extract equivalent to 0.5 sinus gland of *Orconectes limosus*. The *m/z*-values of 11 components, obtained by nanoLC-ESI-MS, are listed in the inset along with their corresponding LC fraction number. Chromatographic conditions were as follows: Pepmap C18 column (75 µm i.d., 15 cm length, LC Packings); elution with a linear gradient from 5-50% solvent B in 1 h at a flow rate of 200 nl/min. Solvent A: 0.1% FA; solvent B: 0.08% FA, 80% acetonitril. Reprinted from[27] with permission.

Direct identification of neuropeptides was achieved by analysis of the desalted tissue extract obtained from the neurosecretory gland of *Orconectes limosus*. The most abundant polypeptides were identified as CHH precursor related peptides A/A* (Fig. 7, LC fractions 3, 4), characterized by their fourfold charged molecular ions at *m/z* 886.63 and 879.88, corresponding to two thirty-three amino acids containing species, respectively (Table 1). By selecting the precursor ions of these already known CPRPs an amino acid sequence coverage of 76% was achieved using the tandem MS fragmentation by low-energy QTOF CID. In Fig. 2 the part of the fragmentation spectrum of Orl-CPRPA* showing the serine-rich region 13-20 is depicted.

Additionally, 4 variants of CPRPs, corresponding to LC fractions 1, 2, 5 and 6 (Fig. 7) were detected as fourfold charged molecular ions by nanoLC-MS at *m/z* 890.62, 883.90, 877.37 and 848.89, respectively (Table 1). Their sequence coverage obtained by this approach of approx. 64% achieved for all variants investigated was not sufficient for determination of all amino acid substitutions and/or modifications. By tandem mass spectrometric investigation only main components were detectable and submitted to MS/MS for sequencing, in analogy to the neuropeptides from C. destructor [49]. Minor components were not ionized under standard conditions, requiring lower pH for protonation.

To inspect this, sinus gland extract has been exposed to a preparative HPLC (data not shown). Fractions, which were potential CHH and MIH candidates, were submitted separately to the nanoESI-MS under different, lower, pH conditions, under which the purified large peptide components were well ionized and identified according to their multicharged molecular ions (Fig. 8).

Table I. m/z values of fragment ions and resulting sequence data of PDH, PDHB and PDHC obtained by low-energy QTOF CID. Identical amino acids are in bold face.

Fragme	PDH	PDHB	PDHC	Fragme	PDH	PDHB	PDHC	PDH	PDHB	PDHC
b1				y18				**N**	**N**	**N**
b2	202.0	202.0	202.0	y17		1758.		**S**	**S**	**S**
b3	331.1	331.1	331.1	y16				**E**	**E**	**E**
b4	444.2	444.2	444.2	y15	1610.	1542.		**L**	**L**	**L**
b5	557.3	557.2	557.2	y14	1497.	1429.	1413.	**I**	**I**	**I**
b6	671.4	671.3	671.3	y13	1384.	1316.	1300.	**N**	**N**	**N**
b7	758.4	742.4	742.3	y12	1270.	1202.	1186.	S	A	A
b8	871.4	855.4	855.4	y11		1131.	1115.	**I**	**I**	**I**
b9	984.6	968.5	968.5	y10	1070.	1018.	1002.	**L**	**L**	**L**
b10	1041.	1025.	1025.	y9	957.5	905.4	889.4	**G**	**G**	**G**
b11	1154.	1112.	1112.	y8	900.5	848.4	832.4	L	S	S
b12		1209.		y7	787.4	761.4	745.3	**P**	**P**	**P**
b13		1310.	1310.	y6	690.4			K	T	T
b14	1478.	1423.	1423.	y5		563.3		V	L	L
b15	1609.	1570.	1554.	y4	463.2	450.2	434.2	M	F	M
b16	1723.		1611.	y3	332.2	303.1	303.1	N	G	G
b17		1756.	1740.	y2	218.1	246.1	246.1	**E**	**E**	**E**
b18				y1		117.1	117.1	A-NH_2	V-MH_2	A-NH_2

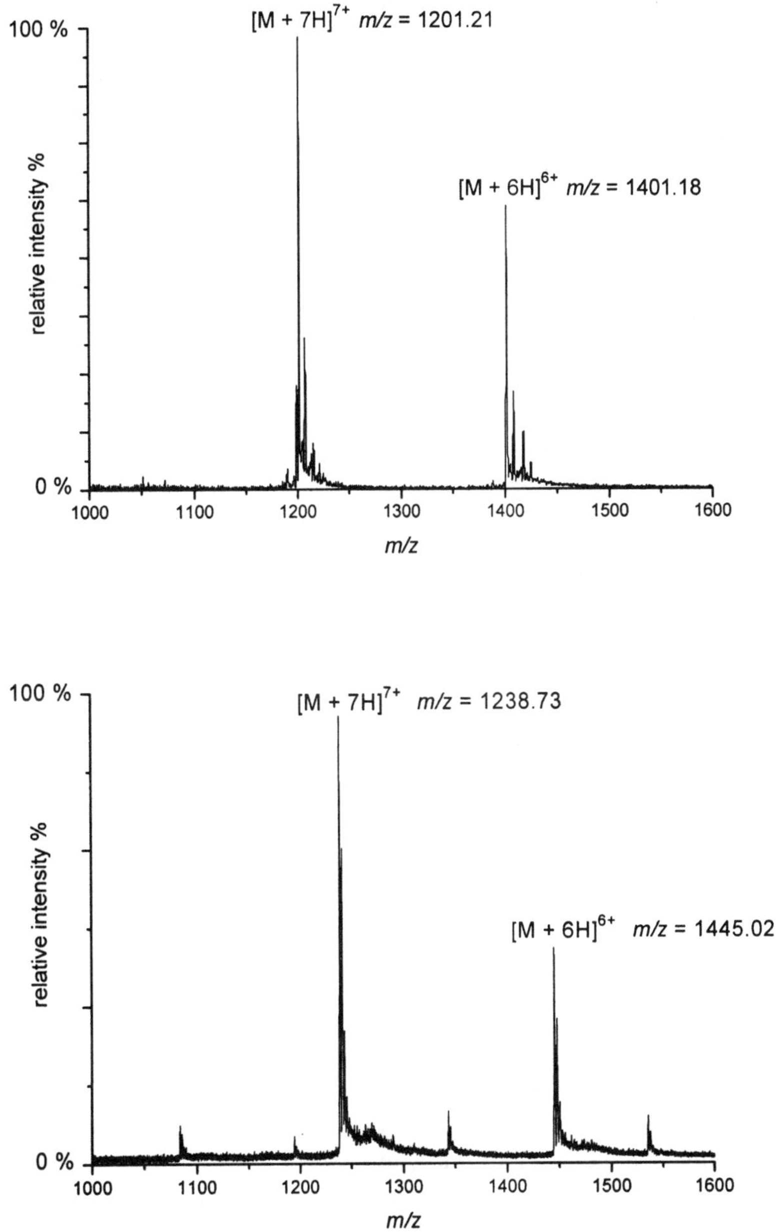

Figure 8. NanoESI-QTOF mass spectra of A: CHH and B: MIH obtained from HPLC separated sinus glands of *Orconectes limosus*. The spectra were recorded in the positive ion mode using methanol/water/acetic acid (50/25/25, v/v/v) as solvent. Reprinted from[27] with permission.

Large peptides belonging to the CHH and MIH family could not be detected under the standard conditions used in our experiments. These peptides were submitted separately to the nanoESI-MS under different, lower, pH conditions, under which these purified large components were well ionized and identified according to their multicharged molecular ions. Possibly the tertiary structure of these native peptides containing disulfide bridges can largely influence the ionization process in the electrospray ion source under particular designed conditions.

High potential of the nanoLC-MS techniques with respect to identification and sequencing of peptides from biological sources without prior purification steps under conditions of high sensitivity can be generally observed. Identification of previously not characterized, novel, peptides from unsequenced genomes can be accomplished by *de-novo* sequencing. The limits of the method by restricted ionizability of peptide components can be overcome by comparison of the LC-UV trace and the TIC in the MS analyzer. For detection of very minor components in complex peptide/protein mixtures, which may be relevant in the full proteome expression analysis, a higher amount of tissue must be prepared and subjected to an array of multidimensional separation techniques, such as multidimensional LC, capillary electrophoresis or two-dimensional gel electrophoresis, and be combined with advanced and highly sensitive mass spectrometric methods.

ACKNOWLEDGEMENTS

The work presented in this review has been performed in the laboratory „Biomedical Analysis", Institute for Medical Physics and Biophysics, Faculty of Medicine, University of Münster, Germany, by (in alphabethical order) Laura Bindila, Dr. Patrick Bulau, Marko Jovanović, Dr. Iris Meisen, and Dr. Alina Zamfir. The contribution of these dedicated collaborators and other members of the group is greatly acknowledged. Financial support was provided by Deutsche Forschungsgemeinschaft (DFG) within the Sonderforschungsbereich 492 "Extracellular Matrix: Biogenesis, Assembly and Cellular Interactions" at the University of Münster (Project Z2 to J.P.-K). Funding of this work by the Deutsche Forschungsgemeinschaft through grants to J.P.-K. (PE 415/15-1,2) and R.K. (Ke 206/17-1,2) is gratefully acknowledged. The QTOF instrument was purchased from a HbfG grant (Land Nordrhein Westfalen) to J. P.-K. The CE instrument was a long-term generous loan by Prof. Dr. Peter Bruckner, Institute for Physiological Chemistry, University of Münster.

REFERENCES

1. Peng, J., Gygi, S.P., 2001, *J. Mass Spectrom* **36:** 1083-1091
2. Manabe, T., 1999, *Electrophoresis* **20:** 3116-3121
3. Gygi, S.P., Rochon, Y., Franza, B.R., Aebersold, R., 1999, *Mol. Cell Biol.* **19**: 1720-1730.
4. O'Farrell, P.H., 1975, *J. Biol. Chem.* **250**: 4007-4021.
5. Klose, J., 1975, *Humangenetik* **26**: 231-243.
6. Edman, P., Begg, G., 1967, *Eur. J. Biochem.* **1**: 80-91.
7. Fenn, J.B., Mann, M., Meng, C.K., Wong, S.K., Whitehouse, V., 1989, *Science* **246**: 64-71.
8. Karas, M., Hillenkamp, F., 1988, *Anal. Chem.* **60**: 2299-2301.
9. Wildgruber, R., Harder, A., Obermaier, C., Boguth, G., Weiss, W., Fey, S.J., Larsen, P.M., Gorg, A., 2000, *Electrophoresis* **21:** 2610-2616.
10. Mann, M., Hendrickson, R.C., Pandey, A., 2001, *Annu. Rev. Biochem* **70:** 437-473.
11. Schevchenko, A., Jensen, O.N., Podtelejnikov, A.V., Sagliocco, F., Wilm, M., Vorm, O., Boucherie, H., Mann, M., 1996, *Proc. Natl. Acad. Sci. USA*, **93:** 14440-14445.
12. Katayama, H., Ishihama, Y., Oda, Y., Asakawa, N., 2000, *Rapid Com. Mass Spectrom.* **14**: 1167-1178.
13. Henzel, W.J., Billeci, T.M., Stults, J.T., Wong, S.C., Grimley, C., Watanabe, C., 1993, *Proc. Natl. Acad. Sci. USA* **90:** 5011-5015.
14. Patterson, S.D., 1994, *Anal. Biochem.* **221:** 1-15.
15. Mortz, E., O'Connor, P.B., Roepstorff, P., Kelleher, N.L., Wood, T.D., McLafferty, F.W., Mann, M., 1996, *Proc. Natl. Acad. Sci. USA*, **93:** 8264-8267.
16. Wilm, M., Neubauer, G., Mann, M., 1996, *Anal. Chem.*, **68:** 527-533.
17. National Research Council Steering Committee: Kenyon, G.L., DeMarini, D.M., Fuchs, E., Galas, D.J., Kirsch, J.F., Leyh, T.S., Moos, W.H., Petsko, G.A., Ringe, D., Rubin, G.M., Sheahan, L.C., 2002, *Mol. Cell. Proteomics*, **1**: 763-780.
18. Pappin, D.J.C., Hojrup, P., Bleasby, A.J., 1993, *Curr. Biol*, **3**: 327-332.
19. Lopez, M.F., Berggren, K., Chernokalskaya, E., Lazarev, A., Robinson, M., Patton, W.F., *Electrophoresis*, **21:** 673-683.
20. Shevchenko, A., Wilm, M., Vorm, O., Mann, M., 1996, *Anal. Chem.*, **68**: 141-143.
21. Wilm, M., Mann, M., 1996, *Anal. Chem.*, **68**: 1-8.
22. Alving, K., Körner, R., Paulsen, H., Peter-Katalinić, J., 1998, *J. Mass Spectrom.*, **33**: 1124-1133.
23. Maček, B., Hofsteenge, J., Peter-Katalinić, J., 2001, *Rapid Commun. Mass Spectrom.*, **15**: 771-777.
24. Mann, M., Wilm, M., 1994, *Anal. Chem.*, **66**: 4390-4399.
25. Roepstorff, P., Fohlman, J., 1984, *Biomed. Mass Spectrom.*, **11:** 601-609
26. Hanisch, F.G., Jovanović, M., Peter-Katalinić, J., 2001, *Anal. Biochem.*, **290**: 47-59.
27. Bulau, P., Meisen, I., Schmitz, T., Keller, R., Peter-Katalinić, J., 2004, *Molec. Cellul. Proteomics*, in press.
28. Shen, Y., Berger, S.J., Anderson, G.A., Smith, R.D., 2000, *Anal. Chem.*, **72**: 2154-2159
29, Waterval, J., Hommels, G., Bestebreurtje, P., Versluis, C., Heck, A.J., Bult, A., Lingeman, H., Underberg, W.J., 2001, *Electrophoresis*, **22**: 2709-2716
30. von Brocke, A., Nicholson, G., Bayer, E., 2001, *Electrophoresis*, **22**: 1251-1266
31. Hearn, M.T., 2001, *Biologicals*, **29:** 159-178
32. Nashabeh, W., Smith, J.T., El Rassi, Z., 1993, *Electrophoresis*, **14:** 407-416
33. Zhang, B., Foret, F., Karger, B.L., 2000, *Anal. Chem.*, **72:** 1015-1022

34. Zamfir, A., Konig, S., Althoff, J., Peter-Katalinić, J., 2000, *J. Chromatogr. A*, **895:** 291-298
35. Zamfir, A., Vukelić, Ž., Peter-Katalinić, J., 2002, *Electrophoresis*, **23:** 2894-2903
36. Zamfir, A., Seidler, D.G., Kresse, H., Peter-Katalinić, J., 2002, *Rapid Commun. Mass Spectrom.*, in press
37. Zamfir, A., Peter-Katalinić, J., 2004, *Electrophoresis*, in press.
38. Huber, C.G., Premstaller, A., Kleindienst, G., 1999, *J. Chromatogr.A* **854**: 141-154
39. Bindila, L., Zamfir, A., Peter-Katalinić, J., 2002, *J. Sep. Sci.*, **25**: 1101-1111.
40. Davis, M.T., Stahl, D.C., Swiderek, K.M., Lee, T.D., 1994, *Methods*, **6**: 304-313.
41. Dongre, A., Eng, J,. Yates III, J.R., 1997, *Trends Biotechnol.*, **15**: 418-425
42. Figeys, D., Aebersold, R., 1998, *Electrophoresis*, **19**: 885-892
43. Gygi, S.P., Corthals, G.L., Zhang, Y., Rochon, Y., Aebershold, R., 2000, *Proc. Natl. Acad. Sci. USA*, **97:** 9390-9395
44. Yates, J.R., 2000, *Trends Genet.*, **16**: 5-8
45. Wagner, K., Racaityte, K., Unger, K.K., Miliotis, T., Edholm, L.E., Bischoff, R., Marko-Varga, G., 2000, *J. Chromatogr. A*, **893**: 293-305.
46. Spahr, C.S., Davis, M.T., McGinley, M.D., Robinson, J.H., Bures, E.J., Beierle, J., Mort, J., Courchesne, P.L., Chen, K., Wahl, R.C., Yu, W., Luethy, R., Patterson, S.D., 2001, *Proteomics*, **1:** 93-107.
47. Cutillas, P.R., Norden, A.G.W., Cramer, R., Burlingame, A.L., Unwin, R.J., 2003, *Clinical Science*, **104**: 483-490.
48. Bulau, P., Meisen, I., Reichwein-Roderburg, B., Peter-Katalinić, J., Keller, R., 2003, Peptides, **24:** 1871-1879.

ABBREVIATIONS:

BGE, background electrolyte
CE, capillary electrophoresis
CHH, crustacean hyperglycemic hormone
CID, collision-induced-dissociation
CIEF-capillary isoelectric focusing
CPRP, CHH precursor related peptide
ESI, electrospray ionization
FA, formic acid
FAPP, fibronectin adhesion promoting peptide
FBIP, fibrinogen binding inhibitor peptide
FF, fibronectin fragment
FTIII, fibronectin type III connecting segment fragment
LE, leucine enkephaline
LF-laminin fragment
MALDI-matrix assisted laser/desorption ionization
MS, mass spectrometry/spectrometer
MS/MS, tandem mass spectrometry/spectrometer
nanoESI, nano-electrospray or nanospray

PDH, pigment-dispersing hormone
QTOF-MS, quadrupole time-of-flight mass spectrometry
SG, sinus gland
TFA, trifluoroacetic acid

SEQUENCE DATA

SWISS-PROT accession numbers:
Pigment-dispersing hormone B (*Orconectes limosus*): P83586
Pigment-dispersing hormone C (*Orconectes limosus*): P83587

A Microscopic Study of Disorder–Order Transitions in Molecular Recognition of Unstructured Proteins: Hierarchy of Structural Loss and the Transition State Determination from Monte Carlo Simulations of $P27^{KIP1}$ Protein Coupled Unfolding and Unbinding

GENNADY M. VERKHIVKER
Pfizer Global Research and Development, La Jolla Laboratories, 10777 Science Center Drive
San Diego CA 921211111, USA

1. INTRODUCTION

1.1 Molecular recognition of unstructured proteins

It has been recently realized that a significant amount of protein domains and even entire proteins can lack intrinsic globular structure under physiological conditions[1-3], suggesting a reappraisal of the conventional structure–function paradigm[4-8]. These proteins, termed as intrinsically unstructured[1,6,7] intrinsically disordered[2-5] or natively unfolded[8-10] can largely comprise of disordered segments in their functional state. The intrinsic plasticity and functional disorder-order folding transitions coupled to binding can provide for these protein systems an important prerequisite for effective molecular recognition, including high specificity coupled with low affinity, the ability to bind with several different targets, a precise control and simple regulation of the binding thermodynamics, and the increased rates of specific macromolecular

Supramolecular Structure and Function 8, Edited by Pifat-Mrzljak
Kluwer Academic/Plenum Publishers, New York 2004

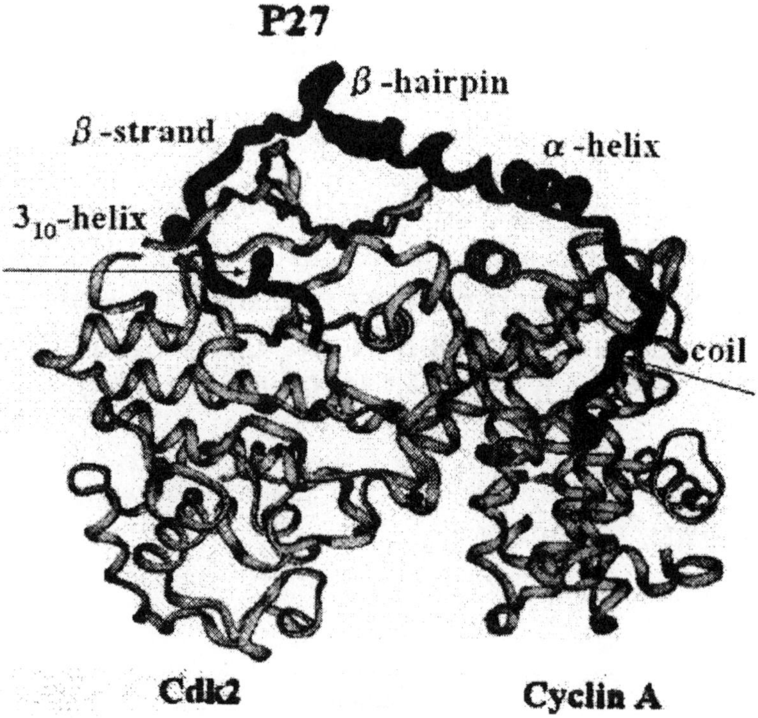

Figure 1. The crystal structure of the tertiary complex with the p27 protein

association[1-12]. Coupling of folding and binding accompanied by a disorder–order transition has been experimentally detected for $p27^{Kip1}$, a member of the Kip/Cip protein family involved in cell cycle regulation[13-18]. The crystal structure of the tertiary complex with the bound 69-amino acid N-terminal inhibitory domain of $p27^{Kip1}$ (residues 25-93)[16] has revealed an ordered conformation of $p27^{Kip1}$ comprising sequentially of the coil (residues from 25 to 34), α-helix (residues from 35 to 60), β–hairpin (residues from 61 to 71), β-strand (residues 75 to 81), and 3_{10}-helix (residues from 85 to 90) (Fig. 1). According to the proposed binding mechanism[16], the tertiary contacts formed between the rigid cyclin and a conserved in the Kip/Cip family Leu—Phe–Gly (LFG) motif of $p27^{Kip1}$ the coil serve as an initial anchor in complex formation. Instead of contributing directly to folding of $p27^{Kip1}$ the hydrophobic residues of coil, β-hairpin, and β-strand form extensive interfacial contacts with the complex, that are further consolidated by stabilizing specific interactions, most notably six backbone–backbone hydrogen bonds, formed by β-strand, and 3_{10}-helix, protruding deeply into the catalytic cleft (Fig. 1).

Circular dichroism spectroscopy experiments have shown that the unbound form of the $p27^{Kip1}$ inhibitory domain is intrinsically disordered, but not entirely unfolded and contains some marginally stable helical structure in the α-helix region[18]. Reducing or increasing the stability of the -helix in the unbound $p27^{Kip1}$ with proline or alanine substitutions has a marginal effect on the binding thermodynamics[18]. However, while the disruption of α-helix in the unbound form by proline mutations does not affect the rate of $p27^{Kip1}$ binding, a marginal stabilization of the α-helix upon alanine substitutions in this region slows down the rate of formation for the inhibited complex by approximately 3–fold[18]. Hence, the intrinsic disorder of α-helix in the unbound form of $p27^{Kip1}$ can result in a kinetic advantage during coupled folding and binding process[18].

1.2 Folding and binding coupling in the energy landscape models

The process of complex formation when one or both binding partners are highly flexible or even completely disordered[19-27] can evolve over a vast configurational space, under the influence of a multitude of conflicting interactions, and on hierarchical length and time scales that makes the complexity of molecular recognition for these systems analogous to the protein folding problem[28-36]. A large number of conformational states available during molecular recognition necessitates the use of a statistical characterization and the energy landscape approach, originally introduced in protein folding[28,31-33] and further developed in molecular recognition[37-53]. Striking similarities in the hierarchy of folding and the transition state ensembles for different, small, and fast–folding protein sequences with a common native topology, discovered based on comprehensive experimental[54-62] and theoretical studies[63-75], have revealed the important role of the native structure in determining protein folding mechanisms. A relatively large entropic cost of forming non–local interactions early in the process can encourage ordering of simple topological motifs with local interactions formed more readily than those with many non–local interactions[59,,63-65,76-81]. Recent experimental data have indicated that proteins with similar topologies can nevertheless fold via different mechanisms, suggesting a more balanced view according to which the interplay between topological characteristics of the native structure and the heterogeneity of specific interactions in the transition state ensemble (TSE) could ultimately explain protein folding mechanisms for a broad range of protein systems[82.84].

The emerging realization that protein folding and molecular recognition phenomena share a number universal aspects, including the existence of a thermodynamically stable native structure, a large number of conformational states available to the system, the complex nature of interactions on the underlying energy landscape[37-53], is manifested by a balance between opposing thermodynamic forces during the process, the loss of conformational entropy and the energy gain upon the native structure formation, that ultimately determines the thermodynamic free energy barrier of the reaction[85]. These physical arguments, that have been fruitful in understanding protein folding mechanisms, have also led to an elegant, fly–casting mechanism based derived from the energy landscape theory and proposed to explain kinetic advantages of unstructured proteins in binding[85]. By analogy with the folding funnel mechanism, the unfolded protein would bind weakly at relatively long distances, followed by coupled folding and binding mechanism as it 'reels in' on the binding partner, utilizing a greater 'capture radius' than a folded protein to provide the enhanced binding rates.

Balancing the desired realism in all–atom computer modeling of folding and binding86-88 with the underlying stochastic nature of these phenomena, that requires averaging over a large number of independent simulations, can be addressed by the employment of equilibrium free energy methods[89-94] and 'non–equilibrium' kinetic temperature–induced unfolding simulations[95-103]. In this work, a microscopic study of the $p27^{Kip1}$ binding mechanism is conducted by simulating hierarchy of structural loss during $p27^{Kip1}$ coupled unfolding and unbinding from the crystal structure of the tertiary complex using high–temperature Monte Carlo simulations. Combined with a simplified, yet all-atom energetic model[44,50,51], these simulations allow to capture both energetic and topological frustration[68] in folding and binding, including the incorporation of secondary structure elements into a tertiary structure of the complex, and the satisfaction of packing constraints at the intermolecular interface. Here, it is proposed that molecular recognition $p27^{Kip1}$ of can proceed through formation of an encounter intermediate corresponding to the ensemble of non–specific, largely unstructured conformations weakly bound to the complex. At the second stage of the molecular recognition reaction this ensemble undergoes a disorder–order transition to the ensemble of well–ordered, native–like states.

Simulating coupled unfolding and unbinding $p27^{Kip1}$ at high temperatures and determination of the populated meta–stable states allows a subsequent mapping of theTSEusing a rigorous kinetic analysis[104-108]. Amicroscopic characterization of the binding mechanism and free energy barrier suggests an atomic picture of coupling between folding and binding that reconciles the initiation binding event with the experimental data, indicating a kinetic advantage for the intrinsically unstructured α-helix in the unbound form. In agreement with the experimental data, it is shown that the topological requirements of the native complex overwhelm any local folding preferences for creating a stable α-helix prior to overcoming the major free energy barrier. Consequently, folding of unstructured proteins upon binding to a given template is largely determined by the requirements to form specific complex that ultimately dictates the folding mechanism.

2. MATERIALS AND METHODS

2.1 Molecular recognition energy model

The knowledge-based simplified energetic model includes intramolecular energy terms for the ligand, given by torsional and nonbonded contributions of the DREIDING force field[109], and intermolecular ligand-protein steric and hydrogen bond interaction terms calculated from a simplified piecewise linear (PL) potential summed over all protein and ligand heavy atoms (Fig. 2a). The parameters of the pairwise potential depend on the following different atom types: hydrogen–bond donor, hydrogen-bond acceptor, both donor and acceptor, carbon–sized nonpolar, sulfur-sized nonpolar, flourine-sized nonpolar, and large nonpolar. The atomic radius is 1.4 °A for fluorine is 1.4 °A and 1.8 °A for carbon, oxygen, and nitrogen atoms. The atomic radius of 2.2 °A is assigned to sulfur and phosphorus, chlorine, and bromine atoms, modeled as large nonpolar atom type. Electronegative atoms with an attached hydrogen are defined to be donors, while oxygen and nitrogen atoms with no bound hydrogens are defined to be acceptors. Sulfur is modeled as being capable of making weak hydrogen bonds which allows for sulfur–donor closer contacts that are seen in some of the crystal structures. Hydroxyl groups are defined in this model to be both donor and acceptor, and carbon atoms are defined to be nonpolar.An empirical desolvation correction is applied to the attractive portion of the interactions between nonpolar and polar atoms. This correction is defined as the ratio between the attractive well depth for nonpolar-polar contacts and the one for nonpolar-nonpolar contacts, and can range between 0 and 1. The parameter is set to 1.0 in this work, thereby imposing a desolvation penalty by disadvantaging the burial of polar groups with the nonpolar atoms.

A hydrogen bond interaction term is assigned to interactions between donor and acceptors, a repulsive interaction contribution is computed for donor-donor and acceptor–acceptor contacts, and a steric intermolecular term is assigned for other contacts. The steric and hydrogen bondlike potentials have the same functional form, with an additional three–body contribution to the hydrogen bond term and the repulsive term for donor-donor and acceptor–acceptor contacts. Both the hydrogen bond interaction energy and the repulsive interaction contribution between donor-donor and acceptor–acceptor close contacts are modulated by an approximate angular dependence (Fig. 2b). These terms are multiplied by the hydrogen bond strength term, which is a function of the angle θ determined by the relative orientation of the protein and ligand atoms (Fig. 2b). The scaling for the repulsive interactions is equivalent to the

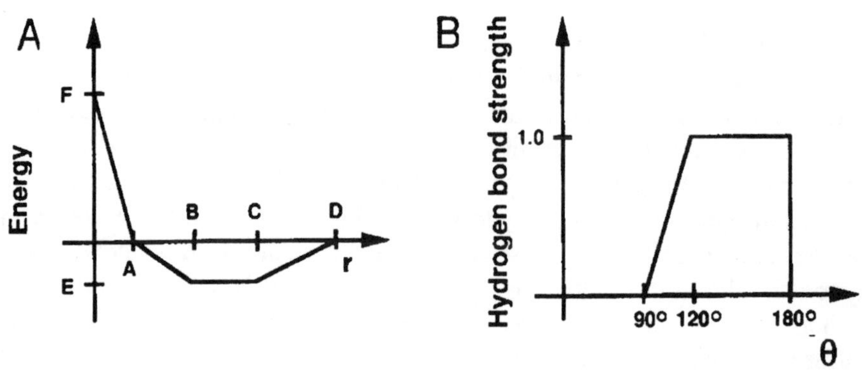

Figure 2. A) The functional form of the ligand–protein interaction energy. For steric interactions, A = 0.93B, C = 1.25B, D = 1.5B, E = -0.4, F = 15.0, and B=r_{ligand}+ $r_{protein}$ is the sum of the atomic radii for the ligand and protein atoms. For hydrogen bond interactions, A = 2.3, B = 2.6, C = 3.1, D = 3.4, E = -4.0, F = 15.0. For sulfur hydrogen bond interactions, A = 2.7, B = 3.0, C = 3.5, D = 3.8, E = -2.0, F = 15.0. For chelating interactions with the metals A = 1.5, B = 1.7, C = 2.5, D = 3.0, E = -10.0, F = 15.0. For repulsive interactions, A = 3.2, E = 0.1, F = 15.0. The repulsive potential is then linearly scaled from E=0.1 to zero between 3.2 °A and 5.0 °A_The units of A, B, C, and D are °A for E and F the units are kcal/mole. B) The hydrogen bond interaction energy and the repulsive term are multiplied by the hydrogen bond strength term, which is a function of the angle θ determined by the relative orientation of the protein and ligand atoms.

dependence used for the hydrogen bond interaction term, but in this cases it implies a maximum penalty when the angle θ is 180 degrees, fading to zero at 90 degrees and below. θ is defined to be the angle between two vectors, one of which points from the protein atom to the ligand atom. For protein atoms with a single heavy atom neighbor, the second vector connects the protein atom with its heavy atom neighbor, while for protein atoms with two heavy atom neighbors, it is the bisector of the vectors connecting the protein atom with its two neighbors.

2.2 Monte Carlo simulations of coupled folding and binding

The protein complex is held fixed in its minimized crystallographic conformation, while rigid body degrees of freedom and a total of 169 rotatable angles of $p27^{Kip1}$ protein are treated as independent variables during simulations. $p27^{Kip1}$ conformations and orientations are sampled in a parallelepiped that encompasses the crystallographic structure of the $p27^{Kip1}$ complex with a large 10.0 °A cushion added to every side of this box surrounding the interface which guarantees a sufficiently unbiased conformational search and allows to model disorder–order transitions. Bonds allowed to rotate in $p27^{Kip1}$ protein include those linking sp^3 hybridized atoms to either sp^3 or sp^2 hybridized atoms and single bonds linking two sp^2 hybridized atoms, resulting in total 169 rotatable bonds. The moves are chosen as follows: a variable is selected at random and then a uniform displacement is given along each rigid body degree of freedom, the ligand is rotated as a rigid body, or a dihedral angle of the ligand is rotated. Monte Carlo simulations allow to dynamically optimize the step sizes at each temperature by taking into account the inhomogeneity of the molecular system[110-112]. The acceptance ratio method is used to update the step sizes every cycle of 1000 sweeps. For all these simulations, we equilibrated the system for 1000 cycles (or one million sweeps), and collected data during 5,000 cycles (or five million sweeps) resulting in 5,000 samples at each temperature. A sweep is defined as a single trial move for each degree of freedom of the system.

A control simulation at T=300K and 20 independent simulations of $p27^{Kip1}$ coupled unfolding and unbinding starting from the crystal structure of the complex have been performed at each of the following temperatures T=400K, 500K, and 600K. The only difference between 20 independent simulations carried out at each temperature was that a different random seed was used in the initial temperature equilibration. Averages within the time regime correspond to what is found experimentally, but any hierarchy in unbinding and unfolding may be obscured because the protein unfolds and unbinds at different times in different simulations. To classify a given conformation one must choose some order parameters that can differentiate the states of the system in the course of the could unfolding and unbinding process. the order parameters that produced a more homogeneous and clear description of the disorder–order transitions during coupled folding and binding were employed : the root mean square deviation (RMSD) of the $p27^{Kip1}$ backbone atoms from the native structure, including the RMSD values for the individual secondary structure elements that can measure unfolding progression, and the total number of intermolecular hydrogen bonds formed by $p27^{Kip1}$ in the complex, that can monitor the extent of unbinding.

2.3 Similarity clustering

The 3D–similarity calculations are based on the spatial proximity of atoms and the atom type. Four types of atoms are distinguished : hydrogen bond donors, hydrogen bond acceptors, hydrogen bond donors and acceptors and nonpolar atoms. The atom type compatibility $a(i,j)$ is assigned a value between 0.0 and 1.0, with the compatibility between two atoms of the same type defined to be 1.0, that between donor and acceptor atom is 0.0, and other combinations of atoms have compatibilities between 0.0 and 1.0.

The spatial proximity between two atoms i and j is evaluated with a Gaussian function $p(i,j) = 10^{(r_{i,j}^2/\sigma^2)}$, where r_{ij} is the distance between atoms i and j, and $\sigma = c^2/\log_{10}(p)$, where c and p denote the cutoff distance and proximity threshold respectively. Both the cutoff distance and the proximity threshold determine the shape of the gaussian function to evaluate spatial proximity of two atoms, with c=°A and p=0.000032.

A descriptor $d(i,j)$ is calculated from the spatial proximity and the atom type compatibility:

$$d(i,j) = p(i,j) * a(i,j) if r(i,j) \leq c \quad (1)$$

$$d(i,j) = 0 if r(i,j) > c \quad (2)$$

An atom descriptor $D_m^n(i)$ for atom in molecule is then calculated by summation over all N atoms in molecule n, $D_m^n(i) = \sum_{j=1}^{N} d_m^n(i,j)$. The intermolecular similarity between molecules and is given by the Tanimoto coefficient[113-115].

$$S(m,n)=\frac{\sum_{i=1}^{M} D_{m(i)}^{m} D_{m}^{n}(i)+\sum_{j=1}^{N} D_{n}^{m}(j) D_{n}^{n}(j)}{\sum_{j=1}^{M} D_{m}^{m}(i)^{2}+\sum_{j=1}^{N} D_{n}^{n}(j)^{2} \sum_{i=1}^{M} D_{m}^{m}(i) D_{m}^{n}(i)-\sum_{j=1}^{N} D_{n}^{m}(j) D_{n}^{n}(j)} \tag{3}$$

Molecules are grouped into clusters by comparing the intermolecular similarity coefficient. The first molecule is assigned to the first cluster. The next molecule is assigned to the cluster in which a cluster member has the highest similarity with the next molecule, if the similarity is above a threshold, chosen to be 0.85. Otherwise, the next molecule is assigned to a new cluster. The first member of the a cluster is called the cluster center. After all molecules are assigned to clusters, the molecules are arranged in new order, starting with the largest cluster and proceeding to the smallest cluster. The reordered set of molecules is subjected to the same clustering procedure. This procedure is iterated until the information entropy converges to a minimum. The clusters with at least 100 members are analyzed. Since conformations which belong to the same cluster are equivalent with 85% structural similarity, different clusters are compared by analyzing cluster centers.

3. SIMULATING DISORDER–ORDER TRANSITIONS IN MOLECULAR RECOGNITION OF UNSTRUCTURED PROTEINS

3.1 Hierarchy of structural loss and the transition state determination from Monte Carlo simulations of $p27^{Kip1}$ protein coupled unfolding and unbinding

A microscopic study of the $p27^{Kip1}$ protein conformational transitions coupled to binding is conducted by simulating hierarchy of structural loss during $p27^{Kip1}$ coupled unfolding and unbinding that is monitored using 100 independent temperature–induced Monte Carlo simulations initiated from the crystal structure of the tertiary complex. The most stable elements of $p27^{Kip1}$

during dynamics at T=300K (Fig. 3a,b) and T=400K (Fig. 4a,b) are the β-hairpin and the β–strand, while the coil portion, α-helix and 3_{10}-helix fluctuate in a considerably broader range. As temperature increases to T=500K, temperature–induced motions produce more distorted $p27^{Kip1}$ conformations that nevertheless are still largely bound to the complex and are characteristic of an expanded form of the native bound structure (Fig. 5a,b). Only at T=600 K coupled unfolding and unbinding can be seen on the simulation time scale, as evident by dramatic and progressively increasing RMSD from the native structure (Fig. 6a) and a pronounced decrease in the total number of hydrogen bonds (Fig. 6b), suggesting that at this temperature $p27^{Kip1}$ undergoes order–disorder transition to the ensemble of non–specific, largely unstructured conformations weakly bound to the complex. At this temperature, the most stable structural elements of the β-hairpin and the β–strand exhibit a gradual increase in RMSD from the native bound orientations. While considerably more mobile coil, α-helix and 3_{10}-helix reveal a very large range of fluctuations, dynamical behavior of these structural elements even at the high temperature is characterized by frequent excursions to the ensemble of short-lived, yet native–like bound conformations.

The differences between $p27^{Kip1}$ solvation free energies in the unbound and bound states, that are computed based on the extent of solvent–accessible surface area, are pronounced at T=300K (Fig. 3c) and T=400K (Fig. 4c) where $p27^{Kip1}$ is largely bound to the Cdk2–cyclin A complex and fluctuates in near–native orientations. Small deviations in the value of gyration radius (Fig. 3d,4d) are also consistent with a thermodynamically stable behavior of the $p27^{Kip1}$ protein at these temperatures. Temperature–induced fluctuations of the $p27^{Kip1}$ protein at higher temperatures (Fig. 5c,6c) produce a rather close overlap in solvation free energies between the unbound and bound states of the protein indicative of unbinding and unfolding of the protein at these temperatures. The values of gyration radius at higher temperatures (Fig. 5d,6d) reveal considerably larger fluctuations, oscillating between smaller values indicative of forming largely unbound and non–specific aggregates and larger values which correspond to more extended $p27^{Kip1}$ bound conformations featuring a native–like shape.

Because the average effect results from a large number of independent coupled folding and binding events, the focus is not on the analysis of highly diverse, individual unfolding/unbinding high–temperature trajectories, but rather on the general features and preserved hierarchy of structural loss observed in the ensemble of 100 trajectories generated at T=600K. The order parameters that

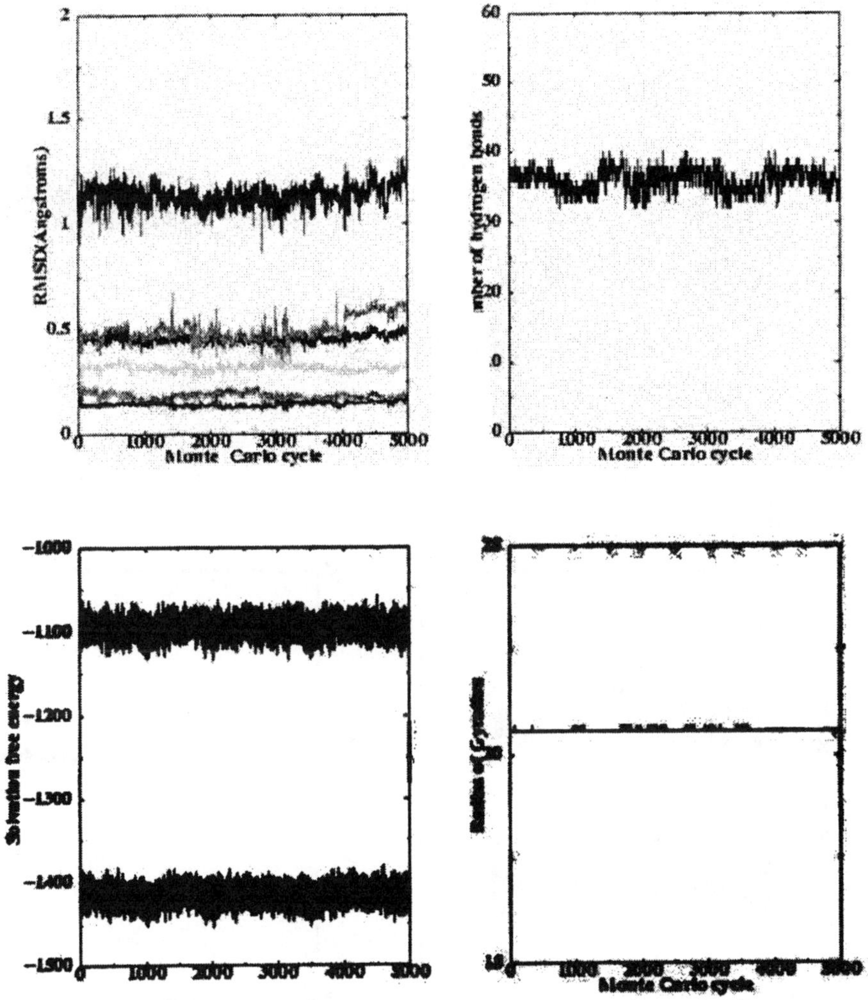

Figure 3. Time–dependent history of (a) the RMSD from the crystal structure during simulations at T=300K for p27 and p27 segments, (b) the total number of the intermolecular hydrogen bonds formed by p27 in the tertiary complex during simulations at T=300K, (c) the solvation free energies in the unbound and bound states of p27 that are computed based on the extent of solvent–accessible surface during simulations at T=300K, (d) the radius of gyration for the p27 protein during simulations at T=300K.

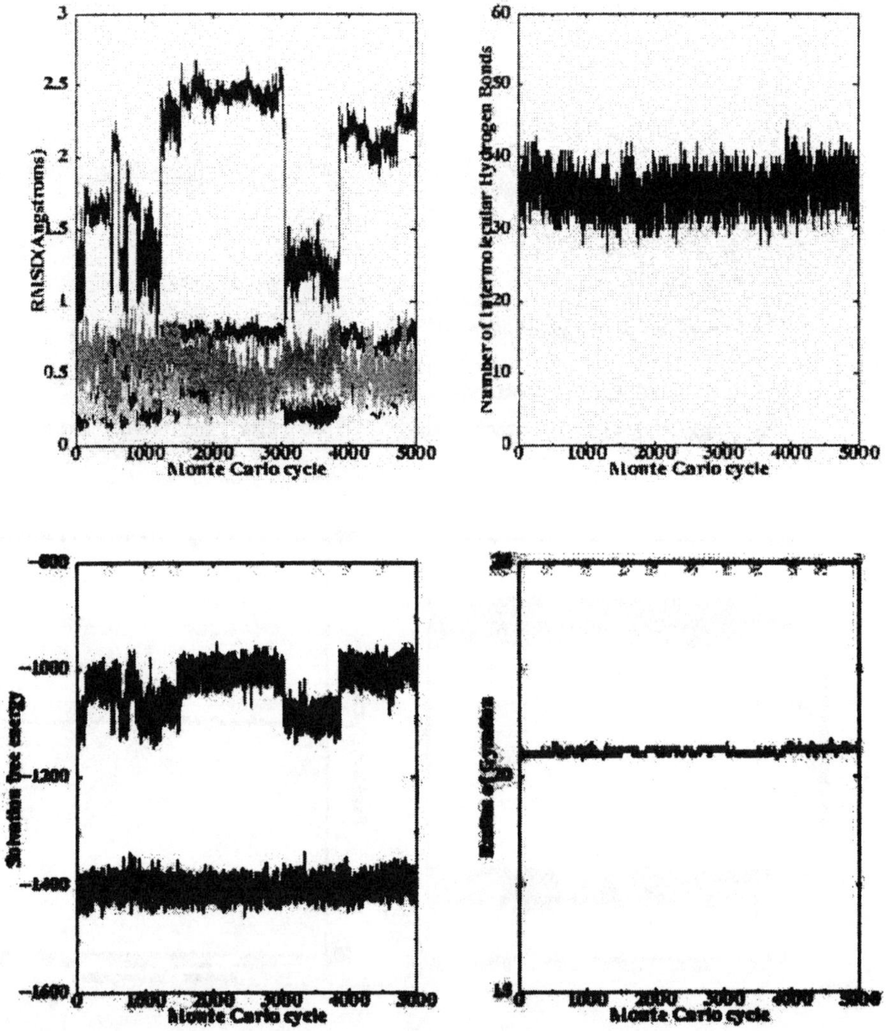

Figure 4. Time–dependent history of (a) the RMSD from the crystal structure during simulations at T=400K for p27 and p27 segments, (b) the total number of the intermolecular hydrogen bonds formed by p27 in the tertiary complex during simulations at T=400K, (c) the solvation free energies in the unbound and bound states of p27 that are computed based on the extent of solvent–accessible surface during simulations at T=400K, (d) the radius of gyration for the p27 protein during simulations at T=400K.

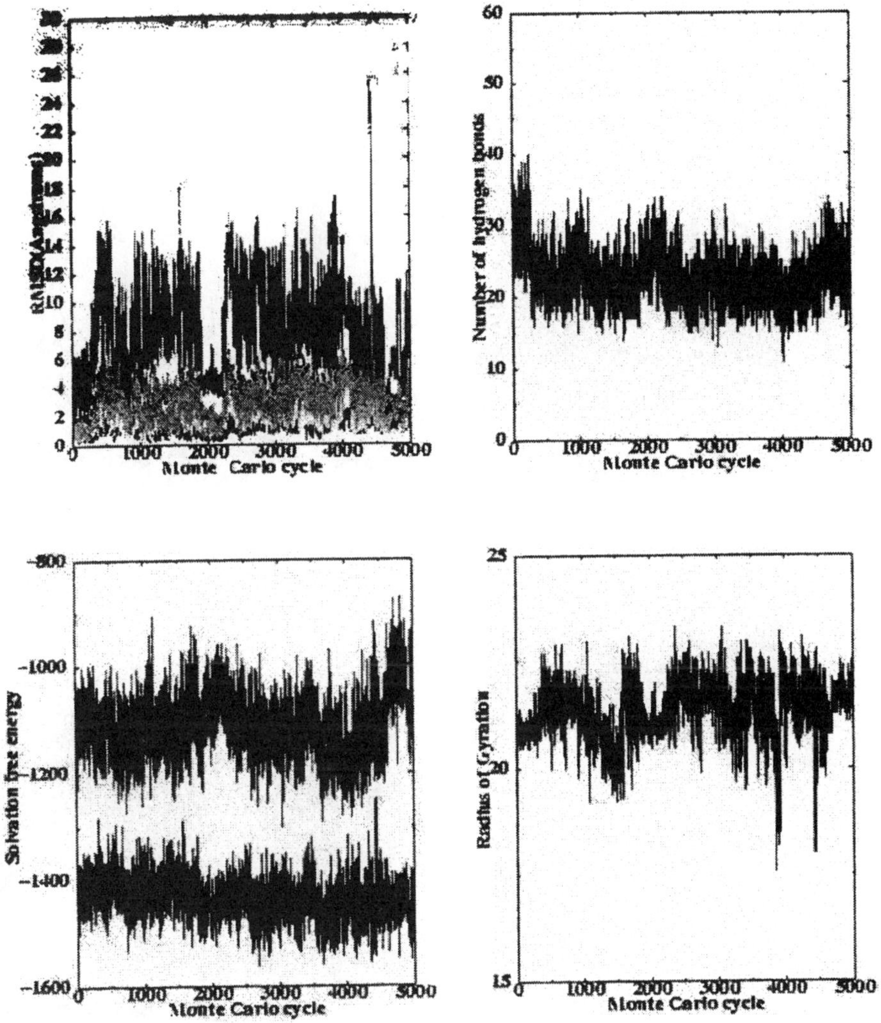

Figure 5. Time–dependent history of (a) the RMSD from the crystal structure during simulations at T=500K for p27 and p27 segments, (b) the total number of the intermolecular hydrogen bonds formed by p27 in the tertiary complex during simulations at T=500K, (c) the solvation free energies in the unbound and bound states of p27 that are computed based on the extent of solvent–accessible surface during simulations at T=500K, (d) the radius of gyration for the p27 protein during simulations at T=500K.

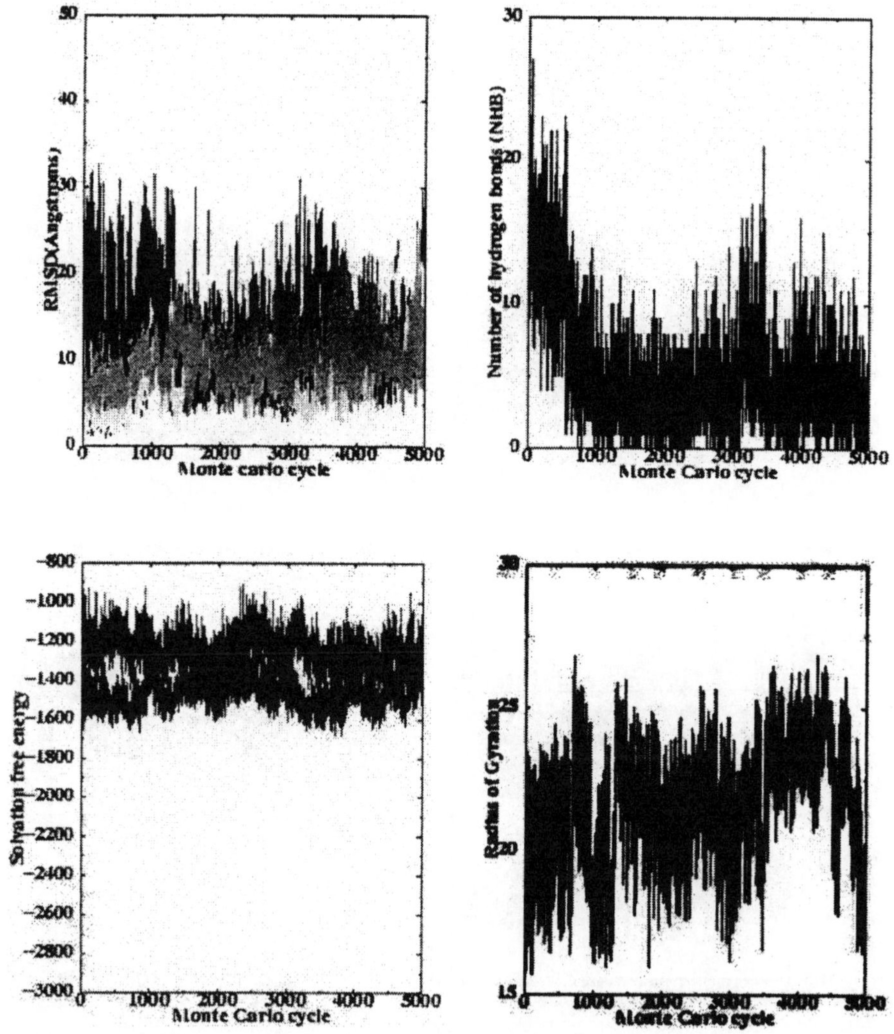

Figure 6. Time–dependent history of (a) the RMSD from the crystal structure during simulations at T=600K for p27 and p27 segments, (b) the total number of the intermolecular hydrogen bonds formed by p27 in the tertiary complex during simulations at T=600K, (c) the solvation free energies in the unbound and bound states of p27 that are computed based on the extent of solvent–accessible surface during simulations at T=600K, (d) the radius of gyration for the p27 protein during simulations at T=600K.

produced a more homogeneous and clear description of the disorder–order transitions and can differentiate the states of the system in the course of simulations are the root mean square deviation (RMSD) of the $p27^{Kip1}$ backbone atoms from the native structure, including the RMSD values for the individual secondary structure elements that can measure unfolding progression, and the total number of intermolecular hydrogen bonds formed by $p27^{Kip1}$ in the complex, that can monitor the extent of unbinding (Fig. 7,8). In the symmetric–funnel limit of a coupled folding and binding event, the occupancies of the intermolecular contacts should decrease in concert with the loss in the native structure. If there is a lack of symmetry in the coupled folding and binding process, the occupancies of some intermolecular contacts should decrease more slowly, and others more rapidly. Despite considerable differences between individual trajectories, the analysis of 20 independent simulations at T=600K initiated from the crystal structure of the tertiary complex has revealed a systematic trend in the hierarchy of structural loss for $p27^{Kip1}$ during coupled unfolding and unbinding (Fig. 7,8). The α–helix and 3_{10}-helix rarely populate native orientations during coupled unfolding and unbinding, that suggests that these structural elements $p27^{Kip1}$ of may be the last ones to be ordered, after overcoming the rate–limiting free energy barrier. The shape of distributions for α–helix and 3_{10}-helix reveals a more symmetric picture, with a concerted loss in the native structure and the total number of intermolecular contacts (Fig. 7,8), which is more consistent with a simple funnel description when the interactions stabilizing the native bound structure also stabilize partially folded conformations. Strikingly, despite large fluctuations of the $p27^{Kip1}$ coil, there is a considerable lack of symmetry in the hierarchy of its structural loss, reflecting frequent transitions between non-specific, largely unbound orientations and native–like conformations (Fig. 7,8).

The most persistent interactions of $p27^{Kip1}$ at the intermolecular interface are formed by the β-hairpin and the β–strand elements that maintain their structural integrity considerably longer during unbinding/unfolding than other $p27^{Kip1}$ elements (Fig. 7). The shape of the distribution for the $p27^{Kip1}$ and its β-hairpin and β-strand components deviates considerably from the concerted picture: as the number of intermolecular hydrogen bonds progressively decreases and $p27^{Kip1}$ gradually dissociates from the bound complex, these structural elements continue to maintain their native bound conformation intact until the late stage of the process. The intermolecular hydrophobic contacts of the β-hairpin conserved residues with the complex and a network of six backbone–backbone hydrogen bonds formed by β-strand at the intermolecular interface determine largely the stability of these elements, that are lost late in the

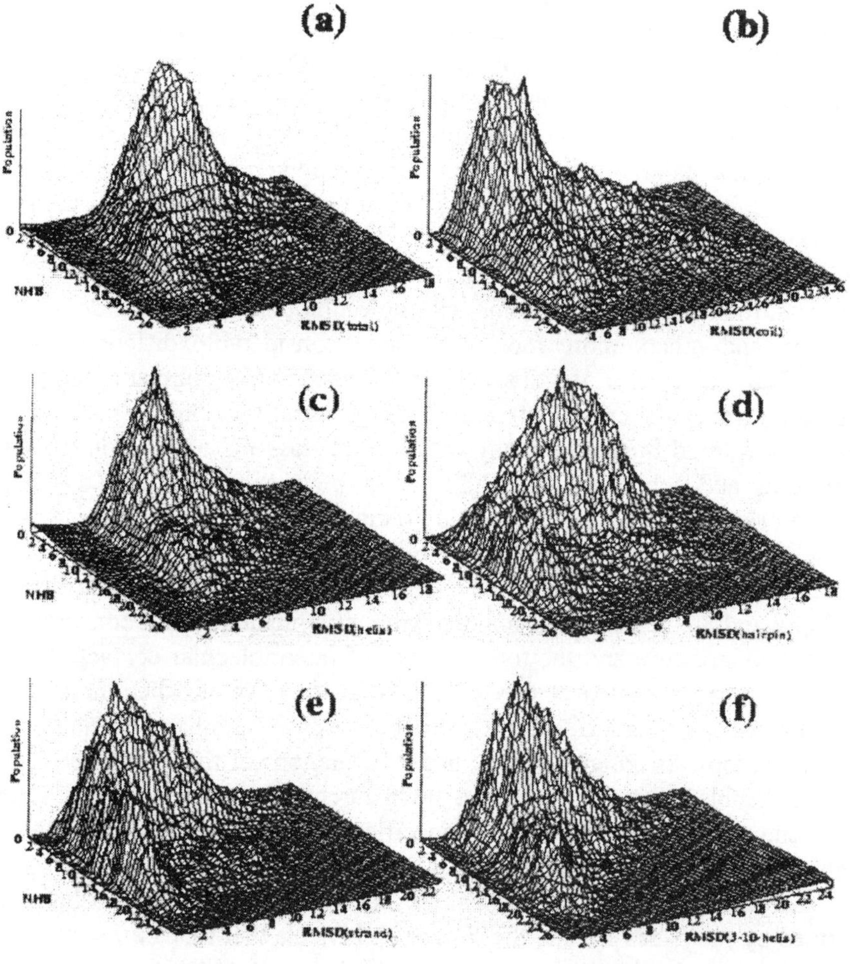

Figure 7. Three-dimensional population histograms generated from 20 independent unfolding/ unbinding simulations at T=600K initiated from the crystal structure of the complex to monitor hierarchy of structural loss for p27 and its segments. The histograms are built as a function of the total number of the intermolecular hydrogen bonds (NHB) formed by the entire p27 in the complex and the RMSD from the crystal structure of the bound p27 in the tertiary complex (as well as the RMSD's from the crystal structure conformation for each of the individual p27 segments).

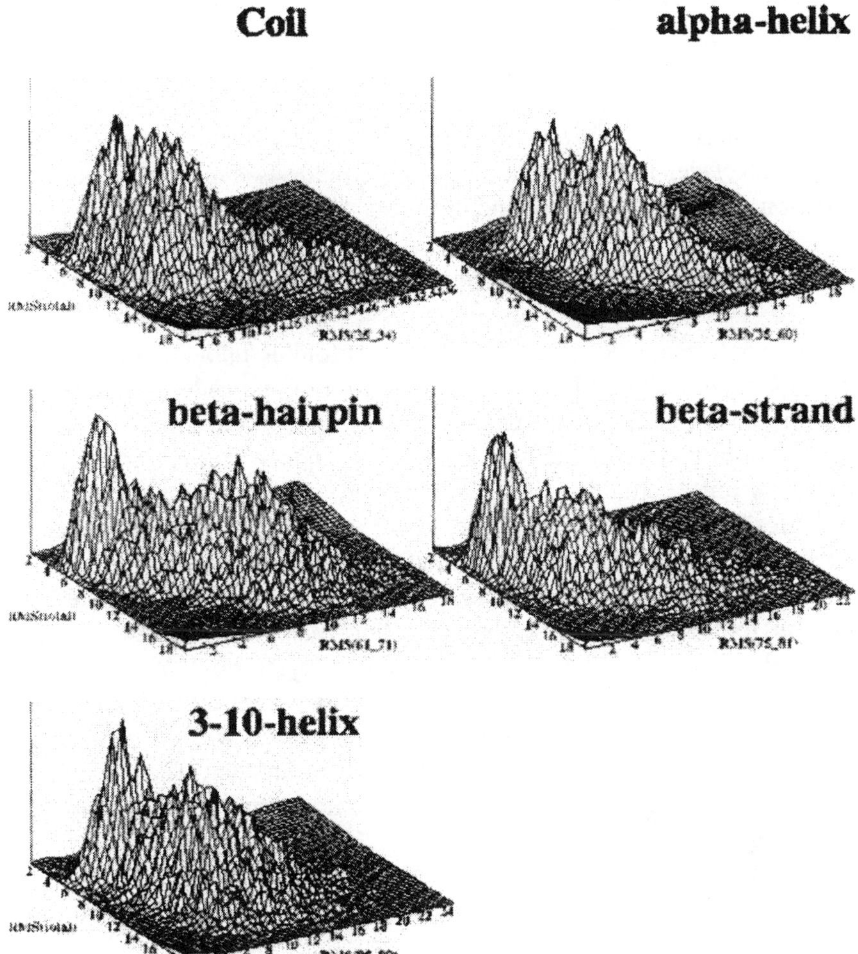

Figure 8. Three-dimensional population histograms generated from 20 independent unfolding/unbinding simulations at T=600K initiated from the crystal structure of the complex to monitor hierarchy of structural loss for p27 and its segments. The histograms are built as a function of the RMSD from the crystal structure for the bound p27 in the tertiary complex and the RMSD's from the crystal structure conformation for each of the individual p27 segments.

unbinding/unfolding process. As the distribution for $p27^{Kip1}$ approaches a dramatic 90 degrees turn (Fig. 7), signaling the loss of the majority of the intermolecular contacts, structural integrity of the β-hairpin and the β-strand elements begins to fade, accompanied by the depletion of the most stable intermolecular specific interactions. Subsequently, $p27^{Kip1}$ undergoes order–disorder transition to the unbound phase and begins to fluctuate in non–specific, largely unstructured conformations that are weakly bound to the cyclin portion of the interface, in agreement with the hypothesis suggesting cyclin A as a initiation docking site during binding.

3.2 Determination of the transition state ensemble

All conformations generated in 20 unfolding/unbinding simulations at T=600K were subjected to structural clustering procedure. As a result, 28 clusters have emerged that generally represent the conformational families of largely unstructured, unbound states, non–specific, weakly bound conformations, partially– unfolded/unbound conformations, and native–like structures. The monitored disorder–order transitions between a meta–stable intermediate, consisting of non–specific, largely unstructured conformations weakly bound to the complex, and the ensemble of the native–like states. In addition to the cluster centers representing major conformational ensembles collected from 20 unfolding/unbinding trajectories at T=600K, 400 conformations were selected from a typical high–temperature unfolding/unbinding trajectory during the time interval when the character of the conformation rapidly changes from one state to another. These conformations were clustered into 3 structurally distinct conformational ensembles that could also represent the putative TS states.

If the free energy profile along a single transition coordinate is complicated, with intermediates, or a well-defined single transition coordinate for a system can not be found, the transition state ensemble can not be determined by sampling the thermal distribution of states in the dominant free energy barrier[105,106]. Because the TS states are inherently unstable with respect to either the non–specific, weakly bound conformations or to the native complex, it is expected short kinetic simulations initiated from a TS conformation will rapidly encounter these states with a similar, close to 50 % probability. These properties of the TS ensemble suggested a computational method for identifying TS conformations in which a collection of short trajectories are initiated from a putative TS conformation, followed by computing the probability that these trajectories reach the native state[106]. This method allows the transition state ensemble to be determined without explicitly knowing the reaction coordinate and regardless of the complexity of the free energy landscape.

20 short kinetic simulations at T=350K were performed from 10 representative conformations of each of the originally determined 28 clusters and additionally selected 3 cluster centers, generated from a single trajectory, to test the putative TS conformations. These kinetic runs are conducted on a short time scale, constituting only 2%of the total simulation time necessary to consistently observe unfolding and unbinding at high–temperature simulations. Transition probability calculations are conducted at lower temperature, because the time it takes for a conformation to diffuse away from the top of a free energy barrier either to the native state or to the non–specific, unbound states should be much faster than the rate at which this barrier can be crossed starting from the initial or final states. The principal caveat of this method is that transitions that are relevant at low temperatures may not appear in high-temperature unfolding trajectories, and therefore one cannot exclude the possibility for alternative pathways that might dominate at lower temperatures.

The TS conformations rapidly converge to the native–like structures in about half of the runs (Fig. 9) and can be described as an expanded form of the native states in the β-hairpin and the β-strand regions of the interface. In contrast, a considerable loss of the native bound structure for the coil, α–helix, and 3_{10}-helix is observed in the transition state conformations. The ensembles of so–called "post-critical" states were also determined, that are defined as the first stable structures that appear immediately after the TS is overcome and lead to a rapid and consistent descend to the native structure[116-118]. A significant fraction of $p27^{Kip1}$ highly populated conformational clusters consistently refolds and rebinds to the native bound structure after short kinetic runs (Fig. 10). The principal difference between the post-critical conformations and the TS ensemble is determined by a considerable structural consolidation of the β-hairpin and the β-strand intermolecular contacts, in particular, by strengthening in six backbone–backbone hydrogen bonds established by β-strand with the complex (Fig. 9,10). The C_{alpha} RMSD from the native structure for representative conformations in these ensembles shown as a function of $p27^{Kip1}$ residue number (residues 25-93) (Fig. 9a,10a) display similar, large fluctuations in the coil and 3_{10}–helix and less pronounced motions in the α-helix. The differences in structural integrity of the β-hairpin and the β-strand are reflected in larger 3.5 °A-4.0 °A RMSD fluctuations from their native structure in the TS ensemble. These results suggest that consolidation of a localized set of specific interactions formed by the β-hairpin and the β-strand at the interface can result

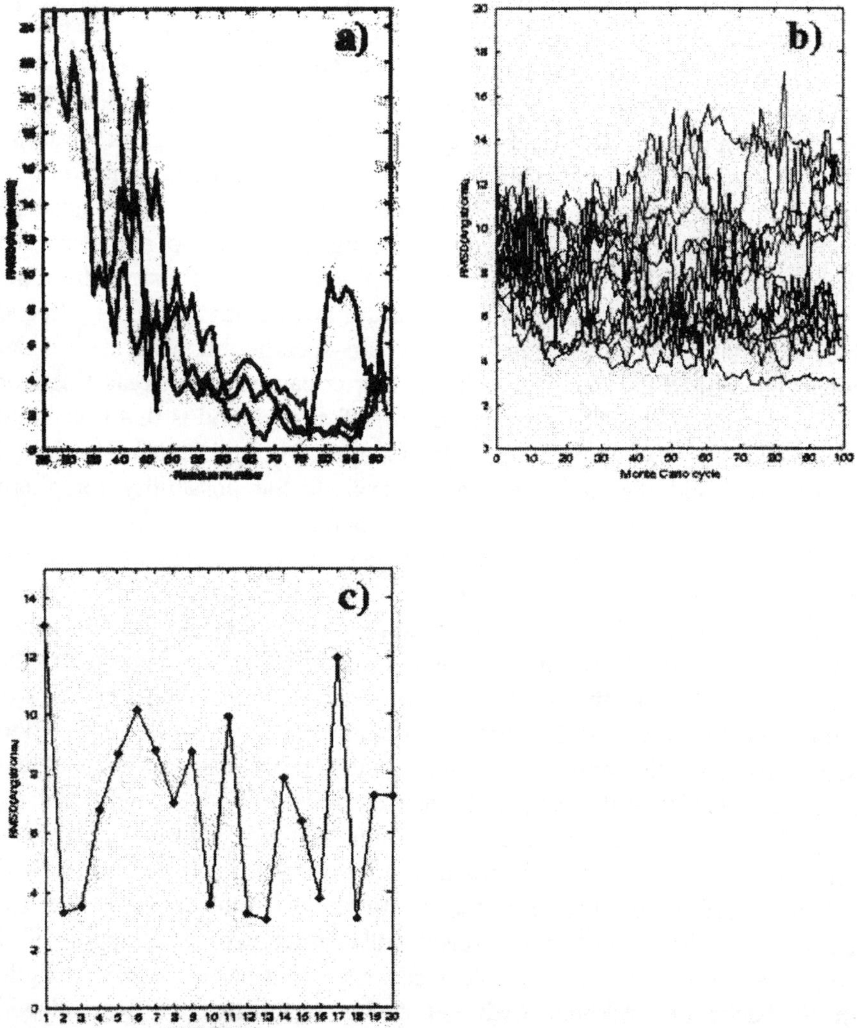

Figure 9. A microscopic characterization of the TS ensemble : a) The RMSD from the native structure for representative conformations in the TS ensemble as a function of p27 residue number (from residue 25 to residue 93) b) Time-dependent history for a panel of 20 short kinetic runs initiated from a typical TS ensemble conformation. The TS conformations rapidly converge to the native–like structures in about half of the runs. c) The RMSD from the native structure for 20 p27 conformations emerging at the end of 20 short kinetic runs initiated from a typical TS ensemble state.

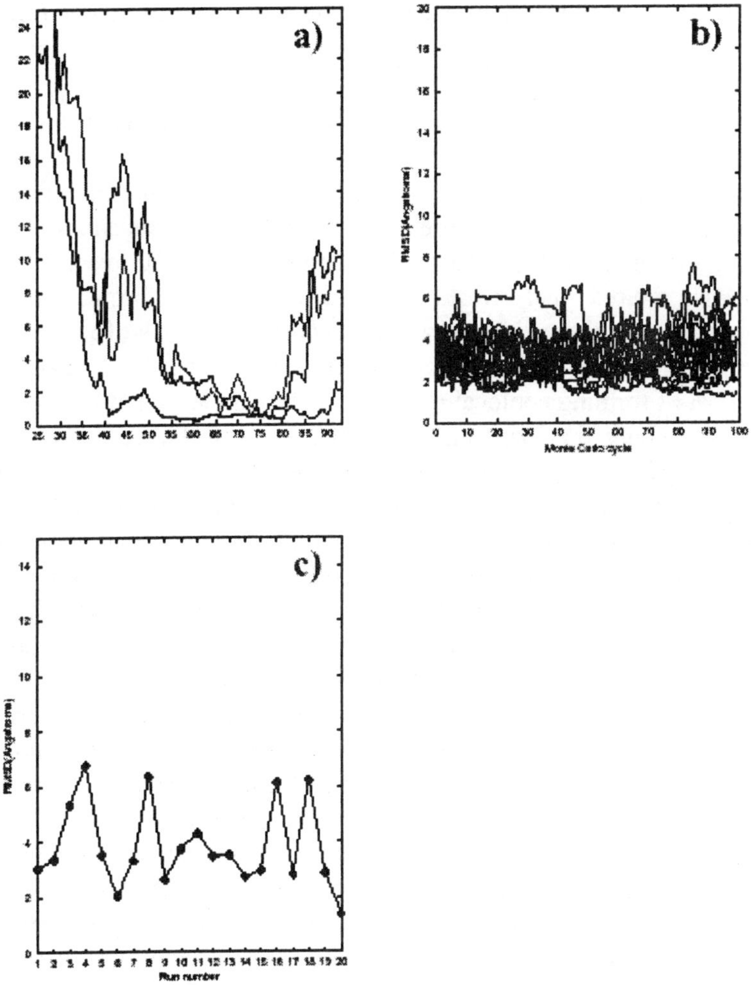

Figure 10. Amicroscopic characterization of the post-critical ensemble : a) TheRMSDfrom the native structure for representative conformations in the post-critical ensemble as a function of p27 residue number (from residue 25 to residue 93) b) Time-dependent history for a panel of 20 short kinetic runs initiated from a post-critical conformation. The post-critical conformations rapidly converge to the native–like structures in most of the runs. c) The RMSD from the native structure for 20 p27 conformations emerging at the end of 20 short kinetic runs initiated from a typical post-critical ensemble state.

in nucleation conformations that are kinetically committed to consistently and rapidly reach the native structure without encountering any major free energy barriers.

3.3 An atomic picture of the binding mechanism

Amicroscopic characterization of the meta–stable states and free energy barrier suggests an atomic picture of the binding mechanism. Initially, the formation of a meta–stable intermediate consisting of entropically favorable, largely unstructured and weakly bound $p27^{Kip1}$ conformations, binding mechanism may be favored by means of frequent collisions of the intrinsically disordered $p27^{Kip1}$ with the rigid cyclin portion of the complex. A relatively large entropic cost of forming nonlocal interactions during the initiation binding event encourages probing the local intermolecular contacts between the conserved LFG motif of $p27^{Kip1}$ coil and recognition groove of cyclin. By analogy with the fly–casting mechanism, the unfolded $p27^{Kip1}$ may anchor weakly in this region to favorably position the remainder of the protein for coupled folding and binding event. The results suggest that the $p27^{Kip1}$ exploits the intrinsic disorder in the α–helix until thermal fluctuations allow the system to wrap around the binding partner, with the β-hairpin starting to form hydrophobic interactions and the β-strand adopting the native structure topology at the intermolecular interface. As $p27^{Kip1}$ protein begins to explore the native–like topology of the interface and intermolecular contacts in these regions, while retaining favorable contacts with cyclin, the entropy loss caused by retaining non-local, intermolecular contacts simultaneously at the remote regions of the interface may not be fully compensated by the lack of sufficient strength in these interactions which are yet to be consolidated. Indeed, the the specific interactions formed with the complex by the β-hairpin and β-strand have considerably stronger orientational requirements than the hydrophobic intermolecular contacts in this region and are not likely to be stabilizing unless they are fully formed. Consequently, $p27^{Kip1}$ may find more advantageous at this stage to partially release contacts with the cyclin in order to optimize the balance between the loss in conformational entropy and the energy gain upon formation of native interactions. At this stage of the process, before the transition state is reached, the conformational entropy dominates and the formation of new native interactions leads to a greater loss in the entropy than the gain in the energy. i.e. the coupled folding and binding reaction proceeds up-hill in free energy terms.

The intrinsic folding preferences to formlocal secondary structure in $p27^{Kip1}$ are discouraged at this stage by insufficient gain in the intermolecular interactions that would not compensate for the loss of entropy, i.e. α–helix formation is found to be kinetically detrimental during the coupled folding and binding process. It appears that only after the major free energy barrier is passed, the α–helix and the 3_{10}-helix elements begin to form their native structure concomitantly with the formation of the native intermolecular contacts. At this stage, the reaction is dominated by the energy gain that is greater then the loss in conformational entropy, and the process rapidly proceeds down-hill to the native complex. These results provide a structural rationale for the experimental data revealing that preorganized native–like local secondary structure in $p27^{Kip1}$ can result in slower binding. Consequently, the observed stabilization of α-helix in the alanine mutants may introduce a requirement for unfolding of α-helix prior to binding leading to the experimentally detected, concomitant decrease in the rate of complex formation.

The emerging structural polarization in the ensemble of unfolding/unbinding trajectories and in the computationally determined TSE are not determined by the folding topological preferences of $p27^{Kip1}$, but rather can be viewed as a consequence of the topological requirements of the intermolecular interface to minimize free energy cost associated with ordering the β-hairpin and β–strand intermolecular contacts which could be critical for nucleating rapid folding and binding. Indeed, structural integrity of the localized set of specific intermolecular hydrogen contacts can result in nucleation conformations that are kinetically committed to attain the native structure. Furthermore, the TSE conformations and post–critical conformations have a similar topology of the intermolecular interface, yet may dramatically diverge during kinetic runs to either the unbound state or to the native structure based on subtle differences in structural consolidation of six backbone–backbone hydrogen bonds formed by β-strand with the complex. Hence, structural polarization of the TSE may also reflect the importance of specific hydrogen bonds in stabilizing the transition state since these interactions are the last to break during coupled unfolding and unbinding and may contribute to the nucleation the refolding and rebinding process. Importantly, significant structural changes in the corresponding portion of the interface upon binding with $p27^{Kip1}$, that are not modeled in our simulations, are localized specifically in the regions interacting with the β-hairpin and β-strand and contribute to consolidation of four intermolecular hydrogen bonds formed by the β-hairpin and six backbone–backbone intermolecular hydrogen bonds made by the β-strand. Hence, it is tempting to suggest that the slow, rate–limiting step of the coupled folding and binding reaction may indeed be determined by the balance between a considerable entropy loss in this region, exacerbated by protein structural changes, and energy gain upon formation of the extensive intermolecular contacts. Universality between protein folding and molecular recognition scenarios are further manifested by a rather broad distribution of conformations in the TSE that are nevertheless share a common and rather localized set of spefic interactions. This suggests an underlying energy landscape of the folding and binding reaction that may reconcile the extreme of a diverse and symmetrical funnel[31-33] with a more "classical"view of folding reaction[34,35].

Adetailed experimental characterization of the binding mechanismand structural mapping of the TSE for $p27^{Kip1}$ and related members of the Kip/Cip family using protein engineering techniques and binding kinetics measurements is required to rigorously assess the validity of computational predictions and quantify the role of specific interactions and topological determinants of folding and binding. Nevertheless, computational characterization of the TSE and the proposed binding mechanism have shown that coupling of folding and binding is not primarily driven by intrinsic folding preferences of $p27^{Kip1}$, but is rather determined by the requirements to form specific complex that ultimately dictates the folding mechanism. In agreement with the experimental data, it was shown in this work that the topology of the native intermolecular interface coupled with the localized, specific interactions formed by $p27^{Kip1}$ with the complex in transition state overwhelms any local folding preferences for creating a stable α-helix prior to overcoming the major free energy barrier.

The results of this study may have some implications for understanding a balance between local and non-local interactions in processes of coupled folding and binding. Similarly to protein folding theory, it may be argued that if local elements of the native structure are rarely populated in the unbound state, but is present in the transition state, such as the native β-hairpin and β-strand segments of $p27^{Kip1}$, then stabilization of these elements should accelerate the reaction and stabilize the complex. Hence, our results provide support a view that a balance between short–range and long-range interactions at the interface may ultimately determine kinetics of the process.

4. CONCLUSIONS

A microscopic characterization of the free energy barrier suggests an atomic picture of the binding mechanism that rationalizes and reconciles the hypothesized initiation binding event with the experimental data, indicating a kinetic advantage for the intrinsically unstructured α-helix in the unbound form. Despite considerable differences between individual trajectories, the analysis of independent simulations at T=600K shows a systematic trend in the hierarchy of structural loss for $p27^{Kip1}$ during coupled unfolding and unbinding. The emerging structural polarization in the ensemble of unfolding/unbinding trajectories and in the computationally determined TSE are not determined by the folding topological preferences of $p27^{Kip1}$, but is interpreted as a consequence of the topological requirements of the intermolecular interface to minimize free energy cost associated with ordering the β-hairpin and β-strand intermolecular contacts which are the last one to disintegrate in unfolding/unbinding and thereby could be important for nucleating rapid folding and binding. In agreement with the experimental data, it has been shown that the topology of the native intermolecular interface coupled with the localized, specific interactions formed by $p27^{Kip1}$ with the complex in transition state overwhelms any local folding preferences for creating a stable α-helix prior to overcoming the major free energy barrier. These results provide a structural rationale for the experimental data revealing that preorganized native–like local secondary structure in $p27^{Kip1}$ can result in slower binding. Hence, folding of unstructured proteins upon binding to a given template is largely determined by the requirements to form specific complex that ultimately dictates the folding mechanism. By using a protein engineering approach similar to the -value analysis developed to characterize the TS ensemble in protein folding, it should be possible to validate the predicted details of the binding mechanism by probing the kinetic consequences of mutations of every residue that makes appreciable interactions in the native state. Synergy of theoretical and experimental advances in the fields of protein folding and binding, based on the increasing amount of information in known structures and mechanisms may provide a fruitful direction for future research and allow to unify interdisciplinary efforts in resolving these key problems in molecular biology.

REFERENCES

1. Wright, P.E., & Dyson, H.J., 1999, Coupling of folding and binding for unstructured proteins. *J. Mol. Biol.***293:** 321-331.
2. Dunker, A.K., Lawson, J.D., Brown, C.J., Williams, R.M., Romero, P., Oh, J.S., Oldfield, C.J., Campen, A.M., Ratliff, C.M., Hipps, K.W., Ausio, J., Nissen, M.S., Reeves, R., Kang, C., Kissinger, C,R., Bailey, R.W., Griswold, M.D., Chiu,W., Garner, E.C., & Obradovic, Z., 2001, Intrinsically disordered protein. *J. Mol. Graph. Model.* **19:** 26-59.
3. Dunker, A.K., & Obradovic, Z., 2001, The protein trinity–linking function and disorder. *Nat. Biotechnol.* **19:** 805-806.
4. Dunker, A.K., Brown, C.J., & Obradovic, Z., 2002, Identification and functions of usefully disordered proteins. *Adv. Protein. Chem.***62**: 25-49.
5. Dunker, A.K., Brown, C.J., Lawson, J.D., Iakoucheva, L.M., & Obradovic, Z., 2002, Intrinsic disorder and protein function. *Biochemistry* **41:** 6573-6582.
6. Dyson, H.J., & Wright, P.E., 2002, Coupling of folding and binding for unstructured proteins. *Curr. Opin. Struct. Biol.* **12:** 54-60.
7. Dyson, H.J., & Wright, P.E., 2002, Insights into the structure and dynamics of unfolded proteins from nuclear magnetic resonance. *Adv. Protein Chem.* **62:** 311-314.
8. Uversky, V.N., 2002, Natively.unfolded proteins: a point where biology waits for physics. *Protein Sci.* **11**: 739-756.
9. Uversky, V.N., Gillespie, J.R., & Fink, A.L., 2000, Why are "natively unfolded" proteins unstructured under physiologic conditions. *Proteins: Struct. Funct. Genet.* **41:** 415-427.
10. Uversky, V.N., 2002, What does it mean to be natively unfolded? *Eur. J. Biochem.* **269:** 2-12.
11. Spolar, R.S., & Record, M.T., 1994, Coupling of local folding to sitespecific binding of proteins to DNA. *Science* **263:** 777-784.
12. Plaxco, K.W. & Gross, M. (1997) The importance of being unfolded. *Nature* **386:** 657-659.
13. Kriwacki, R.W., Hengst, L., Tennant, L., Reed, S.I., & Wright, P.E., 1996, Structural studies of p21Wafl/Cipl/Sdil in the free and Cdk2-bound state: conformational disorder mediates binding diversity. *Proc. Natl. Acad. Sci. USA* **93:** 11504-11509.
14. Hashimoto, Y., Kohri, K., Kaneko, Y., Morisaki, H., Kato, T., Ikeda, K., & Nakanishi, M., 1998,. Critical role for the 310 helix region of p57(Kip2) in cyclin-dependent kinase 2 inhibition and growth suppression. *J Biol Chem* **273:** 16544-16550.
15. Adkins, J.N., Lumb, K., 2002, Intrinsic structural disorder and sequence features of the cell cycle inhibitor p57Kip2. *Proteins: Struct. Funct. Genet.* **46:** 1-7.
16. Russo, A,A,, Jeffrey, P.D., Patten, A.K., Massague, J., & Pavletich, N.P., 1996, Crystal structure of the p27Kipl cyclin-dependent-kinase inhibitor bound to the cyclin A-Cdk2 complex. *Nature* **382:** 325-331.
17. Flaugh, S.L., & Lumb, K.J., 2001, Effects of macromolecular crowding on the intrinsically disordered proteins c-Fos and p27(Kipl). *Biomacromolecules* **2:** 538-540.
18. Bienkiewicz, E.A., Adkins, J.N., & Lumb, K., 2002, Functional consequences of preorganized helical structure in the intrinsically disordered cell-cycle inhibitor p27(Kipl). *Biochemistry* **41:** 752-759.
19. Davis, A.M. & Teague, S.J., 1999, Hydrogen bonding, hydrophobic interactions, and failure of the rigid receptor hypothesis. *Angew. Chem. Int. Ed. Engl.***39:** 736-749.

20. Van Regenmortel, M.H., 1999, Molecular recognition in the postreductionistera. *J. Mol. Recognit.* **12:** 1-2.
21. Carlson, H.A. & McCammon J.A., 2000, Accommodating protein flexibility in computational drug design. *Mol. Pharmacol.* **57:** 213-218.
22. Ma, B.,Wolfson, H.J., & Nussinov, R., 2001, Protein functional epitopes: hot spots, dynamics and combinatorial libraries. *Curr. Opin. Struct. Biol.* **11:** 364-369.
23. Atwell, S., Ultsch, M., De Vos, A.M., & Wells, J.A., 1997, Structural plasticity in a remodeled protein-protein interface. *Science* **278:** 1125-1128.
24. Sundberg, E.J., & Mariuzza, R.A., 2000, Luxury accommodations: the expanding role of structural plasticity in protein-protein interactions. *Structure Fold. Des.* **8:** R137-142.
25. Demchenko, A.P., 2001, Recognition between flexible protein molecules: induced and assisted folding. *J. Mol. Recognit.* **14:** 42-61.
26. DeLano, W.L., Ultsch, M.H., de Vos, A.M., & Wells, J.A., 2000, Convergent solutions to binding at a protein-protein interface. *Science* **287:** 1279-1283.
27. Luque, I., Leavitt, S.A., & Freire E., 2002, The linkage between protein folding and functional cooperativity: two sides of the same coin. *Annu. Rev. Biophys. Biomol. Struct.* **31:** 235-256.
28. Bryngelson, J. D., Onuchic, J. N., Socci, N. D. & Wolynes, P. G., 1995, Funnels, pathways, and the energy landscape of protein folding, a synthesis. *Proteins: Struct. Funct. Genet.: Struct. Funct. Genet.* **21:** 167-195.
29. Dill, K. A., Bromberg, S., Yue, K., Fiebig, K.M., Yee, D.P., Thomas, P.D. & Chan, H.S., 1995, Principle of protein folding–a perspective from simple exact models. *Protein Sci.* **4:** 561-602.
30. Dill, K.A. & Chan, H.S., 1997, From Levinthal to pathways to funnels. *Nat. Struct. Biol.* **4:** 10-19.
31. Onuchic, J.N., Luthey-Schulten, Z., & Wolynes, P.G., 1997, Theory of protein folding: the energy landscape perspective. *Annu. Rev. Phys. Chem.* **48:** 545-560.
32. Onuchic, J.N., Nymeyer, H., Garcia, A.E., Chahine, J., & Socci, N.D., 2000, The energy landscape theory of protein folding: insights into folding mechanisms and scenarios. *Adv. Protein. Chem.* **53:** 87-152.
33. Plotkin, S.S., & Onuchic, J.N., 2002, Understanding protein folding with energy landscape theory. Part I: Basic concepts. *Q. Rev. Biophys.* **35:** 111-167.
34. Shakhnovich, E.I., 1997, Theoretical studies of protein–folding thermodynamics and kinetics. *Curr. Opin. Struct. Biol.* **7:** 29-40.
35. Mirny, L., & Shakhnovich, E., 2001 Protein folding theory: from lattice to all-atom models. *Annu. Rev. Biophys. Biomol. Struct.* **30:** 361-396.
36. Fersht, A.R., & Daggett, V., 2002, Protein folding and unfolding at atomic resolution *Cell* **108:** 573-582.
37. Janin, J., 1996, Quantifying biological specificity: the statistical mechanics of molecular recognition. *Proteins: Struct. Funct. Genet.* **25:** 438-445.
38. Verkhivker, G.M. & Rejto, P.A., 1996, A mean field model of ligand–protein interactions, Implications for the structural assessment of human immunodeficiency virus type 1 protease complexes and receptor–specific binding. *Proc. Natl. Acad. Sci. USA.* **93:** 60-64.
39. Rejto, P.A. & Verkhivker, G.M., 1996, Unraveling principles of lead discovery: from unfrustrated energy landscapes to novel molecular anchors. *Proc. Natl. Acad. Sci. USA.* **93:** 8945-8950.

40. Tsai, C.-J., Xu, D., & Nussinov, R., 1998, Protein folding via binding and vice versa. *Curr. Biol.* **3:** R71-R80.
41. Tsai, C.-J., Ma, B., & Nussinov, R., 1999, Folding and binding cascades: shifts in energy landscapes. *Proc. Natl. Acad. Sci. USA.* **96:** 9970-9972.
42. Sinha, N., & Nussinov, R., 2001, Point mutations and sequence variability in proteins: Redistributions of preexisting populations. *Proc. Natl. Acad. Sci. USA.* **98:** 3139-3144.
43. Ma, B., Shatsky, M., Wolfson, H.J., & Nussinov, R., 2002, Multiple diverse ligands binding at a single protein site: a matter of pre-existing populations. *Protein Sci.* **11:** 184-197.
44. Verkhivker, G.M., Rejto, P,A., Bouzida, D., Arthurs, S., Colson, A.B., Freer, S.T., Gehlhaar, D.K., Larson, V., Luty, B.A., Marrone, T., & Rose, P.W., 1999, Towards understanding the mechanisms of molecular recognition by computer simulations of ligand–protein interactions *J. Mol. Recognit.* **12:** 371-389.
45. Zhang, C., Chen, J., & DeLisi, C., 1999, Protein–protein recognition: exploring the energy funnels near the binding sites. *Proteins: Struct. Funct. Genet.* **34:** 255-267.
46. Vakser, I.A., Matar, O.G., & Lam, C.F., 1999, A systematic study of lowresolution recognition in protein–protein complexes. *Proc. Natl. Acad. Sci.USA.* **96:** 8477-8482.
47. Tovchigrechko, A. & Vakser, I.A., 2001, How common is the funnel-like energy landscape in protein-protein interactions? *Protein Sci.* **10**, 1572-1583.
48. Camacho, C.J., Weng, Z., Vajda, S., & DeLisi, C. (1999) Free energy landscapes of encounter complexes in protein–protein association. *Biophys. J.* **76:** 1166-1178.
49. Camacho, C.J.,Vajda, S., 2001, Protein docking along smooth association pathways. *Proc. Natl. Acad. Sci. USA.* **98:** 10636-10641.
50. Verkhivker, G.M., Bouzida, D., Gehlhaar, D.K., Rejto, P,A., Schaffer L, Arthurs, S., Colson, A.B., Freer, S.T., Larson,V., Luty, B.A., Marrone, T.,& Rose, P.W., 2001, Hierarchy of simulation models in predicting molecular recognition mechanisms from the binding energy landscapes: structural analysis of the peptide complexes with SH2 domains. *Proteins: Struct. Funct. Genet.* **45:** 456-470.
51. Verkhivker, G.M., Rejto, P,A., Bouzida, D., Arthurs, S., Colson, A.B., Freer, S.T., Gehlhaar, D.K., Larson, V., Luty, B.A., Marrone, T., & Rose, P.W., 2001, Navigating ligand–protein binding free energy landscapes: universality and diversity of protein folding and molecular recognition mechanisms. *Chem. Phys. Lett.* **336:** 495-503.
52. Verkhivker, G.M., Bouzida, D., Gehlhaar, D.K., Rejto, P,A., Freer, S.T., & Rose, P.W., 2002, Complexity and simplicity of ligand-macromolecule interactions: the energy landscape perspective. *Curr. Opin. Struct. Biol.* **12:** 197-202.
53. Verkhivker, G.M., Bouzida, D., Gehlhaar, D.K., Rejto, P,A., Freer, S.T., & Rose, P.W., 2002, Monte Carlo simulations of the peptide recognition at the consensus binding site of the constant fragment of human immunoglobulin G: the energy landscape analysis of a hot spot at the intermolecular interface. *Proteins: Struct. Funct. Genet.* **48:** 539-557.
54. Plaxco, K.W., Simons, K.T., & Baker, D., 1998, Contact order, transition state placement and the refolding rates of single domain protein. *J. Mol. Biol.* **277:** 985-994.
55. Grantcharova, V.P., Riddle, D.S., Santiago, J.V., & Baker, D., 1998, Important role of hydrogen bonds in the structurally polarized transition state for folding of the src SH3 domain. *Nat. Struct. Biol.* **5**, 714-720.

56. Riddle, D.S., Grantcharova, V.P., Santiago, J.V., Alm, E., Ruczinski, I., & Baker, D., 1999, Experiment and theory highlight role of native state topology in SH3 folding. *Nat. Struct. Biol.* **6:** 1016-1024.
57. Martinez, J.C.,& Serrano, L., 1999, The folding transition state between SH3 domains is conformationally restricted and evolutionarily conserved. *Nat. Struct. Biol.* **6:** 1010-1016.
58. Plaxco, K.W., Larson, S., Ruczinski, I., Riddle, D.S., Thayer, E.C., Buchwitz, B., & Davidson, A.R., 2000, Evolutionary conservation in protein folding kinetics. *J. Mol. Biol.* **298:** 303-312.
59. Baker, D., 2000, A surprising simplicity to protein folding. *Nature* **405:** 39-42.
60. Plaxco, K.W., Simons, K.T., Ruczinski, I., & Baker, D., 2000, Topology, stability, sequence, and length: defining the determinants of two-state protein folding kinetics. *Biochemistry* **39:** 11177-11183.
61. Larson, S,M,, Ruczinski, I., Davidson, A.R., Baker, D., & Plaxco, K.W. 2002, Residues participating in the protein folding nucleus do not exhibit preferential evolutionary conservation. *J. Mol. Biol.* **316:** 225-233.
62. Fersht, A.R., 2000, Transition-state structure as a unifying basis in proteinfolding mechanisms: contact order, chain topology, stability, and the extended nucleus mechanism. *Proc. Natl. Acad. Sci. USA.* **97:** 1525-1529.
63. Galzitskaya, O.V., & Finkelstein, A.V., 1999, Atheoretical search for folding/unfolding nuclei in three-dimensional protein structures. *Proc. Natl. Acad. Sci. USA.* **96:** 11299-11304.
64. Alm, E., & Baker. D., 1999, Prediction of protein-folding mechanisms from free-energy landscapes derived from native structures. *Proc. Natl. Acad. Sci. USA.* **96:** 11305-11310.
65. Munoz, V., & Eaton W.A., 1999, A simple model for calculating the kinetics of protein folding from three-dimensional structures. *Proc. Natl. Acad. Sci. USA.* **96:** 11311-11316.
66. Tsai, J., Levitt, M., & Baker, D., 1999, Hierarchy of structure loss in MD simulations of src SH3 domain unfolding. *J. Mol. Biol.* **291:** 215-225.
67. Alm, E., & Baker D. (1999). Matching theory and experiment in protein folding. *Curr. Opin. Struct. Biol.* **9:** 189-199.
68. Clementi, C., Nymeyer, H. & Onuchic, J., 2000, Topological and energetic factors: what determines the structural details of the transition state ensemble and en-route intermediates for protein folding? An investigation for small globular proteins. *J. Mol. Biol.* **298:** 937-953.
69. Clementi, C., Jennings, P.A., & Onuchic, J.N., 2000, How native-state topology affects the folding of dihydrofolate reductase and interleukin-1beta. *Proc. Natl. Acad. Sci. USA.* **97:** 5871-5876.
70. Clementi, C., Jennings, P.A., & Onuchic, J.N., 2001, Prediction of folding mechanism for circular-permuted proteins. *J. Mol. Biol.* **311:** 879-890.
71. Vendruscolo, M., Paci, E., Dobson, C.M., & Karplus, M., 2001, Three key residues form a critical contact network in a protein folding transition state. *Nature* **409:** 641-645.
72. Ferrara, P., & Caflisch, A., 2001, Native topology or specific interactions: what is more important for protein folding? *J. Mol. Biol.* **306:** 837-850.
73. Gsponer, J., & Caflisch, A., 2001, Role of native topology investigated by multiple unfolding simulations of four SH3 domains. *J. Mol. Biol.* **309:** 285-298.
74. Gsponer, J., & Caflisch, A., 2002, Molecular dynamics simulations of protein folding from the transition state. *Proc. Natl. Acad. Sci. USA.* **99:** 6719-6724.
75. Dokholyan, N.V., Li, L., Ding, F., & Shakhnovich, E.I., 2002, Topological determinants of protein folding. *Proc. Natl. Acad. Sci. USA.* **99:** 8637-8644.

76. Wolynes, P.G., 1997, Folding funnels and energy landscapes of larger proteins within the capillarity approximation. *Proc. Natl. Acad. Sci. USA.* **94:** 6170-6715.
77. Shoemaker, B.A., & Wolynes, P.G., 1999, Exploring structures in protein folding funnels with free energy functionals: the denatured ensemble. *J. Mol. Biol.* **287:** 657-674.
78. Shoemaker, B.A.,Wang, J., & Wolynes, P.G., 1999, Exploring structures in protein folding funnels with free energy functionals: the transition state ensemble. *J. Mol. Biol.* **287:** 675-694.
79. Portman, J., Takada, S. & Wolynes, P., 2001, Micro-scopic theory of protein folding rates. I. Fine structure of the free energy profile and folding routes from a variational approach. *J. Chem. Phys.* **114**: 5069–5081.
80. Portman, J., Takada, S. & Wolynes, P., 2001, Microscopic theory of protein folding rates. II. Local reaction coordinates and chain dynamics. *J. Chem. Phys.* **114:** 5082–5096.
81. Alm, E,., Morozov, A.V., Kortemme, T., & Baker, D., 2002, Simple physical models connect theory and experiment in protein folding kinetics. *J. Mol. Biol.* **322:** 463-476.
82. McCallister, E.L., Alm, E., & Baker, D., 2000, Critical role of beta-hairpin formation in protein G folding. *Nat. Struct. Biol.* **7:** 669-673.
83. Nauli, S., Kuhlman, B., & Baker, D., 2001, Computer-based redesign of a protein folding pathway. *Nat. Struct. Biol.* **8:** 602-605.
84. Heidary, D.K., & Jennings, P.A., 2002, Three topologically equivalent core residues affect the transition state ensemble in a protein folding reaction. *J. Mol. Biol.* **316:** 789-798.
85. Shoemaker, B.A., Portman, J.J., & Wolynes, P.G., 2000, Speeding molecular recognition by using the folding funnel: the fly-casting mechanism. *Proc. Natl. Acad. Sci. USA.* **97:** 8868-8873.
86. Duan, Y. & Kollman, P.A., 1998, Pathways to a protein folding intermediate observed in a 1–microsecond simulation in aqueous solution. *Science* **282:** 740-744.
87. Zagrovic, B., Snow, C.D., Shirts, M.R., & Pande,V.S., 2002, Simulation of folding of a small alpha-helical protein in atomistic detail using worldwidedistributed computing. *J. Mol. Biol.* **323:** 927-929.
88. Snow, C.D., Nguyen, H., Pande, V.S., & Gruebele, M., 2002, Absolute comparison of simulated and experimental protein-folding dynamics. *Nature* **420:** 102-106.
89. Boczko, E.M., & Brooks III, C.L., 1995, First-principles calculation of the folding free energy of a three-helix bundle protein. *Science* **269:** 393-396.
90. Guo, Z., & Brooks III, C.L. (1997) Thermodynamics of protein folding: a statistical mechanical study of a small all–protein. *Biopolymers*, **42:** 745-757.
91. Guo, Z., Brooks III, C.L., & Boczko E.M., 1997, Exploring the folding free energy surface of a three–helix bundle protein. *Proc. Natl. Acad. Sci. USA* **94:** 10161-10166.
92. Sheinerman, F.B., & Brooks III, C.L., 1998, Molecular picture of folding of a small alpha/beta protein. *Proc. Natl. Acad. Sci. USA* **95:** 1562-1567.
93. Shea, J.E., Onuchic, J.N., & Brooks III, C.L., 1999, Exploring the origins of topological frustration: design of a minimally frustrated model of fragment B of protein A. *Proc. Natl. Acad. Sci. USA* **96:** 12512-12517.
94. Shea, J.E., & Brooks III, C.L., 2001, From folding theories to folding proteins: a review and assessment of simulation studies of protein folding and unfolding. *Annu. Rev. Phys. Chem.* **52:** 499-535.
95. Lazaridis, T., and Karplus, M., 1997, "New view" of protein folding reconciles with the old through multiple unfolding simulations. *Science* **278:** 1928-1931.

96. Sheinerman, F. & Brooks III, C.L., 1998, Calculations on folding of segment B1 of streptococcal protein G. *J. Mol. Biol.* **278:** 439-456.
97. Daggett, V., 2002, Molecular dynamics simulations of the protein unfolding/folding reaction. *Acc. Chem. Res.* **35:** 422-429.
98. De Jong, D., Riley, R., Alonso, D.O., & Daggett, V., 2002, Probing the energy landscape of protein folding/unfolding transition states. *J. Mol. Biol.* **319:** 229-242.
99. Day, R., Bennion, B.J., Ham, S., & Daggett, V., 2002, Increasing temperature accelerates protein unfolding without changing the pathway of unfolding. *J. Mol. Biol.* **322:** 189-203.
100. Mayor, U., Johnson, C.M., Daggett, V., & Fersht, A.R., 2000, Protein folding and unfolding in microseconds to nanoseconds by experiment and simulation. *Proc. Natl. Acad. Sci. USA* **97:** 13518-13522.
101. Daggett, V., 2000, Long timescale simulations. *Curr. Opin. Struct. Biol.* **10:** 160-164.
102. Ladurner, A.G., Itzhaki, L.S., Daggett,V., & Fersht, A.R., 1998, Synergy between simulation and experiment in describing the energy landscape of protein folding. *Proc. Natl. Acad. Sci. USA* **95:** 8473-8478.
103. Brooks III, C.L,. 1998, Simulations of protein folding and unfolding. *Curr. Opin. Struct. Biol.* **8:** 222-226.
104. Pande, V.S., & Rokhsar, D.S., 1999, Molecular dynamics simulations of unfolding and refolding of a beta-hairpin fragment of protein G. *Proc. Natl. Acad. Sci. USA* **96:** 9062-9067.
105. Nymeyer, H., Socci, N.D., & Onuchic, J.N., 2000, Landscape approaches for determining the ensemble of folding transition states: success and failure hinge on the degree of frustration. *Proc. Natl. Acad. Sci. USA* **97:** 634-639.
106. R.Du, V.S., Pande, A.Y., Grosberg, T., Tanaka, E.S., & Shakhnovich, 1998, On the transition coordinate for protein folding. *J. Chem. Phys.* **108:** 334-350.
107. Pande, V.S., & Rokshar, D., 1999, Folding pathway of a lattice model for proteins. *Proc. Natl. Acad. Sci. USA.* **96:** 1273-1278.
108. Verkhivker, G.M., Bouzida, D., Gehlhaar, D.K., Rejto, P,A., Freer, S.T., & Rose, P.W., 2003, Simulating disorder–order transitions in molecular recognition of unstructured proteins : where folding meets binding. *Proc. Natl. Acad. Sci. USA* **100:** 5148-5153.
109. Mayo, S.L., Olafson, B.D., Goddard, W. A. III., 1990, DREIDING: a generic force field for molecular simulation. *J. Phys. Chem.* **94:** 8897-8909.
110. Bouzida, D., Kumar, S. and Swendsen, R.H., 1992, Efficient Monte Carlo methods for the computer simulation of biological molecules. *Phys. Rev. A* **45:** 8894-8901.
111. Bouzida, D., Rejto, P.A., Arthurs, S., Colson, A.B., Freer, S.T., Gehlhaar, D.K., Larson, V., Luty, B.A., Rose, P.W. andVerkhivker, G.M., 1999, Computer simulations of ligand–protein binding with ensembles of protein conformations: A Monte Carlo study of HIV–1 protease binding energy landscapes. *Int. J. Quantum Chem.* **72:** 73-84.
112. Bouzida, D., Rejto, P.A. and Verkhivker, G.M., 1999, Monte Carlo simulations of ligand–protein binding energy landscapes with the weighted histogram analysis method. *Int. J. Quantum Chem.* **73:** 113-121.
113. Willet, P., &Winterman, V., 1986, A Comparison of some measures for the determination of intermolecular structural similarity. *Quant. Struct.-Act. Relat. Pharmacol. Chem. Biol.* **5:** 18-25.
114. Willet, P.,Winterman, V.,& Bawden, D. (1986) Implementation of non–hierarchical cluster analysis methods in chemical information systems: selection of compounds for biological testing and clustering of substructure search output. *J. Chem. Inf. Comput. Sci.* **26:** 109-118.

115. Bawden D., 1888, Browsing and clustering of chemical structures. In *The international language of chemistry*, (W.A. Warr, eds.) Springer–Verlag, Berlin, pp.145-150.
116. Shakhnovich, E.I., 1998, Folding nucleus : specific or multiple ? Insights form lattice models and experiments. *Fold. Des.*3: R108-R111.
117. Thirumalai, D. & Klimov, D.K., 1998, Fishing for folding nuclei in lattice models and proteins. *Fold. Des.* 3: R111-R118.
118. Li, L., Mirny, L.A., & Shakhnovich, E.I., 2000, Kinetics, thermodynamics and evolution oif non-native interactions in a protein folding nucleus. *Nat. Struct. Biol.* 7: 336-342.

Computational Detection of the Binding Site Hot Spot and Predicting Energetics of Ligand Binding at the Remodeled Human Growth Hormone–Receptor Interface Using a Hierarchy of Molecular Docking and Binding Free Energy Approaches

GENNADY M. VERKHIVKER
Pfizer Global Research and Development, La Jolla Laboratories, 10777 Science Center Drive, San Diego CA 921211111, USA

1. INTRODUCTION

1.1 Flexibility in molecular recognition and hot spots at intermolecular interfaces

Understanding mechanismsand fundamental biophysical principles of molecular recognition continues to present a fundamental experimental and theoretical challenge[1-6]. Alanine scanning mutagenesis of protein–protein interfacial residues, combined with structural and thermodynamic studies, have enabled discovery of energetically important hot spot regions at the intermolecular interfaces that are critical in determining binding affinity, i.e. alanine mutation of a hot spot residue in the binding site results in a pronounced drop in binding affinity of the complex[7]. A comprehensive analysis of protein–protein interfaces[8,9] and a survey of hot spots compiled from various protein binding sites have demonstrated a diversity of interaction patterns and a lack of

Supramolecular Structure and Function 8, Edited by Pifat-Mrzljak
Kluwer Academic/Plenum Publishers, New York 2004

general rules for hydrophobicity, polarity, or shape, that can be used to unambiguously predict hot spots at the intermolecular interfaces. Arecent analysis of conserved residues in 11 clustered interface families comprising a total of 97 crystal structures has shown that the composition of hot spots is typically enriched by certain residues, such as Trp, Tyr, Arg, His, Gln, Asn, and Pro and can be surrounded by a shell of less important residues[10-12]. The discovery of hot spots appears to be broadly relevant in a variety of protein–protein recognition events where, despite a typically large size of intermolecular interface, binding affinity and specificity may be determined by a functional epitope consisting of only a small fraction of the interfacial residues[7,10]. An overlap between flexible consensus binding sites and energetically critical hot spots, discovered through a combination of structural and mutagenesis studies, have triggered experimental and computational studies aiming to understand the nature of structural flexibility and functional diversity in molecular recognition[7-13].

Molecular recognition between proteins and flexible target molecules, including other proteins and small molecules is often accompanied by a considerable flexibility of the protein binding sites and structural rearrangements upon binding between the associated partners[14-18]. Protein dynamics at the intermolecular interfaces can have a profound effect in determining binding thermodynamics, kinetics and consequently in modulating binding affinity and specificity of molecular recognition[19-22]. Protein binding interfaces can be not only structurally flexible, but also functionally adaptive, with a diverse repertoire of protein systems capable of binding with high affinity to ligands, different from their natural binding partners in composition, size, and shape[23,24]. Evolution can often find convergent solutions to stable intermolecular interfaces by using structural flexibility and plasticity of the hot spot residues in the binding sites to allow accommodation to different binding partners[12,23,24].This notion was manifested in discovery of a combinatorially selected high–affinity synthetic peptide that can mimic specific interactions of much larger natural proteins with the hot spot residues of the constant fragment of human immunoglobulin G protein, while using the interacting groups from a different structural scaffold[23]. Computational alanine scanning and binding free energy calculations conducted for this system have indicated that while a specific set of the conserved hot spot protein residues provides thermodynamic stability of the native structure, different hot spot residues contribute decisively to the binding affinity of the peptide–protein complex[25].

1.2 Structural and functional studies of human growth hormone–receptor binding

Structural and functional studies of human growth hormone–receptor binding between hGH and hGHbp have provided a comprehensive view and atomic details of the binding thermodynamics that is determined by a small fraction of residues of a large intermolecular interface[26-28]. In a series of pioneering biochemical studies, homolog-scanning and alanine scanning[29] mutagenesis[30] strategies had allowed an initial mapping of the binding determinants in the hGH and hGHbp molecules even before the high-resolution structures of the complex and individual components became available. These studies determined a relatively small set of residues that resulted in 4-time lower binding affinity, when mutated to alanine, with many of these residues being altered in the corresponding nonbinding homologs. Using a variety of biophysical methods, including titration calorimetry, the stoichiometry of the hGH—hGHbp complex was determined, revealing two hGHbp receptor molecules bound to a single hGH hormone molecule[31]. The crystals composition of the hGH–hGHbp complex confirmed that biologically significant dimerization of the growth hormone receptorwas indeed mediated through a single hormone molecule[32]. The solution of the X-ray structure of the 1:2 hGH–hGHbp complex[33] has revealed not only the stoichiometry of the complex and the structures of the complex components, but has also identified the critical contact residues of the hormone–receptor interface, thereby permitting a direct structural interpretation of the mutagenesis and biophysical studies. The structure showed that two receptor molecules employed the same binding determinants to interact with two considerably dissimilar sites on the opposite sites of the hormone, that required the receptor binding surfaces to undergo local conformational changes[33,34]. Furthermore, essentially the same set of binding determinants was used by hGH to bind to both hGHbp and prolactin receptor, despite only 28 % homology between the complementary binding sites of these receptors[33-35].

A detailed study of hormone binding determinants in the hGHbp receptor was conducted by systematically replacing side chains with alanine and measuring the binding affinities of the alanine hGHbp mutants with the hormone[36]. This mutational strategy probed the importance of 49 charged surface residues, 9 aromatic residues, and 26 neighboring residues. A similar analysis was performed on the hormone side, in which the role of 31 buried hGH residues in binding with the hGHbp receptor was determined by converting each of these residues to alanine and measuring their effect on modulating binding affinity and kinetics of the reaction in the site 1 of hGHbp[37]. Alanine substitutions of only 8 residues among the studied set (K41, L45, P61, R64, K172, T175, F176, and R178), forming two small patches on the protein interface, accounted for approximately 85 % of the binding energy. Moreover, a significant increase in the off–rates was detected for six mutants P61A, R64A, K172A, T175A F176A and R178A, while only subtle effects on the on–rates were observed. These data have indicated that the hormone and the receptor associate by rapid diffusion and the primary role of the intermolecular hot spot residues is to produce a tightly bound complex after the hormone has reached the binding site[37]. In a subsequent widely recognized work[38], the crystal structure of the hGH-hGHbp complex was used to perform alanine scanning of 33 buried side–chains on the receptor. Despite a large hGH:hGHbp intermolecular interface, that covers 1350 $Å^2$ and involves 33 receptor side–chain residues, the discovered hot spot of binding energy comprises of only 9 of these residues (R43, E44, I103, W104, I105, P106, I164, D165, W169), providing virtually all binding affinity. The critical interactions in the hot spot are primarily hydrophobic and the functiona epitope is dominated by contributions fromtwo hydrophobic residues W104 and W169 contributing approximately 4.5 kcal/mol, i.e. more than three-quarters of the total binding free energy[38]. Importantly, empirical parameters such as the extent of side-chain burial surface area or van derWaals interactions served as pure predictors of the relative effects on binding affinity, when the alanine scanning was performed. However, these correlations were considerably better when only the burial of well-packed and dominating W104 and W169 hot spot residues was considered.

Mutational studies have shown that more than two-thirds of the contact hGH–hGHbp interfacial residues have practically no effect on binding affinity when converted to alanine[36-38]. Alanine-shaving, which is the process of making multiple simultaneous alanine mutations, has been used to experimentally test the role of affinity–inert residues[39,40]. Strikingly, a less than 10–fold drop in affinity was detected when 16 energetically unimportant residues were simultaneously mutated to alanine[39]. Multiple alanine scanning of peripheral to the functional epitope contact hGH residues F25, Y42, and Q46 have shown only local structural changes near sites of mutation and a little effect on binding kinetics and thermodynamics, resulting from large but compensating changes in the enthalpy and entropy of binding[40] Hence, large cumulative reductions on protein–protein contacts can be tolerated with minimal changes in binding affinity for peripheral, functionally–inert residues in the hGH—hGHbp interface. These studies have underscored the inherent difficulties of inferring binding free energy changes from local contacts at the flexible regions in the interface and have suggested the possibility for more dynamic regions to be less critical for binding affinity[40]. The 2.6 Å resolution crystal structure of the G120R designed mutant of hGH that binds only a single molecule of hGHbp has guided further a detailed survey of the structural and functional basis for hormone–receptor recognition[41]. This analysis was primarily conducted to distinguish direct binding energy contributions of the hot spot receptor residues from indirect effects that may be mediated through peripheral affinity–inert contacts. The revised alanine scan of the hGHbp residues based on the 1:1 complex showed that only 11 residues affected significantly binding affinity with hGH (R43, E44, I193, W104, I105, P106, D126, E127, D164, I165, and W169)[41].The true functional epitope is assembled cooperatively and even more localized with only six hGHbp side-chains reflecting major direct contributions to binding affinity[41].

The computed changes in solvent-accessible surface area using the reported alanine shaves[41] resulted in the largest change of only 4.2 $Å^2$ for the shaved mutant of the hGH receptor[10]. Interestingly, truncation to alanine of all 46 affinity-inert side-chains led to only 56 % (60.6 $Å^2$) increase in the calculated solvent-accessible surface for the hot spot[10]. The failure of alanine shaving to significantly increase the solvent accessibility of the hot spot was suggested as a rationale for similar binding affinities of alanine shaved mutants compared to those of wild-type proteins. Computational alanine scanning of the hGH-hGHbp complex has shown an excellent agreement between the relative binding free energy changes upon mutation of W104 and W169 residues and experimental data, indicating that it may be energetically costly to significantly change nearby residues to compensate for the loss of critical interactions[42]. However, for charged and polar residues R43, D164, and R217 in hGHbp, which are less important in binding affinity, the agreement with experiment was only qualitative, suggesting that appreciable local side-chain rearrangements of the surrounding residues may take place to restore these defects[42].

1.3 Rational remodeling of the human growth hormone–receptor interfaces

The structural and functional data of the hGH–hGHbp binding have suggested that the discovered hot spot could present a template for rational remodeling of the interfaces and design of alternative scaffolds and small molecule mimics[38,41]. Remodeling of the interface between the hGH and the hGHbp receptor was studied by mutating to alanine a critical W104 residue in the receptor and selecting a pentamutant (K168R, D171T, K172Y, E174A, and F176Y) of hGH by phage display that fills the created cavity and largely restores binding affinity[43]. A 2.1 Å resolution x-ray structure of the W104A mutant of the hGHbp receptor bound to the hGH mutant containing five mutations (K168R, D171T, K172Y, E174A, and F176Y) showed that the receptor cavity was filled by selected hydrophobic mutations of hGH and was accompanied by considerable structural rearrangements occurred in the interface at sites distant from the mutations[43]. This pioneering study has shown how structural plasticity at the remodeled hGH–hGHbp interface can rescue large functional changes and largely restore binding affinity of the wild-type hGH–hGHbp complex by a limited number of mutations, thereby providing a mechanism for mutations to be accommodated during coevolution of high affinity binding partners[43].

The hypothesis that functional versatility of hormone—receptor binding may be an evolutionary consequence of selecting a set of structurally diverse binding partners was probed by employing phage display mutagenesis and designing novel hGH variant molecules that are different from the wild type hormone[43,44]. The structure of the complex between the phage display optimized high–affinity to site 1 hGH bound with hGHbp molecules was determined at 2.6 Å resolution and revealed that 15 mutations introduced in the designed variant of hGH result in significant structural changes in the hormone–receptor interface[45]. A tighter binding in site 1 was accompanied by 15 % smaller and 20 % more hydrophobic interfacial surface area compared to the wild type complex. Strikingly, structural plasticity of the hormone–receptor recognition leads to a novel binding solution at the site 2, with a structurally entirely distinct conformation of the receptor molecule obtained in the absence of any selection presure[45]. These results have indicated that structural elements of the hGH—hGHbp system may have an inherent structural flexibility and functional plasticity that enables them to evolve and bind to a variety of binding surfaces.

An elegant protein engineering solution to the specificity problem often implies creating an artificial cavity that can be introduced into the ligand-binding site of the protein by truncating side chains using site-directed mutagenesis. A complementary modification can then be designed on the ligand, generating a new 'bumped' compound that no longer binds the wild-type protein but interacts specifically with the mutant. Rational redesign of protein–protein and ligand–protein interfaces and generating modified molecules that interact only with specifically mutated proteins has become a powerful approach for manipulating remodeled protein interfaces and generating ligands with new, orthogonal specificities[46-50]. A recently reported approach for generating synthetic molecules directly modulating specific interactions at the hGH—

hGHbp interface involved a cavity formation, introduced at the interface by mutating to glycine the hot spot residues T175 from the hormone and W104 from the receptor[51]. While binding affinity of the W104A mutant of hGHbp is reduced by more than 2500-fold relative to the wild-type complex, T175A mutation results in only 25-fold reduction in binding free energy. The double mutant W104G/T175G reduces binding affinity of the wild–type complex by a factor of and results in a formation of a sufficiently large cavity in the interface that can accommodate small molecules complementing this defect and restoring the affinity of the complex. A library of 200 indole analogs and derivatives of structurally related 5- and 6-membered fused aromatic heterocycles was then screened for ligands that complement this defect and restore binding affinity of the wild-type complex. A significant recovery of phage was detected only in the presence of ligands containing the benzimidazole core motif; the individually assayed representative molecules that largely restored binding affinity were reported with their respective dissociation constants[51].

We have previously conducted initial simulations with the remodeled hGH–hGHbp complex for a panel of potent benzimidazole-containing inhibitors, that can restore the binding affinity of the wild-type complex, and for a set of known non–active small molecules that contain different heterocyclic motifs[52]. Despite presence of numerous pockets on the protein surface for the mutant 7 hGH–hGHbp complex, the binding site cavity can be distinguished as the energetically most favorable hot spot for the benzimidazole–containing inhibitors, while for a set of non–active molecules the lowest energy ligand conformations do not necessarily bind in the engineered cavity[52].

Here, we study microscopic details of ligand binding in the engineered cavity created at the hGH-hGHbp remodeled interface with a panel of small molecules restoring the binding affinity of the wild-type complex. We extend a hierarchical strategy, proposed originally for a rapid screening of ligand conformations in a given binding site, to a broader problem of screening pockets and cavities on the protein surface and identifying the binding site hot spot at the intermolecular interface of the remodeled hGH–hGHbp complex. Multiple docking simulations are performed using an evolutionary algorithm and simplified, knowledge-based energy function to rapidly explore a large conformational space and locate families of low–energy ligand conformations binding to the protein surface. Structural clustering of the ligand bound conformations is followed for the low–energy cluster representatives by a detailed binding free energy analysis using a molecular mechanics AMBER force field combined with the solvation energy computed using a continuum solvation model. We show that the proposed protocol provides a robust exploration of low–energy conformational clusters at various locations on the protein surface and allows to detect the binding site hot spot at the intermolecular interface of the remodeled hGH–hGHbp complex.

Since phage recovery ratio does not quantitatively correspond to the binding affinity enhancement of the mutant complex in the presence of a given ligand[51], we study energetics of ligand binding for five reported compounds with the measured dissociation constants (Fig. 1) to compare the computed binding free energies with the experimental data. Despite the absence of the crystal structures of the bound inhibitors with the remodeled hGH—hGHbp complex, the precise presentation of the ligand elements that could mimic the indole rings of W104 and W169 is known to be the essential prerequisite for the molecules to restore the binding affinity of the wild-type hGH–hGHbp complex[51]. As a result, the inhibitors must bind in the engineered cavity of the redesigned interface to fulfill the observed biological function. In addition, equilibrium simulations of ligand–protein binding dynamics at the intermolecular hot spot are performed with the aid of parallel simulated tempering technique revealing a considerable structural diversity of the energetically favorable ligand binding modes contributing to equilibrium dynamics during molecular recognition. However, the low–energy

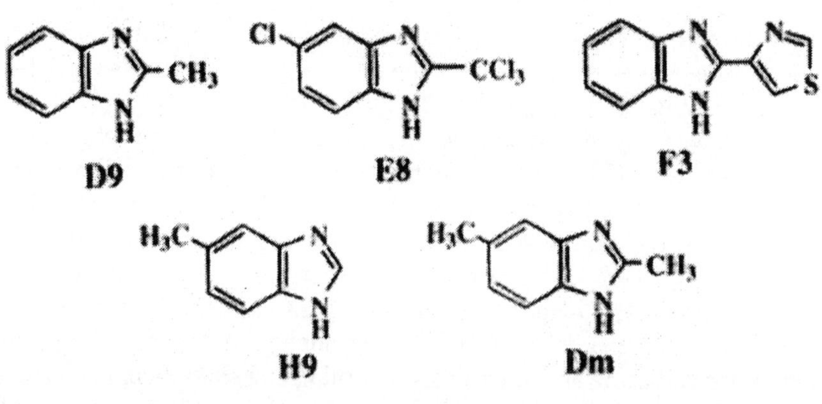

Figure 1. Chemical structures for a panel of studied inhibitors serving as molecular switches and restoring binding affinity of the hGH-hGHbp remodeled interface. The notations of the compounds from the original work[51] are kept and correspond to the group nomenclature in the initial library screening (see for more details[51])

ligand binding modes, generated from equilibrium simulations in the binding site cavity, are generally compatible with the structural orientation of the benzimidazole core motif that mimics the location of the W104 indole ring in the crystal structure of the wild-type complex. Binding free energy calculations are then performed for all conformations in the low–energy structural clusters structural clusters featuring this structural positioning in the core motif of the inhibitors. We show that this protocol allows an accurate analysis of binding thermodynamics for studied inhibitors and can provide a plausible rationale of the experimental data by clarifying the role of key energetic contributions in restoring binding affinity of the hGH–hGHbp complex.

2. METHODS

2.1 Simplified molecular recognition energy model

Simplified energy models can faithfully describe a multitude of the available binding modes for the complex and reproduce relative thermodynamic stability of the native complex with respect to intermediate complexes and alternative binding modes[53-56]. A multi–stage strategy[57,58] with a hierarchy of different energy functions is pursued in this work to achieve a synergy of robust conformational sampling and accurate estimation of binding energetics. The simplified energy function[58,59] is used in conjunction with evolutionary search and Monte Carlo simulations[59-61] to sample the conformational space and and adequately describe the multitude of the low–energy states available to the system.

The knowledge–based simplified energetic model includes intramolecular energy terms for the ligand, given by torsional and nonbonded contributions of the DREIDING force field[62], and intermolecular ligand–protein steric and hydrogen bond interaction terms calculated from a simplified piecewise linear (PL) potential summed over all protein and ligand heavy atoms (Fig. 2A). The parameters of the pairwise potential depend on the following different atom types: hydrogen-bond donor, hydrogen-bond acceptor, both donor and acceptor, carbon-sized nonpolar, sulfur-sized nonpolar, flourine-sized nonpolar, and large nonpolar. The atomic radius is 1.4 Å for fluorine is 1.4 Å and 1.8 Å for carbon, oxygen, and nitrogen atoms. The atomic radius of 2.2 Å is assigned to sulfur and phosphorus, chlorine, and bromine atoms, modeled as large nonpolar atom type. Electronegative atoms with an attached hydrogen are defined to be donors, while oxygen and nitrogen atoms with no bound hydrogens are defined to be acceptors.

Sulfur is modeled as being capable of making weak hydrogen bonds which allows for sulfur-donor closer contacts that are seen in some of the crystal structures. Crystallographic water molecules and hydroxyl groups are defined in this model to be both donor and acceptor, and carbon atoms are defined to be nonpolar. An empirical desolvation correction is applied to the attractive portion of the interactions between nonpolar and polar atoms. This correction is defined as the the ratio between the attractive well depth for nonpolar–polar contacts and the one for nonpolar–nonpolar contacts, and can range between 0 and 1. The parameter is set to 1.0 in this work, thereby imposing a desolvation penalty by disadvantaging the burial of polar groups with the nonpolar atoms.

A hydrogen bond interaction term is assigned to interactions between donor and acceptors, a repulsive interaction contribution is computed for donor–donor and acceptor–acceptor contacts, and a steric intermolecular term is assigned for other contacts. The steric and hydrogen bond-like potentials have the same functional form, with an additional three-body contribution to the hydrogen bond term and the repulsive term for donor–donor and acceptor–acceptor contacts. Both the hydrogen bond interaction energy and the repulsive interaction contribution between donor–donor and acceptor–acceptor close contacts are modulated by an approximate angular dependence (Fig. 2B). These terms are multiplied by the hydrogen bond strength term, which is a function of the angle θ determined by the relative orientation of the protein and ligand atoms (Fig. 2B). The scaling for the repulsive interactions is equivalent to the dependence used for the hydrogen bond interaction term, but in this cases it implies a maximum penalty when the angle θ is 180 degrees, fading to zero at 90 degrees and below. θ is defined to be the angle between two vectors, one of which points from the protein atom to the ligand atom. For protein atoms with a single heavy atom neighbor, the second vector connects the protein atom with its heavy atom neighbor, while for protein atoms with two heavy atom neighbors, it is the bisector of the vectors connecting the protein atom with its two neighbors.

For molecular docking simulations, the energy landscape must be relatively smooth for robust structure prediction of ligand–protein complexes and oftening the potentials is a way to smooth the force field and enhance sampling of the conformational space while retaining adequate description of the binding energy landscape[53-58]. The PL energy function has no singularities at interatomic distances and can effectively explore large conformational spaces. While the PL energy function is proven to be more adequate for sampling non–polar and hydrogen bonds patterns, this simplified energy model does not

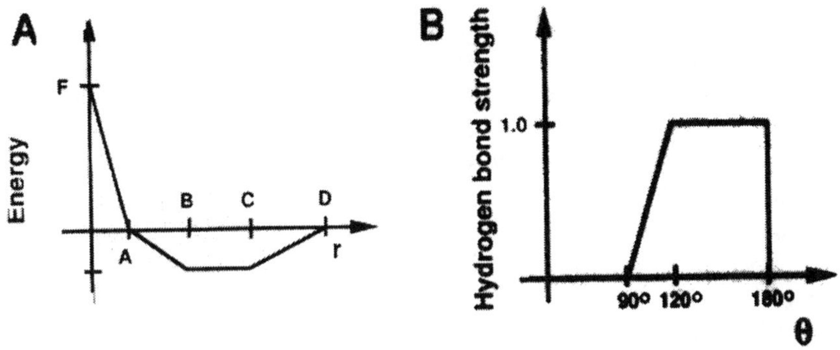

*Figure 2.*A) The functional form of the ligand–protein interaction energy. For steric interactions, A = 0.93B, C = 1.25B, D = 1.5B, E = -0.4, F = 15.0, and B= $r_{ligand} + r_{protein}$ is the sum of the atomic radii for the ligand and protein atoms. For hydrogen bond interactions, A = 2.3, B = 2.6, C = 3.1, D = 3.4, E = -4.0, F = 15.0. For sulfur hydrogen bond interactions, A = 2.7, B = 3.0, C = 3.5, D = 3.8, E = -2.0, F = 15.0. For chelating interactions with the metals A = 1.5, B = 1.7, C = 2.5, D = 3.0, E = -10.0, F = 15.0. For repulsive interactions, A = 3.2, E = 0.1, F = 15.0. The repulsive potential is then linearly scaled from E=0.1 to zero between 3.2 Å and 5.0 Å. The units of A, B, C, and D are °A for E and F the units are kcal/mole. B) The hydrogen bond interaction energy and the repulsive term are multiplied by the hydrogen bond strength term, which is a function of the angle θ determined by the relative orientation of the protein and ligand atoms.

include a direct electrostatic component and therefore may be less accurate in detecting the exact energetics of the binding modes, especially when extensive networks of electrostatic interactions are present in the crystal structure. This function is less accurate in detecting the exact location and energetics of the native state because of the inaccuracy in quantifying the magnitude of protein–protein and ligand–protein interactions.

2.2 Binding free energy calculations

The conformational states generated with the PL energy function and docking and equilibrium simulations are evaluated with a more detailed binding free energy model, which includes the molecular mechanics AMBER force field[63] and the solvation energy term based on continuum generalized Born and solvent accessible surface area (GB/SA) solvation model[64-70]. This procedure is conceptually similar to the MM/PBSA (molecular mechanics Poisson-Boltzmann surface area) approach[71-78], and replaces time-consuming PB continuum calculations with less demanding GB solvation calculations, correlating well with the PB results[79]. The MM/PBSA approach uses molecular dynamics simulations of the system to generate a thermally averaged ensemble of conformations. Based on this set of structures, the total free energy of the system is evaluated as a sum of the polar solvation energy, which is computed using a finite–difference Poisson-Boltzmann (PB) approach, the nonpolar solvation term derived from the solvent-accessible surface area (SA), and solute entropy contribution. The molecular mechanical energy of the molecule includes the electrostatic, van der Waals contributions and internal strain energy. The ensemble of structures for the uncomplexed protein and ligand are generated in the MM/PBSA approach by using the molecular dynamics trajectory of the complex, and simply separating the protein and ligand coordinates, followed by an additional minimization of the unbound protein and unbound ligand. This methodology has been successfully applied in various applications of protein–protein and ligand–protein recognition[72-78] and recently in the computational alanine scanning of the hGH–hGHbp complex[42].

The average total free energy of the molecule G is evaluated as follows:

$$G_{molecule} = G_{solvation} + E_{MM} - TS_{solute} \tag{1}$$

$$G_{solvation} = G_{cavity} + G_{vdw} + G_{pol} \tag{2}$$

In the GB/SA model, the G_{cavity} and G_{vdw} contributions are combined together *via* evaluating solvent-accessible surface areas:

$$G_{SA} = G_{cavity} + G_{vdw} = \sum_i \sigma_i SA_i \tag{3}$$

where G_{SA} is the nonpolar solvation term derived from the solvent-accessible surface area (SA).

$$G_{pol} = -166.0(1 - \frac{1}{\varepsilon}) \sum_i \sum_j \frac{q_i q_j}{(r_{ij}^2 + \alpha_{ij}^2 \exp(-D_{ij}))^{0.5}} \tag{4}$$

where G_{pol} is the polar solvation energy which is computed using the GB/SA solvation model. S_{solute} is the vibrational entropy of the molecule. E_{MM} is the molecular mechanical energy of the molecule summing up the electrostatic E_{es} interactions, van der Waals contributions E_{vdw}, and the internal strain energy E_{int}:

$$E_{MM} = E_{es} + E_{vdw} + E_{\mathrm{int}} \tag{5}$$

Using these equations, the binding free energy of the ligand–protein complex is computed as follows:

$$\Delta G_{bind} = G_{complex} - G_{protein} - G_{ligand} \tag{6}$$

From this equation, one can determine contributions of the ligand–protein interaction energy ΔG_{MM}, strain energy ΔG_{strain}, and solvation energy $\Delta G_{GB/SA}$ to the total binding free energy.

$$\Delta G_{bind} = \Delta G_{interaction} + \Delta G_{strain} + \Delta G_{solvation} \tag{7}$$

$$\begin{aligned}\Delta G_{interaction} = {} & E_{MM}^{complex} - E_{MM}^{boundprotein} - E_{MM}^{boundligand} = \\ & E_{es}^{complex} - E_{es}^{boundprotein} - E_{es}^{boundligand} + \\ & E_{vdw}^{complex} - E_{vdw}^{boundprotein} - E_{vdw}^{boundligand} + \\ & E_{int}^{complex} - E_{int}^{boundprotein} - E_{int}^{boundligand}\end{aligned} \tag{8}$$

$$\Delta G_{strain} = (E_{MM}^{boundprotein} - E_{MM}^{freeprotein}) + (E_{MM}^{boundligand} - E_{MM}^{freeligand}) \tag{9}$$

$$\Delta G_{solvation} = G_{solvation}^{complex} - G_{solvation}^{freeprotein} - G_{solvation}^{freeligand} \tag{10}$$

Binding free energy evaluations are performed for a representative from each cluster, a cluster center. The structures for the uncomplexed protein and ligand are generated by using the corresponding samples of the ligand–protein complex in the cluster centers and separating the protein and ligand coordinates, followed by an additional minimization of the unbound protein and unbound ligand.

The crystal structure of the wild–type hGH—hGHbp hormone–receptor complex was used in computational modeling and was subjected to glycine mutation at the W104 position on the hGHbp and T175 position on the hormone. The mutated chimera structure was minimized using 200 iteration of conjugate gradient minimization to relieve any strain contacts. The energy of each ligand–protein complex is subjected to the conjugate gradient minimization as implemented in the version 7.0 of the MacroModel molecular modeling software package[68]. All protein residues within 3 Å radius sphere from the ligand are treated as flexible during minimization. All protein residues within 2 Å radius from the flexible shell form a first shell of restrained atoms with the force constant 100.0 $kJ/molA^2$. A second shell of restrained atoms with the force constant 200 $kJ/molA^2$ consists of the residues within 2 Å radius from the first shell and finally, the third shell of restrained atoms is generated by the residues which reside within 2 Å from the second shell and they are restrained with the force constant 300 $kJ/molA^2$. The remaining protein atoms are treated as frozen atoms and do not move during the minimization procedure. The interactions between frozen atoms and restrained atoms, and frozen atoms and flexible atoms are included in the total energy value. A residue–based cutoff of 8 Å is set for computing non-bonded van der Waals interactions and 20 Å residue–based cutoff is used for computing electrostatic interactions. The structures of the studied ligands were built and minimized with the MNDO atomic charges calculated using MOPAC program. The protein atoms have been assigned the AMBER force field charges.

The normal mode analysis and the vibrational entropy calculations are done using classical statistical mechanics formula with the AMBER module *nmode* for the energy minimized structures of the complex, the free protein, and free ligand without water molecules. Because of significant variances in computing solute entropy using the MM/GBSA approach, this term was not included in the total binding free energy value.

2.3 Computer simulations of ligand–protein docking

The protein structure of the remodeled hGH–hGHbp mutant complex is held fixed in its minimized conformation, while rigid body degrees of freedom and rotatable angles of the ligands are treated as independent variables. Multiple docking simulations are performed using an evolutionary search algorithm[54,55] and the simplified, knowledge-based energy function to rapidly explore a large conformational space and locate families of low–energy ligand conformations binding to the protein complex. During docking simulations, ligand conformations and orientations are sampled in a parallelepiped that encompasses the binding site obtained from the crystallographic structure of the wild-typehGH— hGHbp complex with a large 25.0 Å cushion added to every side of this box surrounding the interface which guarantees a sufficiently unbiased conformational search to locate the enginnered binding cavity. Bonds allowed to rotate include those linking sp^3 hybridized atoms to either sp^3 or sp^2 hybridized atoms and single bonds linking two sp^2 hybridized atoms. The moves are chosen as follows: a variable is selected at random and then a uniform displacement is given along each rigid body degree of freedom, the ligand is rotated as a rigid body, or a dihedral angle of the ligand is rotated.

Evolutionary algorithm, a stochastic optimization technique based on the ideas of natural selection, was used previously in ligand–protein docking simulations[54,55]. During the evolutionary search, a population of candidate ligand conformers competes for survival against a fixed number of opponents randomly selected from the remainder of the population. A win is assigned to the competitor with the lowest energy and the number of competitions that a member wins determines the survival probability to the next generation. All surviving members produce offspring, subject to a constant population size. In the population of ligand conformers, each member represents an encoded vector consisting of the rigid body coordinates and the torsional angles about the rotatable bonds. The initial ligand conformations are generated by randomizing the encoded vector, where the center of mass of the ligand is restricted to the rectangular parallelepiped that defines the binding site. The three rigid–body rotational degrees of freedom, as well as the torsional angles for all rotatable bonds are uniformly initialized between 0 and 360 degrees. In simulations with multiple protein conformations, each member of the initial population represents a ligand conformation with a randomly assigned protein conformation from the given ensemble. During the search, the surviving members of the population with the lowest energy represent the ligand conformation with the corresponding protein conformation. The protein conformation of the winner is preserved when offspring is produced, otherwise a new randomly selected protein conformation is assigned to a population member. For each docking simulation, the evolutionary search was performed for a total of 120 generations with a population size of 1200 members. To provide a necessary level of diversity, each member competes against three opponents at each generation. The size of the standard deviation for the Gaussian mutation in the process of generating offsprings is varied adaptively using selection pressure. As a result, large mutations are encouraged early in the simulation to facilitate rapid search, while smaller mutations are made as the simulation progresses to refine solutions near to the global energy minimum. The minimized best member of the final generation defines the predicted structure for the ligand–protein complex.

2.4 Monte Carlo equilibrium simulations of ligand–protein binding

Equilibrium simulations of ligand binding with the remodeledhGH—hGHbp mutant complex are performed in a parallelepiped that encompasses the engineered cavity with a 5.0 Å cushion added to every side of this box to accurately reproduce the equilibrium distribution between ligand binding modes in the binding site cavity. We have carried out equilibrium simulations using parallel simulated tempering dynamics[80-87] with 50 replicas of the ligand–protein system attributed respectively to 50 different temperature levels that are uniformly distributed in the range between 5300K and 300K. Independent local Monte Carlo moves are performed independently for each replica at the corresponding temperature level, but after a simulation cycle is completed for all replicas, configuration exchanges for every pair of adjacent replicas are introduced. The m-th and n–th replicas, described by a common Hamiltonian $H(X)$ are associated with the inverse temperatures β_m and β_n, and the corresponding conformations X_m X_n. The exchange of conformations between adjacent replicas m and n is accepted or rejected according to Metropolis criterion with the probability $p = \min(1, \exp[-\delta]$, where $\delta = [\beta_n - \beta_m][H(X_m) - H(X_n)]$. Starting with the highest temperature, every pair of adjacent temperature configurations is tested for swapping until the final lowest value of temperature is reached. This process of swapping configurations is repeated 50 times after each simulation cycle for all replicas whereby the exchange of conformations presents an improved global update which increases thermalization of the system and overcomes slow dynamics at low temperatures on rough energy landscapes, thereby permitting regions with a small density of states to be sampled accurately. During simulation, each replica has a non–negligible probability of moving through the entire temperature range and the detailed balance is never violated which guarantee each replica of the system to be equilibrated in the canonical distribution with its own temperature[80-87].

Monte-Carlo simulations allowto dynamically optimize the step sizes at each temperature by taking into account the inhomogeneity of the molecular system[88]. The acceptance ratio method is used to update the step sizes every cycle of 1000 sweeps. For all these simulations, we equilibrated the system for 1000 cycles (or one million sweeps), and collected data during 5,000 cycles (or five million sweeps) resulting in 5,000 samples at each temperature. A sweep is defined as a single trial move for each degree of freedom of the system. A key parameter is the acceptance ratio which is the ratio of accepted conformations to the total number of trial conformations. At a given cycle of the simulation, each degree of freedom can change randomly throughout some prespecified range determined by the acceptance ratio obtained during the previous cycle. This range varies from one degree of freedom to another because of the complex nature of the energy landscape. At the end of each cycle, the maximum step size is updated and used during the next cycle.

Simulations are arranged in cycles, and after a given cycle i, where the average acceptance ratio for each degree of freedom j is $\langle P_j \rangle^i$, the step sizes

σ_j^i

for each degree of freedom are updated for cycle i+1 according to the formula

$$\sigma_i^{j+1} = \sigma_i^j \frac{\ln\left[a\langle P_{ideal}\rangle + b\right]}{\ln\left[a\langle P_j\rangle^i + b\right]} \tag{11}$$

where $\langle P_{ideal}\rangle$ is the desired acceptance ratio, chosen to be 0.5. The parameters a and b are used to ensure that the step sizes remain well–behaved when the acceptance ratio approaches 0 or 1. They are assigned so that the ratio σ^{i+1}/σ^i is scaled up by a constant value s for $\langle P_j\rangle^i = 0$, and down by the same constant for $\langle P_j\rangle^i = 1$. Solving the equations

$$s^{-1} = \frac{\ln\left[a\langle P_{ideal}\rangle + b\right]}{\ln[b]} \tag{12}$$

$$s = \frac{\ln\left[a\langle P_{ideal}\rangle + b\right]}{\ln[a + b]} \tag{13}$$

with $s = 3$yields a = 0.673 and b = 0.065.

2.5 Similarity clustering

The 3D-similarity calculations are based on the spatial proximity of atoms in a binding site and the atom type. Four types of atoms are distinguished :hydrogen bond donors, hydrogen bond acceptors, hydrogen bond donors and acceptors and nonpolar atoms. The atom type compatibility *a(i,j)* is assigned a value between 0.0 and 1.0, with the compatibility between two atoms of the same type defined to be 1.0, that between donor and acceptor atom is 0.0, and other combinations of atoms have compatibilities between 0.0 and 1.0.

The spatial proximity between two atoms i and j is evaluated with a Gaussian function $p(i,j)=10^{(-r_{i,j}^2/\sigma^2)}$, where r_{ij} is the distance between atoms i and j, and $\sigma=-c^2/\log_{10}(p)$, where c and p denote the cutoff distance and proximity threshold respectively. Both the cutoff distance and the proximity threshold determine the shape of the gaussian function to evaluate spatial proximity of two atoms, with c=3.0 Å and p =0.000032.

A descriptor d(i,j) is calculated from the spatial proximity and the atom type compatibility:

$$d(i,j)=p(i,j)*a(i,j) if r(i,j)\leq c \tag{14}$$

$$d(i,j)=0 if r(i,j)>c \tag{15}$$

An atom descriptor $D_m^n(i)$ for atom in molecule is then calculated by summation over all N atoms in molecule n, $D_m^n(i)=\sum_{j=1}^{N} d_m^n(i,j)$. The intermolecular similarity between molecules m and n is given by the Tanimoto coefficient[89-91]:

$$S(m,n)=\frac{\sum_{i=1}^{M} D_m^m(i)D_m^n(i)+\sum_{j=1}^{N} D_n^m(j)D_n^n(j)}{\sum_{i=1}^{M} D_M^M(i)^2+\sum_{j=1}^{N} D_n^n(j)^2-\sum_{i=1}^{M} D_m^m(i)D_m^n(i)-\sum_{j=1}^{N} D_n^m(j)D_n^n(j)} \tag{16}$$

Molecules are grouped into clusters by comparing the intermolecular similarity coefficient. The first molecule is assigned to the first cluster. The next molecule is assigned to the cluster in which a cluster member has the highest similarity with the next molecule, if the similarity is above a threshold, chosen to be 0.85. Otherwise, the next molecule is assigned to a new cluster. The first member of the a cluster is called the cluster center. After all molecules are assigned to clusters, the molecules are arranged in new order, starting with the largest cluster and proceeding to the smallest cluster. The reordered set of molecules is subjected to the same clustering procedure. This procedure is iterated until the information entropy converges to a minimum. The clusters with at least 100 members are analyzed. Since conformations which belong to the same cluster are equivalent with 85% structural similarity, different clusters are compared by analyzing cluster centers.

3. RESULTS AND DISCUSSION

A large size of the surface of the hGH—hGHbp complex and the existence of various pockets and cavities, including the deeply buried engineered binding cavity, presents a challenge for molecular docking and scoring methods to detect the buried engineered cavity among other 'attractive alternatives' on the protein surface. Of many pockets and cavities on the protein surface, only the engineered binding cavity can bind the inhibitors in order to restore specific protein-protein interactions. The engineered cavity is deeply buried at the intermolecular interface of the remodeled hGH—hGHbp complex that makes this region inherently difficult to sample in the course of docking simulations. Multiple docking simulations are performed for studied ligands to locate families of low–energy ligand conformations that bind to the hGH–hGHbp remodeled complex. As a result of 1000 independent flexible docking simulations for each ligand, numerous clusters of low-energy bound conformations were found. A pattern of structural clusters, that cover a significant fraction of the protein surface and visit some buried regions, is common for all studied ligands and is presented for the E8 inhibitor (Fig. 3a,b). To describe the shape of the generated binding energy landscapes, the scatter plots between the PL energy values and the root mean square deviations (rmsd) from the the predicted lowest energy structure were generated (Fig. 4). A relatively frustrated character of the binding energy surface can be observed with many energetically similar but structurally different local minima

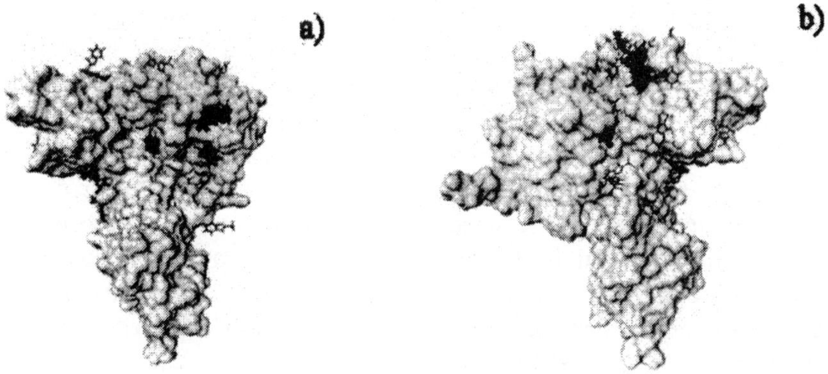

Figure 3. (a) 'Front' and (b) 'back' views of the crystal structure for the 1:1 hGH–hGHbp complex with the computationally engineered double mutation W104G/T175G. The 1000 docked bound conformations of the E8 inhibitor on the protein surface are shown. These conformations form structurally distinct clusters on the protein surface and visit buried regions, including the buried binding site cavity.

populated families of similar conformations, owing primarily to inherent difficulties of adequately sampling buried regions of the hGH—hGHbp redesigned interface. To analyze the energetics of ligand binding modes obtained from docking simulations, all clusters, including clusters containing a single structurally different conformation, were ranked according to their PL energy. We find that the low–energy binding modes correspond to structurally different ligand orientations that often occupy 'attractive ' alternative pockets on the protein surface, as evident from visual inspection and indicated by the high rmsd values that may range from rmsd = 20 Å to 30 Å from the lowest energy ligand conformation in the binding site cavity (Figs. 5a–9a).

The absence of a pronounced energy gap between alternative low–energy locations on the protein surface and the lowest energy conformational cluster, that luckily occupies the binding site cavity, is a signature of a frustrated energy landscape. Uncorrelated, rugged character of the underlying binding energy landscapes obtained with the PL energy function hinders a direct identification of the engineered cavity, even though buried regions of the interface are sampled during docking simulations. A marginal thermodynamic stability of the lowest energy cluster in this energetic model inhibits its robust kinetic accessibility in docking simulations and prevents a reliable detection of the binding site cavity.

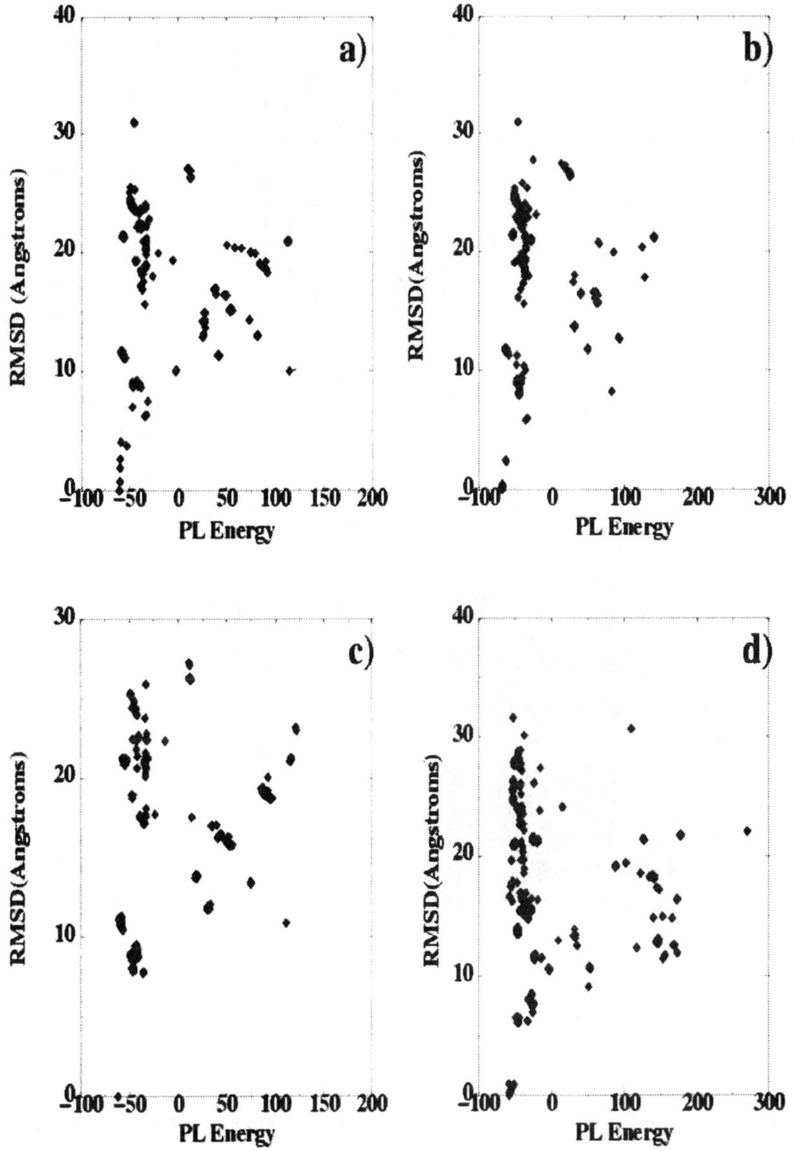

Figure 4. The scatter plots between the PL energies and the rmsd from the the predicted in docking simulations lowest energy conformations for D9 (a), Dm (b), H9 (c) and E8 (d).

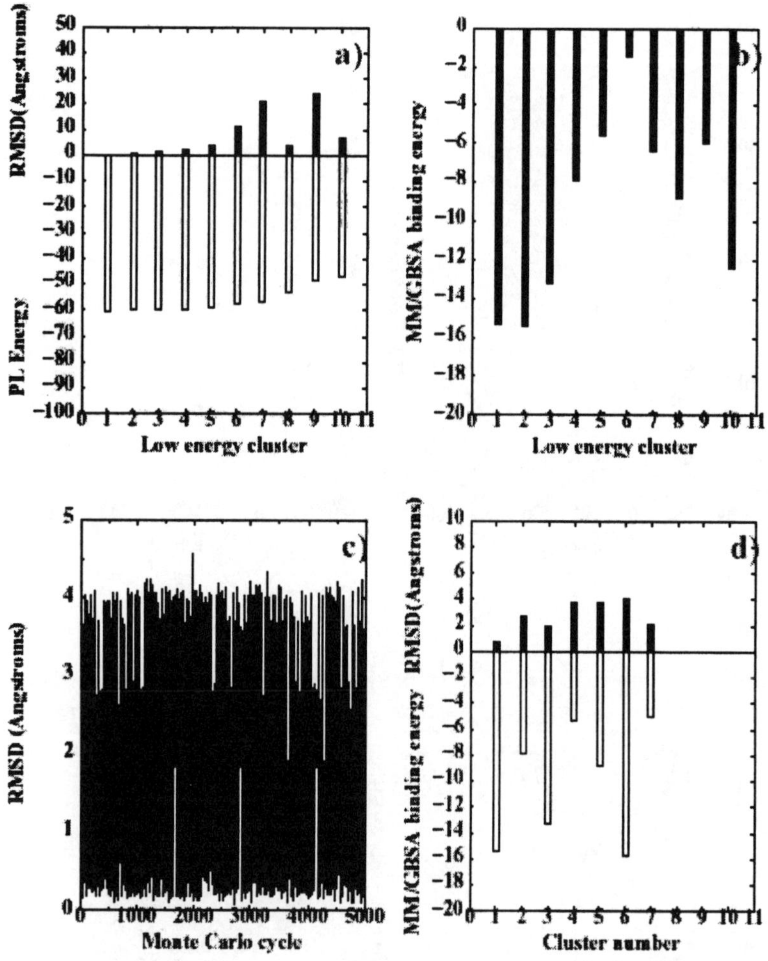

Figure 5. (a) The PL energies and the corresponding rmsd differences of the low–energy cluster centers for the D9 inhibitor from the reference lowest energy conformation determined in docking simulations. (b) MM/GBSA binding free energies for the 10 low–energy conformational families of D9 generated in docking simulations. (c) Time–dependent equilibrium history of D9 binding in the binding site cavity at T=300K. (d) MM/GBSA binding free energies of conformational clusters generated in equilibrium simulations at T=300K for D9 and the rmsd differences of the corresponding cluster centers from the reference lowest energy conformation of D9 determined in docking simulations.

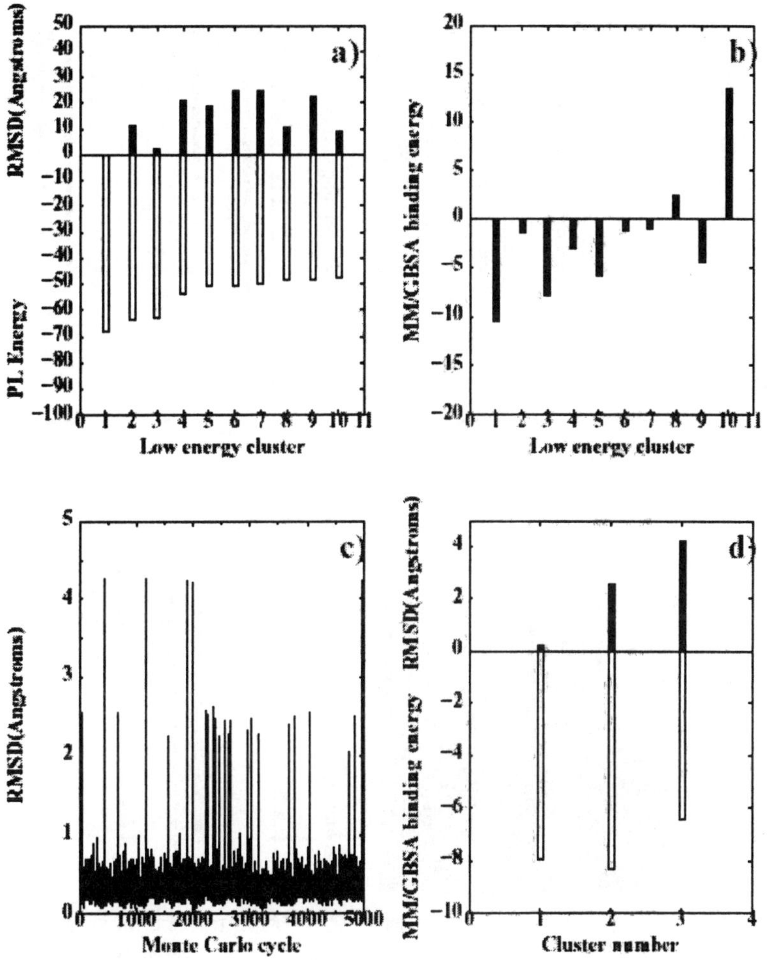

Figure 6. (a) The PL energies and the corresponding rmsd differences of the low–energy cluster centers for theDminhibitor from the reference lowest energy conformation determined in docking simulations. (b)MM/GBSAbinding free energies for the 10 low–energy conformational families of Dm generated in docking simulations. (c) Time–dependent equilibrium history of Dm binding in the binding site cavity at T=300K. (d) MM/GBSA binding free energies of conformational clusters generated in equilibrium simulations at T=300K for Dm and the rmsd differences of the corresponding cluster centers from the reference lowest energy conformation of Dm determined in docking simulations.

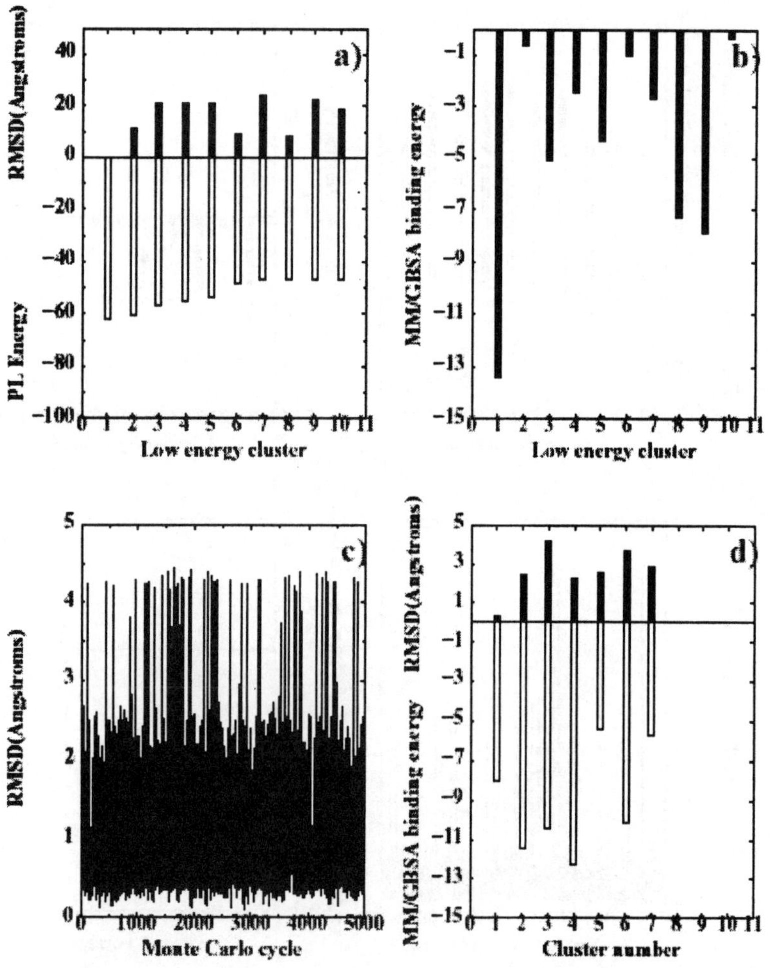

Figure 7. (a) The PL energies and the corresponding rmsd differences of the low–energy cluster centers for the H9 inhibitor from the reference lowest energy conformation determined in docking simulations. (b) MM/GBSA binding free energies for the 10 low–energy conformational families of H9 generated in docking simulations. (c) Time–dependent equilibrium history of H9 binding in the binding site cavity at T=300K. (d) MM/GBSA binding free energies of conformational clusters generated in equilibrium simulations at T=300K for H9 and the rmsd differences of the corresponding cluster centers from the reference lowest energy conformation of H9 determined in docking simulations.

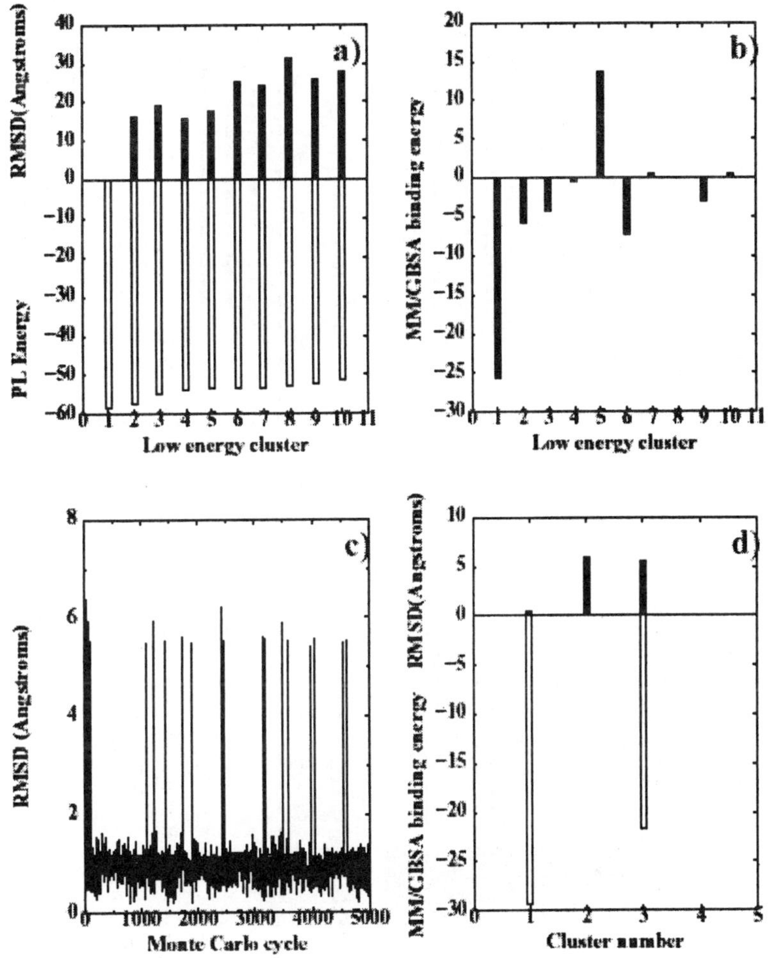

Figure 8. (a) The PL energies and the corresponding rmsd differences of the low–energy cluster centers for the E8 inhibitor from the reference lowest energy conformation determined in docking simulations. (b) MM/GBSA binding free energies for the 10 low–energy conformational families of E8 generated in docking simulations. (c) Time–dependent equilibrium history of E8 binding in the binding site cavity at T=300K. (d) MM/GBSA binding free energies of conformational clusters generated in equilibrium simulations at T=300K for E8 and the rmsd differences of the corresponding cluster centers from the reference lowest energy conformation of E8 determined in docking simulations.

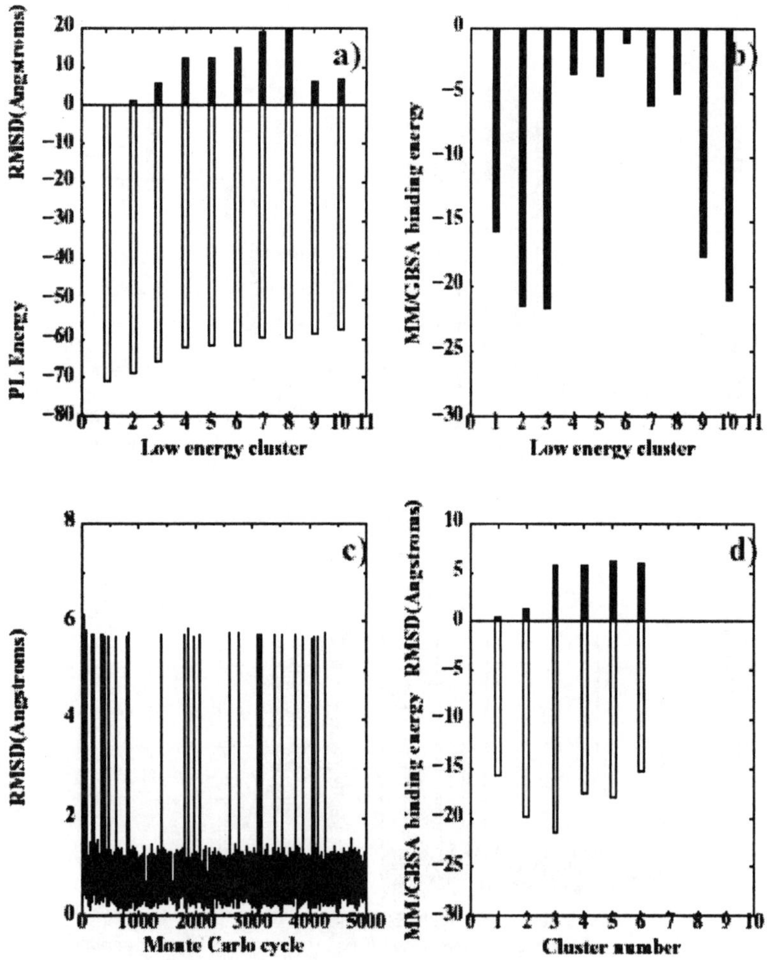

Figure 9. (a) The PL energies and the corresponding rmsd differences of the low–energy cluster centers for the F3 inhibitor from the reference lowest energy conformation determined in docking simulations. (b) MM/GBSA binding free energies for the 10 low–energy conformational families of F3 generated in docking simulations. (c) Time–dependent equilibrium history of F3 binding in the binding site cavity at T=300K. (d) MM/GBSA binding free energies of conformational clusters generated in equilibrium simulations at T=300K for F3 and the rmsd differences of the corresponding cluster centers from the reference lowest energy conformation of F3 determined in docking simulations.

These results suggest that the inability to accurately reproduce subtle differences in the relative ligand binding energies, rather than inherent difficulties in sampling buried cavities, is the bottleneck of this simple model. A sensitive and thermodynamically more accurate energetic model is necessary to distinguish between 'correct' and 'incorrect' binding sites. These results underscore that the desired synergy of the exhaustive sampling of the conformational space and accurate evaluation of the energetics is difficult to achieve in the framework of a single energy model.

The employment of the MM/GBSA energy model to compute binding free energies of the 10 low–energy conformational families generated in docking simulations dramatically changes the frustrated energetics of the ligand binding modes. A considerably better separation is achieved now for the lowest energy cluster of ligand conformations that occupy the binding site cavity and emulate the position of W104 residue by the appropriate positioning of the benzimidazole ring (Figs. 5b-9b). A rather small difference between energetically similar cluster centers for D9 ligand with the binding energies ranging from 15.3 kcal/mol to -13.2 kcal/mol results in concomitant minor deviations in rmsd from·the lowest energy cluster corresponding to 0.7 Å and 1.8 Å respectively (Fig.5b). As the rmsd values from the lowest energy cluster increase to 2.6 Å and 4.0 Å for other conformational families, the binding free energies drop considerably to -7.8 and -5.5 kcal/mol. Hence, insignificant structural departures from the lowest energy cluster cause in this model only minor energy deviations, resulting in a more correlated character of the energy landscape. For Dm (Fig. 6b), H9 (Fig. 7b), and E8 (Fig. 8b) ligands, a single favorable low–energy cluster of ligand conformations, binding in the designed cavity, is separated by an appreciable energy gap from other structurally different families of low–energy bound conformations. For the F3 inhibitor, the binding free energy assessment using MM/GBSA model changes the ranking of the low–energy clusters, leading to lower energy values for cluster centers 2,3, and 10, which differ by rmsd = 1.2 Å , 5.7 Å , and 6.5 Å from the predicted from docking simulations lowest energy conformation (Fig. 9b). The observed structural differences between low–energy binding modes of F3 are caused by a 180 degrees ligand flip in the binding site cavity; the benzimidazole core in this binding mode still has some overlap with the position of the W104 indole ring in the wild-type hGH–hGHbp complex.

The proposed protocol which utilizes a hierarchy of energy models of different accuracy and complexity, provides robust exploration of low–energy conformations that bind to various pockets on the protein surface and allows to detect the binding site cavity at the intermolecular interface of the remodeled hGH–hGHbp complex. The favorable binding free energies for the low–energy clusters of ligand conformations, that bind in the engineered cavity, are primarily determined by the favorable van der Waals interactions and nonpolar solvation contribution. These results agree with computational alanine scanning of the hGH–hGHbp complex, in which alanine mutation for the W104 and W169 residues resulted in a considerable reduction of the van der Waals contacts at the binding interface. In the absence of the crystal structures for studied inhibitors bound at the remodeled hGH—hGHbp interface, a direct comparison with the experimental data is possible only with the binding affinities of the ligand–protein complexes. Structural and mutagenesis data have indicated that the precise presentation of elements mimicking the indole rings of W104 and W169 is an essential feature for ligands to restore the binding affinity of the wild-type complex, and therefore, the panel of studied inhibitors must bind in the hot spot of the intermolecular interface.

To further understand a structural diversity of ligand bound conformations in the binding site cavity that could fulfill an important functional requirement of mimicking critical mutated residues in the hot spot, equilibrium simulations are performed for studied ligands using simulated tempering in conjunction with the simplified PL energy function. During equilibrium simulations in the binding site cavity, we monitor whether putative ligand binding modes result in a unique structural orientation of the benzimidazole core motif that emulates the indole ring of W104 and whether the requirement of structural compatibility with the mutated hot spot residues is a unique structural attribute for the low–energy ligand conformations to bind in the engineered cavity.

Parallel simulated tempering dynamics and subsequent energetic refinement of the major structurally different conformational clusters reveals a considerable diversity of low–energy ligand binding modes in the binding site cavity for all active benzimidazole-containing inhibitors dated (Figs. 5-9). The functionally important hot spot residues at the intermolecular interface of the wild–type hGH—hGHbp complex are located near the center of the contact epitope, with the key interactions represented by alkyl–aromatic stacking contacts (Fig. 10a). The aliphatic groups of charged residues D171, K172, and T175 from the hGH pack tightly against the W104 and W169 receptor residues; the alkyl portion of R43 resides beneath the indole ring of W169 and makes direct electrostatic interactions with D171 and T175 residues. We find that a sufficiently large cavity at the remodeled interface created upon double W104G/T175G mutation (Fig. 10b) is capable of accommodating structurally different ligand binding modes for the benzimidazole-containing inhibitors which can complement this defect and restore most of the lost interactions.

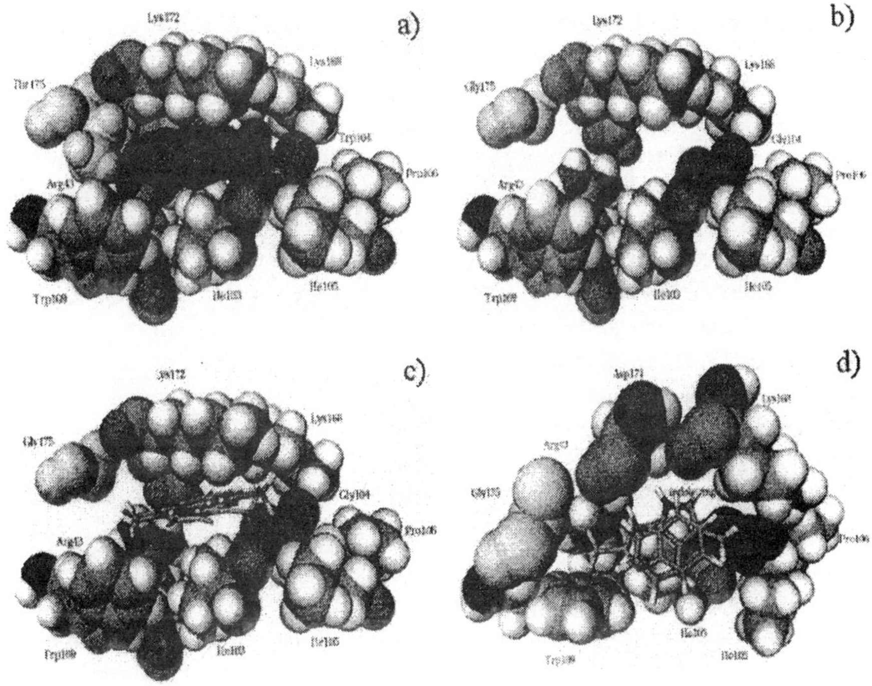

Figure 10. Space–filling CPK models for the functionally important hot spot residues at the intermolecular interface of the crystal structure wild–type hGH—hGHbp complex. (b) Space–filling CPK models for the hot spot residues in the remodeled hGH–hGHbp mutant interface. The double mutation W104G/T175G results in a formation of a sufficiently large cavity in the interface that can accommodate small molecules complementing this defect. (c) The low–energy cluster centers generated from equilibrium simulations for D9 inhibitor at T=300K in the binding site cavity. An overlap is seen with the indole ring of W104 residue from the wild-type complex. (d) The predicted orientation of the D9 inhibitor in the lowest energy cluster obtained from equilibrium simulations.

For D9 inhibitor, at least three low–energy binding modes are well populated in equilibrium simulations at room temperature (Fig. 5c,d). These binding modes deviate by rmsd = 0.8 Å , 1.88 Å , and 4.0 Å from the lowest energy structure (Fig. 5c,d) that serves as the reference state in calculations. However, these seemingly alternative binding modes differ by a 180 degrees ligand flip in the binding cavity, whereas still intimately mimicking the position of the W104 hot spot residue from the wild-type complex (Fig. 10c,d). The predicted orientation of the benzimidazole core in the largest low–energy cluster of D9 closely overlaps W104 side–chain with the 2-methyl group off the benzimidazole protruding further into the cavity and filling the position of the side–chain of T175 in the wild-type complex (Fig. 5c,d).

Similar results are obtained for Dm(Fig. 6c,d) and H9 inhibitors (Fig. 7c,d). Energetically comparable low–energy clusters are observed that differ by 4.25 °A from the reference Dm structure (Fig. 6c,d) and by 4.12 Å from the H9 reference structure (Fig. 7c,d). The low–energy ligand binding modes in the engineered cavity are presented for Dm (Fig. 11a,b) and H9 (Fig. 11c,d). While the lowest energy binding modes for Dm and H9 are very similar, they differ by about 180 degrees flip from the lowest energy cluster determined for D9 inhibitor. Nevertheless, the benzimidazole core mimics even more precisely the position of W104 residue in energetically favorable binding modes of Dm and H9 with the 5-methyl group coming off the benzimidazole ring and occupying a vacancy in the binding cavity left by T175 side–chain in the wild-type complex (Fig. 11). The 2-methyl group in Dm interacts with the alkyl portion of the K168 and K172 side–chains, which is consistent with the observed pattern in the wild-type complex where key interactions in the hGH–hGHbp hot spot are alkyl–aromatic stacking interactions, particularly between the alkyl portion of K172 of hGH and W104 of the hGHbp. The observed structural variations in the binding modes of D9, Dm, and H9 inhibitors do not interfere with the critical recognition requirement to mimic the W104 indole ring and provide a favorable entropic contribution to the binding free energies of these inhibitors. The discovered flexibility of the low-energy ligand conformations may also reflect the ability of hydrophobic interactions to confer similar affinity in a nondirectional way by tolerating minor packing imperfections.

We find that structurally dissimilar and low-energy binding modes that differ by 180 degrees flip may be allowed even for larger and more potent E8 and F3 inhibitors (Fig. 12). Interestingly, structural orientation of the benzimidazole core motif for these inhibitors in the lowest energy clusters also overlaps with the indole ring position of the W104 residue of the wild-type complex. The bulky trichloromethyl group of E8 and the thiazole group of F3 in the 2-position are pushed somewhat outwards in this binding mode to avoid possible short contacts with G175 and interacts favorably with P61, I103, I105 and the aliphatic group of K172 (Fig. 12). In the second largest low energy cluster, that represents a structurally different binding mode for E8 inhibitor, the trichloromethyl group of the 5-chloro-2-trichloromethyl benzimidazole attempts to fill the vacancy in the binding site cavity left by T175 side–chain in the wild-

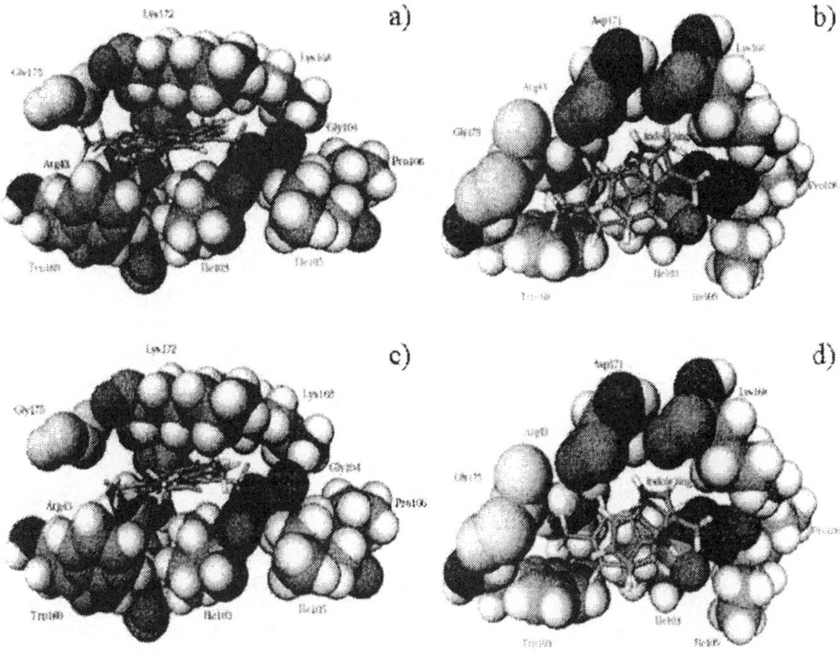

Figure 11. (a) The low–energy cluster centers generated from equilibrium simulations for Dm inhibitor at T=300K in the binding site cavity. An overlap is seen with the indole ring of W104 residue from the the wild-type complex. (b) The predicted orientation of the Dm inhibitor in the lowest energy cluster obtained from equilibrium simulations. (c) The low–energy cluster centers generated from equilibrium simulations for H9 inhibitor at T=300K in the binding site cavity. An overlap is seen with the indole ring of W104 residue from the the wild-type complex. (d) The predicted orientation of the H9 inhibitor in the lowest energy cluster obtained from equilibrium simulations.

type complex. This low–energy binding mode utilizes favorable van der Waals contacts in the binding site cavity with W169 and adds to the bulk of critical nonpolar contacts mildly stabilizing electrostatic interactions with R43 residue (Fig. 12a,b). It is quite possible that local protein conformational changes produced during minimization procedure of the complexes in the MM/GBSA model are not entirely sufficient to capture the degree of structural rearrangements necessary to accommodate large trichloromethyl group of E8 and the thiazole group of F3 in the optimal proximity of the G175 residue.

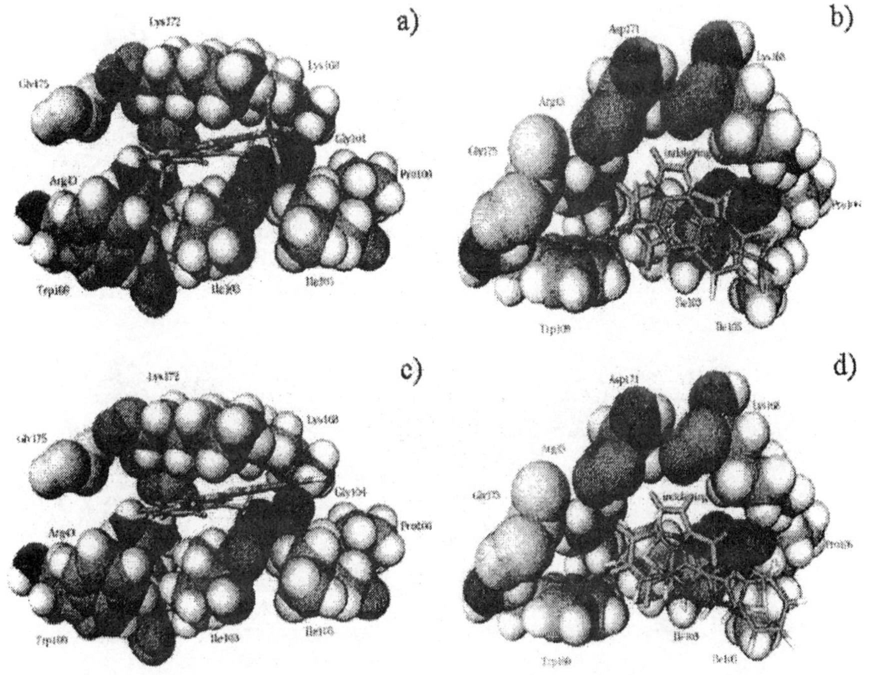

Figure 12. (a) The low–energy cluster centers generated from equilibrium simulations for E8 inhibitor at T=300K in the binding site cavity. An overlap is seen with the indole ring of W104 residue from the the wild-type complex. (b) The predicted orientation of the E8 inhibitor in the lowest energy cluster obtained from equilibrium simulations. (c) The low–energy cluster centers generated from equilibrium simulations for F3 inhibitor at T=300K in the binding site cavity. An overlap is seen with the indole ring of W104 residue from the the wild-type complex. (d) The predicted orientation of the F3 inhibitor in the lowest energy cluster obtained from equilibrium simulations.

Nevertheless, the diversity of low–energy ligand binding modes for studied inhibitors are generally compatible with the structural orientation of the benzimidazole core motif that mimics the location of the W104 indole ring in the crystal structure of the wild-type complex.

A convincing correlation between the predicted and experimental binding affinities is obtained based on binding free energy calculations of a single lowest energy conformational cluster determined after screening the results of docking simulations (Fig. 13a). By comparing the computed binding free energies determined from equilibrium simulations with the experimentally observed dissociation constants (Fig. 13b), a similar agreement is seen. As a result, we infer that there may be a larger diversity of low energy conformations in the binding cavity than was originally suggested in docking simulations. However, these ligand binding modes are still compatible with the position of W104 residue. Structural orientations of the ligands that considerably overlap with W104 position, primarily by using the benzimidazole core motif, is a prerequisite for the low–energy ligand conformations binding in the engineered cavity. The results point to a critical role of the intermolecular van der Waals interactions in restoring binding affinity of the remodeled hGH–hGHbp mutant complex. The bulk of the ligand-protein interactions is determined by the appropriate positioning of the benzimidazole core motif in place of mutated W104 residue. The total electrostatics contribution is mildly destabilizing for the majority of the inhibitors, except the most potent E8 ligand, where the electrostatic intermolecular interactions are not fully offset by the unfavorable polar solvation.

There are a number of limitations of the proposed hierarchical approach in its current implementation, including only a partial account of the protein flexibility in the MM/GBSAmodel and lack of the solute entropy contribution that was not included in the model due to typical significant variances in its values. The entropy change of proteins upon binding, that can be estimated by either the normal mode analysis or the quasi-harmonic approximation, is extremely difficult to evaluate for the hGH—hGHbp system because of a large size in the all-atom model. In the computational alanine scanning analysis of the hGH—hGHbp system, the results of MM/PBSA simulations without inclusion of the solute entropy term were in a good agreement with the experiment[42]. Another, more serious concern could arise from the assumption of only local conformational changes during binding of the inhibitors. In the framework of the MM/GBSA approach, the ensembles of structures for the uncomplexed protein and ligand are generated by using all conformations from a given cluster. Separation into the protein and ligand structures is followed by minimization of the complexes as well as the unbound protein and unbound ligand for each member of the cluster. This generally allows to capture local protein conformational changes in response to motions of the ligand binding mode. However, in the absence of the crystal structures of the complexes with the inhibitors, this still leaves a considerable ambiguity regarding a possibility of a large global conformational change, observed for instance in the remodeled W104A mutant interface in the presence of functional changes in residues

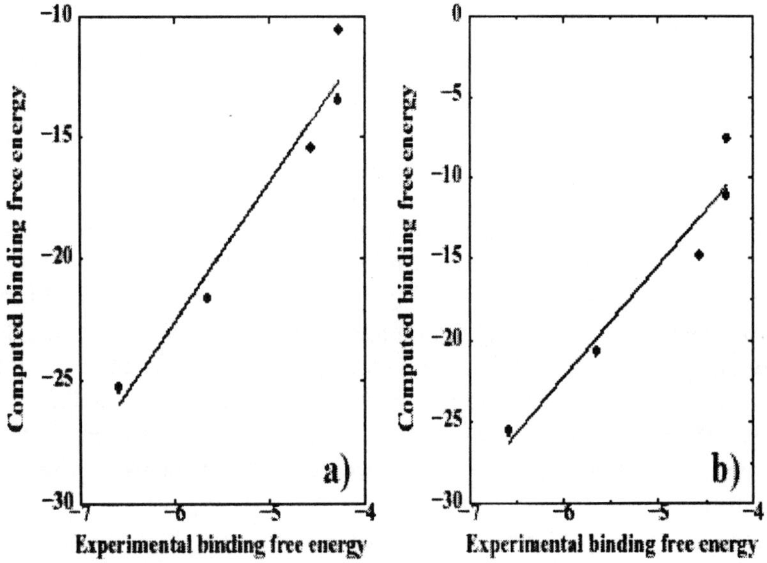

Figure 13. (a) Correlation between the computed and experimental binding free energies based on binding free energy calculations over a single lowest energy conformational cluster determined after screening the results of docking simulations. The value of correlation coefficient R= 0.97 with the regression coefficient (slope) of 5.77. (b) Correlation between the computed and experimental binding free energies determined from equilibrium simulations based on binding free energy contributions averaged over all conformations families of low–energy binding modes that are characterized by an overlap of the benzimidazole coremotif with the indole ring of the key W104 residue of the receptor. The value of correlation coefficient R = 0.96 with the regression coefficient (slope) of 6.85.

K168R, D171T, K172Y, E174A, and F176Y. It is quite clear that there no current modeling techniques that could have predicted or could have adequately simulated the details of these large structural changes. Obviously, only the solution of the crystal structures for the W104G/T175G mutant complex with the panel of studied inhibitors can ultimately validate or disprove some structural suggestions made in our study. Arguably, in the absence of evolutionary pressure to mutate neighboring hot spot residues, as was the case in the phage display remodeling experiments, it may be energetically disadvantageous to rearrange the entire interface to compensate for the loss of interactions, and only small conformational changes near the mutational site may occur. Similar arguments were also presented to explain a good agreement with the experimental changes in binding affinity in computational alanine scanning of W104 and W169 residues[42]. The successful prediction of the relative binding energetics for a panel of studied ligands suggests that the determined ligand binding modes and accompanied local protein conformational changes captured in the MM/GBSA approach may represent a reasonable structural model for ligand binding at the remodeled hGH–hGHbp interface that allows to quantitatively reproduce differences in binding affinity.

4. CONCLUSIONS

A hierarchy of docking and scoring approaches is used to study binding of small molecules that act as molecular switches and restore binding affinity at the engineered cavity of the hGH-hGHbp remodeled interface. We have extended a computational strategy, proposed originally for a rapid screening of ligand conformations in a given binding site, to a problem of screening pockets and cavities on the protein surface and identifying the binding site hot spot at the intermolecular interface. Equilibrium simulations of ligand binding in the binding site cavity of the hGH–hGHbp mutant complex have revealed a certain diversity of binding modes that considerably overlap with the indole ring of W104 residue in the wild-type complex. Structural orientations of the ligands that considerably overlap with W104 position, primarily by using the benzimidazole core motif, is a prerequisite for the low–energy ligand conformations binding in the engineered cavity.

The presented approach combines equilibrium dynamics, structural clustering and binding free energy evaluations of the low–energy conformational clusters using MM/GBSA model and allows an adequate analysis of binding thermodynamics for studied inhibitors. Binding energetics is predicted in a good agreement with the experimental data for a panel of benzimidazole-containing compounds that complement this defect at the redesigned interface and restore the binding affinity of the wild-type hGH–hGHbp complex. The results have revealed an important role of the intermolecular van der Waals interactions in restoring binding affinity of the remodeled hGH–hGHbp mutant complex. The total electrostatic contribution is mildly destabilizing for all studied inhibitors, except the most potent E8 ligand, where the electrostatic intermolecular interactions are not fully offset by the unfavorable polar solvation. The results of this work agree with the computational alanine scanning analysis for the hGH–hGHbp complex and a comprehensive analysis of electrostatic effects in the hGH—hGHbp interface.

The proposed approach can be also applied to computational mapping of binding sites and finding optimal locations for organic solvents on protein surfaces, that was recently addressed in a conceptually similar hierarchical docking approach[92]. This hierarchical protocol could also present a convenient complementary strategy to computational mapping of potential binding sites in proteins and locating putative binding points using multiple copy simultaneous search (MCSS) approach[93-96]. A recruitment of unoccupied pockets near the active sites may be another potential application of the presented method, that could be useful in complementing experimental mapping techniques, such as the multiple solvent crystal structures method that can locate and characterize consensus binding sites capable of binding small molecules[97,98]. Understanding structural and energetic characteristics of the hot spots in protein–protein and ligand–protein interfaces by using experimental and computational approaches can ultimately facilitate rational and combinatorial inhibitor design[99].

REFERENCES

1. McCammon, J.A., 1998, Theory of biomolecular recognition. *Curr. Opin. Struct. Biol.* **8**: 245-249.
2. Muller-Dethlefs, K., Hobza, P., 2000, Noncovalent interactions: a challenge for experiment an theory. *Chem. Rev*, **100**: 143-167.
3. Gane, P.G., Dean, P.M., 2000, Recent advances in structure-based rational drug design. *Curr. Opin. Struct. Biol.* **10**: 401-404.
4. Davis, A.M., Teague, S.J., 1999, Hydrogen bonding, hydrophobic interactions, and failure of the rigid receptor hypothesis. *Angew Chem. Int. Ed. Engl.* **38**: 736-749.
5. Van Regenmortel, M.H., 1999, Molecular recognition in the post-reductionist era. *J. Mol. Recognit.* **12**: 1-2.
6. Carlson, H.A., McCammon, J.A., 2000, Accommodating protein flexibility in computational drug design. *Mol. Pharmacol.***57**: 213-218.
7. DeLano, W.L., 2002, Unraveling hot spots in binding interfaces : progress and challenges. *Curr. Opin. Struct. Biol.* **12**: 14-20.
8. Jones, S., Thornton, J.M., 1996, Principles of protein–protein interactions. *Proc. Natl. Acad. Sci. U S A* **93**: 13-20.
9. Lo Conte, L., Chothia, C., Janin, J., 1999, The atomic structure of protein-protein recognition sites. *J. Mol. Biol.* **285**: 2177-2198.
10. Bogan, A.A., Thorn, K.S., 1998, Anatomy of hot spots in protein interfaces. *J. Mol. Biol.* **280**: 1-9.

11. Hu, Z., Ma, B., Wolfson, H., 2000, Nussinov R. Conservation of polar residues as hot spots at protein interfaces. *Proteins* **39**: 331-342.
12. Ma, B., Wolfson, H.J., Nussinov, R., 2001, Protein functional epitopes: hot spots, dynamics and combinatorial libraries. *Curr. Opin. Struct. Biol.* **11**: 364-369.
13. Sundberg, E.J., Mariuzza, R.A., 2000, Luxury accommodations: the expanding role of structural plasticity in protein-protein interactions. *Structure Fold Des.* **8**: R137-R142.
14. Frauenfelder, H., Leeson, D.T., 1998, The energy landscape in non-biological and biological molecules. *Nat. Struct. Biol.* **5**: 757-759.
15. Rejto, P.A., Freer, S.T., 1996, Protein conformational substates from X-ray crystallography. *Prog. Biophys. Mol. Biol.* **66**: 167-196.
16. Tsai, C.J., Ma, B., Nussinov, R., 1999, Folding and binding cascades: shifts in energy landscapes. *Proc. Natl. Acad. Sci. U S A* **96**: 9970-9972.
17. Kumar, S., Ma, B., Tsai, C.J., Sinha, N., Nussinov, R., 2000, Folding and binding cascades: dynamic landscapes and population shifts. *Protein Sci.* **9**: 10-19.
18. Demchenko, A.P., 2001, Recognition between flexible protein molecules : induced and assisted folding. *J. Mol. Recognit.* **14**: 42-61.
19. Forman-Kay, J.D., 1999, The 'dynamics' in the thermodynamics of binding. *Nature Struct .Biol.* **6**: 1086-1087.
20. Lee, A.L., Kinnear, S.A., Wand, A.J., 2000, Redistribution and loss of side chain entropy upon formation of a calmodulin-peptide complex. *Nat. Struct. Biol.* **7**: 72-77.
21. Zidek, L., Novotny, M.V., Stone, M.J., 1999, Increased protein backbone conformational entropy upon hydrophobic ligand binding. *Nat. Struct. Biol.* **6**: 1118-1121.
22. Cavanagh, J., Akke, M., 2000, May the driving force be with you–whatever it is. *Nat. Struct. Biol.* **7**: 11-13.
23. DeLano, W.L., Ultsch, M.H., de Vos, A.M., Wells, J.A., 2000, Convergent solutions to binding at a protein-protein interface. *Science* **287**: 1279-1283.
24. Tondi, D., Slomczynska, U., Costi, M.P., Watterson, D.M., Ghelli, S., Shoichet, B.K., 1999, Structure-based discovery and in-parallel optimization of novel competitive inhibitors of thymidylate synthase. *Chem. Biol.* **6**: 319-331.
25. Verkhivker, G.M., Bouzida, D., Gehlhaar, D.K., Rejto, P.A., Freer, S.T., Rose, P.W., 2002, Monte Carlo simulations of the peptide recognition at the consensus binding site of the constant fragment (Fc) of human immunoglobulin G : the energy landscape analysis of a hot spot at the intermolecular interface. *Proteins* **48**: 539-557.
26. Wells, J.A., 1991, Systematic mutational analyses of protein–protein interfaces. *Methods Enzymol.* **202**: 390-411.
27. Wells, J.A., de Vos, A.M., 1993, Structure and function of human growth hormone: implications for the hematopoietins. *Annu. Rev. Biophys. Biomol. Struct.* **22**: 329-51
28. Wells, J.A., 1996, Binding in the growth hormone receptor complex. *Proc. Natl. Acad. Sci. U S A* **93**: 1-6.
29. Cunningham, B.C., Jhurani, P.,Ng, P.,Wells, J.A., 1989, Receptor and antibody epitopes in human growth hormone identified by homolog-scanning mutagenesis. *Science* **243**: 1330-1336.
30. Cunningham, B.C., Wells, J.A., 1989, High-resolution epitope mapping of hGH-receptor interactions by alanine-scanning mutagenesis. *Science* **244**: 1081-1085.
31. Cunningham, B.C., Ultsch, M., De Vos, A.M., Mulkerrin, M.G., Clauser, K.R., Wells, J.A., 1991, Dimerization of the extracellular domain of the human growth hormone receptor by a single hormone molecule. *Science* **254**: 821-825.

32. Ultsch, M., de Vos, A.M., Kossiakoff, A.A., 1991, Crystals of the complex between human growth hormone and the extracellular domain of its receptor. *J. Mol. Biol.* **222**: 865-868.
33. de Vos, A.M., Ultsch, M., Kossiakoff, A.A., 1992, Human growth hormone and extracellular domain of its receptor: crystal structure of the complex. *Science* **255**: 306-312.
34. Kossiakoff, A.A., Somers, W., Ultsch, M., Andow, K., Muller, Y.A., de Vos, A.M., 1994, Comparison of the intermediate complexes of human growth hormone bound to the human growth hormone and prolactin receptors. *Protein Sci.* **3**: 1697-1705.
35. Somers,W., Ultsch, M., de Vos, A.M., Kossiakoff, A.A., 1994, The X-ray structure of a growth hormone–prolactin receptor complex. *Nature* **372**: 478-481.
36. Bass, S.H., Mulkerrin, M.G., Wells, J.A., 1991, A systematic mutational analysis of hormone-binding determinants in the human growth hormone receptor. *Proc. Natl. Acad. Sci. U S A* **88**: 4498-44502.
37. Cunningham, B.C., Wells, J.A., 1993, Comparison of a structural and a functional epitope. *J. Mol. Biol.* **234**: 554-563.
38. Clackson, T.,Wells, J., 1995, A hot spot of binding energy in a hormone–receptor interface. *Science* **267**: 383-386.
39. Jin, L.,Wells, J.A., 1994, Dissecting the energetics of an antibody-antigen interface by alanine shaving and molecular grafting. *Protein Sci.* **3**: 2351-2357.
40. Pearce, K.H.Jr., Ultsch, M.H.,Kelley, R.F., de Vos, A.M.,Wells, J.A., 1996, Structural and mutational analysis of affinity-inert contact residues at the growth hormonereceptor interface. *Biochemistry* **35**: 10300-10307.
41. Clackson, T., Ultsch, M.H., Wells, J.A., de Vos, A.M., 1998, Structural and functional analysis of the 1:1 growth hormone:receptor complex reveals the molecular basis for receptor affinity. *J. Mol. Biol.* **277**: 1111-1128.
42. Huo, S., Massova, I.,Kollman, P.A., 2002, Computational alanine scanning of the 1:1 human growth hormone-receptor complex. *J. Comput. Chem.* **23**: 15-27.
43. Atwell, S., Ultsch, M., De Vos, A.M., Wells, J.A., 1997, Structural plasticity in a remodeled protein-protein interface. *Science* **278**: 1125-1128.
44. Lowman, H.B., Wells, J.A., 1993, Affinity maturation of human growth hormone by monovalent phage display. *J. Mol. Biol.* **234**: 564-578.
45. Schiffer, C., Ultsch, M., Walsh, S., Somers, W., de Vos, A.M., Kossiakoff, A., 2002, Structure of a phage display-derived variant of human growth hormone complexed to two copies of the extracellular domain of its receptor: evidence for strong structural coupling between receptor binding sites. *J. Mol. Biol.* **316**: 277-289.
46. Bishop, A., Buzko, O., Heyeck-Dumas, S., Jung, I., Kraybill, B., Liu, Y., Shah, K., Ulrich, S., Witucki, L., Yang, F., Zhang, C., Shokat, K.M., 2000, Unnatural ligands for engineered proteins: new tools for chemical genetics. *Annu. Rev. Biophys. Biomol. Struct.* **29**: 577-606.
47. Bishop, A.C., Buzko, O., Shokat, K.M., 2001, Magic bullets for protein kinases. *Trends Cell. Biol.* **11**: 167-172.
48. Shogren-Knaak, M.A., Alaimo, P.J., Shokat, K.M., 2001, Recent advances in chemical approaches to the study of biological systems. *Annu. Rev. Cell Dev. Biol.* **17**: 405-433
49. Alaimo, P.J., Shogren-Knaak, M.A., Shokat, K., 2001, Chemical genetic approaches for the elucidation of signaling pathways. *Curr. Opin. Chem. Biol.* **5**: 360-367.
50. Specht, K.M., Shokat, K.M., 2002,The emerging power of chemical genetics. *Curr. Opin. Cell Biol.* **14**: 155-159.

51. Guo, Z., Zhou, D., Schultz, P.G., 2000, Designing small-molecule switches for protein-protein interactions. *Science* **288**: 2042-2045.
52. Verkhivker, G.M., Bouzida, D., Gehlhaar, D.K., Rejto, P.A., Freer, S.T., Rose, P.W., 2003, Computational detection of the binding site hot spot at the remodeled human growth hormone–eceptor interface. *Proteins* **53**: 201-219.
53. Rejto, P.A., Verkhivker, G.M., 1996, Unraveling principles of lead discovery: from unfrustrated energy landscapes to novel molecular anchors. *Proc. Natl. Acad. Sci. USA* **93**: 8945-8950.
54. Verkhivker, G.M., Rejto, P.A., 1996, A mean field model of ligand–protein interactions, Implications for the structural assessment of human immunodefi-ciency virus type 1 protease complexes and receptor–specific binding. *Proc. Natl. Acad. Sci. USA* **93**: 60-64.
55. Verkhivker, G.M., Rejto, P.A., Gehlhaar, D.K., Freer, S.T., 1996, Exploring energy landscapes of molecular recognition by a genetic algorithm, analysis of the requirements for robust docking of HIV–1 protease and FKBP–12 complexes. *Proteins* **25**: 342-353.
56. Verkhivker, G.M., Bouzida, D., Gehlhaar, D.K., Rejto, P.A., Schaffer, L., Arthurs, S., Colson, A.B., Freer, S.T., Larson, V., Luty, B.A., Marrone, T., Rose, P.W., 2001, Hierarchy of simulation models in predicting molecular recognition mechanisms from the binding energy landscapes. Structural analysis of the peptide complexes with SH2 domains. *Proteins* **45**: 456-470.
57. Verkhivker, G.M., Bouzida, D., Gehlhaar, D.K., Rejto, P.A., Arthurs, S., Colson, A.B., Freer, S.T., Larson, V., Luty, B.A., Marrone, T., Rose, P.W., 2000, Deciphering common failures in molecular docking of ligand–protein complexes. *J. Comput. Aided Mol. Des.* **14**: 731-751.
58. Verkhivker, G.M., Bouzida, D., Gehlhaar, D.K., Rejto, P.A., Schaffer, L., Arthurs, S., Colson, A.B., Freer, S.T., Larson, V., Luty, B.A., Marrone, T., Rose, P.W., 2002, Hierarchy of simulation models in predicting structure and energetics of the Src SH2 domain binding to the tyrosyl phosphopeptides. *J. Med. Chem.* **45**: 72-89.
59. Bouzida, D., Rejto, P.A., Arthurs, S., Colson, A.B., Freer, S.T., Gehlhaar, D.K., Larson, V., Luty, B.A., Rose, P.W., Verkhivker, G.M., 1999, Computer simulations of ligand–protein binding With ensembles of protein conformations: a Monte Carlo study of HIV–1 protease binding energy landscapes. *Int. J. Quantum. Chem.* **72**: 73-84.
60. Bouzida, D., Rejto, P.A.,Verkhivker, G.M., 1999, Monte Carlo simulations of ligand–protein binding energy landscapes With the weighted histogram analysis method. *Int. J. Quantum. Chem.* **73**: 113-121.
61. Verkhivker, G.M., Rejto, P.A., Bouzida, D., Arthurs, S., Colson, A.B., Freer, S.T., Gehlhaar, D.K., Larson, V., Luty, B.A., Marrone, T., Rose, P.W., 1999, Towards understanding the mechanisms of molecular recognition by computer simulations of ligand-protein interactions. *J. Mol. Recognit.* **12**: 371-389.
62. Mayo, S.L., Olafson, B.D., Goddard, W.A., 1990, III. DREIDING: A generic force field for molecular simulation. *J. Phys. Chem.* **94**: 8897-8909.
63. Weiner, S.J., Kollman, P.A., Case, D.A., Singh, U.C., Chio, C., Alagona, G., Profeta, S., Weiner, P., 1984, A new force field for molecular mechanical simulation of nucleic acids and proteins. *J. Am. Chem. Soc.* **106**: 765-784.
64. Still, W.C., Tempczyk, A., Hawley, R.C., Hendrickson, T., 1990, Semianalytical treatment of solvation for molecular mechanics and dynamics. *J. Am. Chem. Soc.* **112**: 6127-6129.
65. Mohamadi, F., Richards, N.G.J., Guida, W.C., Liskamp, R., Lipton, M., Caufield, C., Chang, G., Hendrickson, T., Still, W.C., 1990, MacroModel-an integrated software system for

modeling organic and bioorganic molecules using molecular mechanics. *J. Comput. Chem.* **11**: 440-467.

66. Qiu, D., Shenkin, P.S., Hollinger, F.P., Still, W.C., 1997, The GB/SA continuum model for solvation. A fast analytical method for the calculation of approximate born radii. *J. Phys. Chem.* **A 101**: 3005-3014.
67. Weiser, J., Weiser, A.A., Shenkin, P.S., Still, W.C., 1998, Neighbor-list reduction: optimization for computation of molecular van der Waals and solventaccessible surface areas. *J. Comput. Chem.* **19**: 797-808.
68. Weiser, J., Weiser, A.A., Shenkin, P.S., Still, W.C., 1998, Erratum: Neighbor-list reduction: optimization for computation of molecular van der Waals and solvent-accessible surface areas. *J. Comput. Chem.* **19**: 1110.
69. Weiser, J., Shenkin, P.S., Still, W.C., 1999, Approximate atomic surfaces from linear combinations of pairwise overlaps (LCPO). *J. Comput. Chem.* **20**: 217-230.
70. Weiser, J., Shenkin, P.S., Still, W.C., 1999, Fast, approximate algorithm for detection of solvent-inaccessible atoms. *J. Comput. Chem.* **20**: 586-596.
71. Srinivasan, J., Cheatham, T.E., Cieplak, P.,Kollman, P.A., Case, D.A., 1998, Continuum solvent studies of the stability of DNA, RNA and phosphoramidate-DNA helices. *J. Am. Chem. Soc.* **120**: 9401-9409.
72. Massova, I., Kollman, P.A., 1999, Computational alanine scanning to probe proteinprotein interactions: a novel approach to evaluate binding free energies. *J. Am. Chem. Soc.* **121**: 8133-8143.
73. Chong, L.T., Duan, Y., Wang, L., Massova, I., Kollman, P.A., 1999, Molecular dynamics and free-energy calculations applied to affinity maturation in antibody 48G7. *Proc. Natl. Acad. Sci. USA* **96**: 14330-14335.
74. Kuhn, B., Kollman, P.A., 2000, A ligand that is predicted to bind better to avidin than biotin: insights from computational fluorine scanning. *J. Am. Chem. Soc.* **122**: 3909-3916.
75. Kuhn, B., Kollman, P.A., 2000, Binding of a diverse set of ligands to avidin and streptavidin: an accurate quantitative prediction of their relative affinities by a combination of molecular mechanics and continuum solvent models. *J. Med. Chem.* **43**: 3786-3791.
76. Lee, M.R; Duan, Y.; Kollman, P.A., 2000, Use of MM-PB/SA in Estimating the Free Energies of Proteins: Application to Native, Intermediates, And Unfolded Villin Headpiece. *Proteins* **39**: 309-316.
77. Wang, W., Kollman, P.A., 2000, Free energy calculations on dimer stability of the HIV protease using molecular dynamics and a continuum solvent model. *J. Mol. Biol.* **303**: 567-582.
78. Kollman, P.A., Massova, I., Reyes, C., Kuhn, B., Huo, S., Chong, L., Lee, M., Lee, T., Duan, Y.,Wang, W., Donini, O., Cieplak, P., Srinivasan, J., Case, D.A., Cheatham, T.E., 2000, III. Calculating structures and free energies of complex molecules: combining molecular mechanics and continuum models. *Acc. Chem. Res.* **33**: 889-897.
79. Tsui, V., Case, D.A., 1992, Molecular dynamics simulations of nucleic acids with a generalized Born solvation model. *J. Amer. Chem. Soc.* **122**: 2489-2498.
80. Marinari, E., Parisi, G., 1992, Simulated tempering : a new Monte Carlo scheme. *Europhys. Lett.* **19**: 451-458.
81. Hukushima, K., Nemoto, K., 1996, Exchange Monte Carlo method and application to spin glass simulations. *J. Phys. Soc. (Jap)* **65**: 1604-1607.
82. Hansmann, U.H.E., Okamoto, Y., 1996, Monte Carlo simulations in generalized ensemble : multicanonical algorithm versus simulated tempering. *Phys. Rev.* **E 54**: 5863-5865.

83. Hansmann, U.H.E., Okamoto, Y., 1997, Generalized-ensemble Monte Carlo method for systems with rough energy landscape. *Phys. Rev.* **E 56**: 2228-2233.
84. Hansmann, U.H.E., Okamoto, Y., 1997, Numerical comparisons of three recently proposed algorithms in the protein folding problem. *J. Comput. Chem.* **18**: 920-933.
85. Hansmann, U.H.E., 1997, Parallel tempering algorithm for conformational studies of biological molecules. *Chem. Phys. Lett.* **281**: 140-150.
86. Sugita, Y., Okamoto, Y., 1999, Replica-exchange molecular dynamics method for protein folding. *Chem. Phys. Lett.* **314**: 141-151.
87. Verkhivker, G.M., Rejto, P.A., Bouzida, D., Arthurs, S., Colson, A.B., Freer, S.T., Gehlhaar, D.K., Larson, V., Luty, B.A., Marrone, T., Rose, P.W., 2001, Navigating ligand–protein binding free energy landscapes: universality and diversity of protein folding and molecular recognition mechanisms. *Chem. Phys. Lett.* **336**: 495-503.
88. Bouzida, D., Kumar, S., Swendsen, R.H., 1992, Efficient Monte Carlo methods for the computer simulation of biological molecules. *Phys. Rev.* **A 45**: 8894-8901.
89. Willet, P., Winterman, V., 1986, A Comparison of some measures for the determination of intermolecular structural similarity. *Quant. Struct-Act. Relat. Pharmacol. Chem. Biol.* **5**: 18-25.
90. Willet, P., Winterman, V., Bawden, D., 1986, Implementation of non–hierarchical cluster analysis methods in chemical information systems: selection of compounds for biological testing and clustering of substructure search output. *J. Chem. Inf. Comput. Sci.* **26**: 109-118.
91. Bawden, D., 1988, Browsing and clustering of chemical structures. In: *The international language of chemistry*. (W.A. Warr, ed.), Berlin:Springer–Verlag; p.145-150.
92. Dennis, S., Kortvelyesi, T., Vajda, S., 2002, Computational mapping identifies the binding sites of organic solvents on proteins. *Proc. Natl. Acad. Sci. USA* **99**: 4290-4295.
93. Caflisch, A., Miranker, A., Karplus, M., 1993, Multiple copy simultaneous search and construction of ligands in binding sites: application to inhibitors of HIV-1 aspartic proteinase. *J. Med. Chem.* **36**: 2142-2167.
94. Stultz, C.M., Karplus, M., 1999, MCSS functionality maps for a flexible proteins. *Proteins* **37**: 512-529.
95. Joseph-McCarthy, D., Tsang, S.K., Filman, D.J., Hogle, J.M,, Karplus, M., 2001, Use of MCSS to design small targeted libraries: application to picornavirus ligands. *J. Am. Chem. Soc.* **123**: 12758-12766.
96. Bitetti-Putzer, R., Joseph-McCarthy, D., Hogle, J.M., Karplus, M., 2001, Functional group placement in protein binding sites: a comparison of GRID and MCSS. *J. Comput. Aided. Mol. Des.* **15**: 935-960.
97. Mattos, C., Ringe, D., 1996, Locating and characterizing binding sites on proteins. *Nat. Biotechnol.* **14**: 595-599.
98. Ringe, D., Mattos, C., 1999, Analysis of the binding surfaces of proteins. *Med. Res. Rev.* **19**: 321-331.
99. Powers, R.A., Shoichet, B.K., 2002, Structure-based approach for binding site identification on AmpC beta-lactamase. *J. Med. Chem.* **45**: 3222-3234.

Molecular and Cellular Levels of Biological Evolution

MIROSLAV RADMAN
INSERM U 571, Faculté de Médécine – Necker Enfants Malades, Université René Descartes, 156 rue de Vaugirard, 75015 Paris, France

1. INTRODUCTION

Evolution is differential survival, or selection, of self-replicating entities. What is selected by given environmental conditions are specific traits, or phenotypes, that correspond to specific genotypes that assure the heritability ("memorization") of the selected traits. This is how Darwinian evolution works. The efficacy of Darwinian evolution depends on the repertoire of variability: the more diversity the higher the chance that a well-adapted variant will already be present within the repertoire. The wealth of genetic variation is directly proportional to the size of the population and to the rate of genetic change (mutation) in the course of replication. Mutations are genetic alterations driving the biological evolution and causing many, if not all, diseases. Mutation originates *via* two mechanisms: "vertical" variation is *de novo* change of one or few genomic (DNA or RNA) bases transmitted to the progeny, whereas "horizontal" variation occurs by genetic recombination creating new mosaics of already pre-existing sequences. The latter process is a true innovation whereas the former is useful for the fine-tuning of specificities or efficacies. This is how genomes of all life forms were evolving over millions of years, but this is also how the immune system creates hundreds of millions of different antibody encoding genes within a single human life in a small cellular compartment of B and T lymphocytes.

Long-term survival of natural populations relies on their capacity to adapt to unpredictable environmental changes *via* genetic variability that is

Supramolecular Structure and Function 8, Edited by Pifat-Mrzljak
Kluwer Academic/Plenum Publishers, New York 2004

generated by mutation and recombination. Until recently, it was generally assumed that the mutation rates are the lowest possible given the cost of anti-mutator mechanisms. The dogma of molecular and evolutionary biology was that mutation is an unavoidable stochastic event due to the limits of the precision of biological processes - a trade-off between the genetic fidelity and its cost. Evolutionary biologists have considered that the pre-existing biodiversity provides sufficient raw material for any selection, in particular for large bacterial populations, and that adaptive evolution works via selection from the pre-existing variability. The process of mutagenesis was not supposed to be affected by the selective pressures. This may be so when considering the global biodiversity, but selection usually acts locally on limited size populations, and there, the diversity is often limited, in particular for populations having experienced a recent selection (bottleneck) (Arjan, et al. 1999). The genetic « arms race » between parasites and hosts is particularly revealing. Viruses and bacteria mutate at high rates resulting in the generation of otherwise rare variants that can escape the host's immune system. In turn, the immune system mutates to create and test antibodies that recognize these new variants. Clearly, these and similar adaptive processes (e.g., the somatic evolution of tumours) are limited by the supply of mutations.

When adaptation is limited by the available genetic diversity, genetically unstable bacterial populations are expected to adapt, at least short term, more rapidly than non-mutators because of increased supply of mutations (Taddei, et al. 1997). Two classes of genes are known to accelerate mutation and/or recombination rates in bacterial populations: stress-inducible wild type genes, usually part of the SOS regulon, and genes whose functional loss, or down regulation, increases the rate of genetic variability (mutator and/or hyper-rec mutants). The coexistence of these two evolution-driving strategies, one acting at the level of individual cells, the other at the population level (Radman M., et al. 2000), will be discussed here as well as the molecular mechanisms involved.

2. GENETIC MUTATORS

Bacteria find sometimes their evolutionary success through a biochemical failure in their DNA replication or repair processes, a genetic defect that helps them not only to adapt but to become adaptable. Such mutants have lost some specific function, this loss resulting in increased rates of mutation (mutators) and/or recombination (hyper-rec mutants) everywhere in their genome. Clearly, most newly arising mutations either have no effect (silent mutations) or are harmful, whereas only very rare specific mutations are favorable under particular selective conditions. The ratio of deleterious to favorable mutations in random mutagenesis may be as high as four to five orders of magnitude (Taddei, et al. 1997). It came therefore as a surprise that high mutation rates are favored during adaptation in spite of the high cost incurred by the generation of numerous deleterious mutations (Sniegowski, et al. 1997; Taddei, et al. 1997). It is however not surprising that when adaptation is achieved, low mutation rates are favored (Tröbner and Piechocki 1984). Computer simulations (Taddei, et al. 1997; Tenaillon, et al. 1999; Tenaillon, et al. 2000) and experiments (Chao and Cox 1983; Tröbner and Piechocki 1984; Sniegowski, et al. 1997) are consistent with this paradigm.

It is possible to select, in the laboratory, bacterial mutants with increased fidelity of protein or DNA biosynthesis (Fijalkowska, et al. 1993) and references therein). Therefore, it is clear that there was no durable selective pressure in nature for maximal fidelity. It becomes apparent that selection for genetic fidelity is balanced by selection for efficiency and robustness. All approaches hint to trade-offs between the cost and benefit of mutations in the course of adaptive evolution. High mutation rates are costly, but when this is the only way to survive selection, then such cost is readily paid: successive selections for specific adaptive mutations will consistently lead to the second order selection of mutator mutants (Mao, et al. 1997; Taddei, et al. 1997; Tenaillon, et al. 1999). However, in spite of their short-term success, the cost of mutators is such (see below) that they are most likely the evolutionary dead ends (Funchain, et al. 2000), and the only long-term beneficiaries of the mutator activity may be the genes/alleles (or operons) created by them and then transferred to genetically stable cells (Tenaillon, et al. 2000; Denamur et al., 2000).

Theory and experiments show that successive selections for any particular mutations, or new alleles, will indirectly select also for high mutation or recombination rates (Mao, et al. 1997; Taddei, et al. 1997). This phenomenon is named second-order selection. How does the second-order selection work? The fraction of spontaneously arising mutator mutations in a bacterial laboratory culture is around 10^{-5} (Boe, et al. 2000). Strong mutators, such as the mismatch repair deficient mutators, increase the mutation rate by 100 to 1000-fold (Glickman and Radman 1980) and the interspecies recombination rate by about 1000-fold (Rayssiguier, et al. 1989). A million-fold enrichment for mutators, making them the predominant population, will occur after three successive selections for specific mutations (Mao, et al. 1997). This can be regarded as a short-term evolution of the capacity to evolve: by selecting for particular genetic events one co-selects for the general mechanism which produces such events, hence the second order selection. Metaphorically, one could say that, in terms of adaptive evolution, bacteria have now « learned how to learn » but, mechanistically, mutators increase in frequency bacause of their genomic association (hitch-hiking) with favorable mutations generated by the mutator activity. Therefore, sex and recombination are expected to lower the probability of mutator fixation (Tenaillon, et al. 2000).

Thus, for natural populations, one can expect that - depending on the way the new selected traits have been acquired (vertically, by mutation, or horizontally, by recombination) – mutator and/or hyper-rec mutants could be locally greatly enriched for. For example, new metabolic operons, or antibiotic resistance genes, can be acquired horizontally, but new metabolic specificities and some antibiotic resistance alleles can be generated by point mutations. Indeed, natural populations of bacteria bear much higher frequency of genetic mutators than one would expect from the rate of their emergence and the decreased fitness of individual mutator cells (see below).

3. THE COST OF GENETIC MUTATORS

In vitro (Chao and Cox 1983; Mao, et al. 1997; Sniegowski, et al. 1997) and in vivo (Giraud et al, 2001) experiments, and the computer simulations (Taddei, et al. 1997; Tenaillon, et al. 1999; Tenaillon, et al. 2000), all suggest that bacteria are better off being mutators while adapting and non-mutators when adapted. Funchain, et al. (2000) have shown that some mutator clones can go to extinction following successive bottlenecks. In terms of supply of mutations, the effect of increasing by 100-fold the mutation rate is equivalent to a 100-fold increase in population size (the amount of new mutations will be equivalent). But, unlike mutators, the adapted wild type does not bear the cost of the ongoing deleterious mutation load. Therefore, considering the "economy" of adaptation, bacterial populations would be better off being mutators while adapting and non-mutators when adapted. There is now compelling evidence that bacteria are endowed with such inducible mutator capacity [reviews, (Radman 1999; Radman et al. 2000)]

4. DNA REPAIR SYSTEMS

Aside from heredity and evolution, i.e., the fine balance between the maintenance of the memory of evolution (DNA sequence) and its variability necessary for adaptive evolution, DNA repair systems' prime role is the maintenance of the physical integrity of genomic DNA strands. Neither the dividing nor the non-dividing cells can function with DNA strand discontinuities, or chemical blocks, because neither replication nor transcription can fully operate. Presumably, the biological responses to DNA damage are so important for life because of this fundamental problem. Cells require the signaling systems to monitor faithfully any DNA strand discontinuity or non-readable DNA damage. First such cellular response to DNA damage was discovered in bacteria, the SOS response, which shares all principal hallmarks of the eukaryotic responses to DNA damage, i.e., the arrest of DNA replication and cell division and induction or enhancement of DNA repair activities until the DNA strands become clean and continuous. SOS response will be revisited below.

How is the stability of DNA sequence maintained? A plethora of proteins has evolved whose action, at many levels and through diverse strategies, results in the observed stability of genetic information (Friedberg, et al. 1995). The double-stranded structure is essential in allowing DNA protection and repair proteins to maintain the original nucleotide sequence in spite of frequent chemical insults to the fine DNA structure and the intrinsic imperfection of DNA replication machinery.

The basic strategies to conserve the DNA sequence involve : (a) The maintenance of the chemical purity of ingredients for DNA replication, i.e., the repair of DNA templates and the enzymatic chemical purification and equilibration of dNTP pools ; (b) The high fidelity DNA copy machine, i.e., efficient selection of correct dNTPs by DNA polymerase and the immediate removal of the mistake by the proofreading exonuclease and (c) The quality control of new strands, i.e., the postreplicative mismatch repair of the new strands to match the sequence of the template. The latter system lowers the frequency of spontaneous mutations by100 to 1000-fold.

5. EDITING OF DNA REPLICATION AND RECOMBINATION BY MISMATCH REPAIR

It is remarkable that the ubiquitous mismatch repair system not only edits DNA replication but also the homologous recombination, which itself is one of the most versatile DNA repair mechanisms for establishing DNA strand continuity (Friedberg, et al. 1995). Mechanistically, this double editing is fairly well understood. All mismatch repair systems appear to use the transient strand discontinuity (nick) to trigger and direct the repair event. In a small fraction of bacteria (those endowed with the Dam methylase for methylating adenine residues in the 5'GATC sequence) such nick is assured by the MutH protein which cuts the unmethylated (newly synthesized) strand at the GATC sequence proximal to the mismatch. The nicking occurs only when the MutS protein has detected the mismatch and attracted the MutL partner [review, (Modrich and Lahue 1996)]. The helicaseII in *E. coli* peels-off the strand from the free ends and the peeled-off strand is degraded by exonucleases. It may be that all newly synthesized DNA is discontinuous *in vivo* [see (Varlet, et al. 1991)] and that the ligase activity gives the impression of the leading/lagging strand asymmetry? Perhaps only the MutH utilizing bacteria can allow themselves to have a hyperactive DNA ligase? Alternatively, the direct interaction of mismatch repair proteins with the DNA polymerase complex may allow them to distinguish the template from the copy. The strand direction is one of the least understood details of mismatch repair for most of the bacterial and all eukaryotic species.

The mismatch-stimulated (as detected by MutS and MutL binding) localized helicase-catalysed melting of DNA may account for the editing of both DNA replication and recombination: in replication, the mismatched Okazaki fragment would be aborted, and in recombination, the mismatched invading strand would be aborted [(Rayssiguier, et al. 1989; Stambuk and Radman 1998)]. Hence, when MutS or MutL are not functional, the mutation rate goes up and the recombination between similar sequences is allowed [review, (Radman M. , et al. 1995; Modrich and Lahue 1996)]. Thus, we speak of the mutator phenotype and the relaxed genetic barrier between the species (plus the chromosomal rearrangements) phenotype. The latter phenotype allows for efficient horizontal transfer of genes among diverged related species (Denamur et al. 2001), which is often referred to as the hyper-rec phenotype which, however, does not apply to recombination between genetically identical partners (Vulic, et al. 1997).

6. MUTATION RATES IN NATURE

Mutation rates in microbes cultivated in the laboratory are generally very low (Drake, et al. 1998), but in natural populations of *Escherichia coli* and *Salmonella enterica*, strains with high mutation rates (mutators) often exceed a frequency of 1% (Treffers, et al. 1954; Jyssum 1960; Gross and Siegel 1981; Tröbner and Piechocki 1984; LeClerc, et al. 1996; Matic, et al. 1997). The fraction of strong mutators among Pseudomonas isolated from the lung of cystic fibrosis patients is about 20% (Oliver, et al. 2000).

Most of the strong mutator mutants that have emerged in nature and in the laboratory are mismatch repair deficient mutants. There does not appear to be any special reason for their abundance as compared with other mutator mutants : the large target size of the four mismatch repair genes, in particular the *mutS*, can account for their abundance (Bregeon, et al. 1999). Other potential mutator genes are either small (e.g., *mutT*) or essential for the viability (e.g., genes encoding replicative DNA polymerase).

However, the long-term robustness of mutator cells is compromised by the continuous production of deleterious mutations (Funchain, et al. 2000). It is likely that the high fraction of mutators reflects a recent, or ongoing, selective pressure. *In fine*, most mutator populations may well go to extinction but their selected favorable alleles may survive in other genomes following a horizontal transfer. Furthermore, mutator bacteria may be able to regain lower mutation rate once adaptation has been achieved (Denamur, et al. 2000).

7. THE SOS SYSTEM: 30TH ANNIVERSARY OF ITS PUBLICATION

In November 1970, I wrote and privately circulated a memo among the leading experts in the field of DNA repair and mutagenesis proposing that "*E. coli* possesses an inducible DNA repair system. This hypothetical repair, which we call SOS repair, is manifested only following damage to DNA and requires *de novo* protein synthesis. Mutagenesis by ultraviolet light is observed only under conditions of functional SOS repair: we therefore suspect that this is a mutation-prone repair." Because of the resistance from the scientific community, this was published four years later (Radman M. 1974). In fact, the initial term was « SOS replication » to imply that it is a trans-lesion DNA replication rather than a mechanism for removal of DNA damage [see [Friedberg, 1997] for an historical account]. The memo was one step forward from an earlier paper (Defais, et al. 1971) which showed that phage lambda UV induced mutagenesis, repair and lysogenic induction are apparently co-expressed. To this list, the memo has further added the already published inhibition of cell division (filamentous growth or filamentation) and prophage induction (Witkin 1967), and (as the key ingredient) the cellular mutagenesis, in a scenario of a co-regulated global cellular response to DNA damage. To honor Weigle's discovery of enhanced reactivation of UV irradiated phage lambda in UV irradiated host cells (which he called UV-reactivation (Weigle 1953) – a condition *sine qua non* to obtain phage mutagenesis – the phenomenon was renamed W-reactivation. In addition to the notion of coordinated regulation of metabolically unrelated genes, already present in Witkin's paper, the iconoclast aspect of the SOS hypothesis was the notion that bacterial mutagenesis is not an unavoidable stochastic even but a well regulated, genetically programmed process. The apparent smell of Lamarckism (bacteria mutate when needed!) was apparently scary at that time. Much later came the excitement, and controversy, of the Cairns et al (1988) paper claiming the gene-directed inducible mutagenesis that is independent of SOS [Cairns, 1991]. Neither claim, except the inducibility, resisted the experimental tests. Today, thanks to the discovery of two SOS induced mutagenic DNA polymerases and the studies of the effect of mutator alleles on the fitness of bacterial populations, it appears that SOS mutagenesis is a device to render the Darwinian selection more efficient. Eventually, the mutagenic SOS response to bacterial starvation (or aging) on plates, involving also the metabolic cAMP regulon, was discovered (Taddei, et al. 1995) and the role of SOS in adaptive mutagenesis established (McKenzie, et al. 2000). Hence, much of the inducible adaptive mutagenesis is an SOS phenomenon. The idea of « inducible evolution » (Radman 1980; Echols 1981) is gaining experimental support.

8. TRANS-LESION SYNTHESIS BY SOS POLYMERASES: A MECHANISM FOR SURVIVAL AND EVOLUTION

Damages, or lesions, in DNA templates that are non-instructive for base-pairing interrupt the DNA replication process by blocking the elongation of nascent chains. TLS is the DNA copying process that can overcome such blockage and allow the completion of DNA replication in spite of the presence of lesions. TLS was proposed to produce « targeted » mutations, i.e., mutations occurring at the site of DNA damage [review, Friedberg, 1995]. Initially, it was proposed that TLS is an error-prone DNA synthesis as a key component of the stress-inducible cellular SOS response in bacteria (Radman 1974). TLS was first demonstrated *in vivo,* by a physical analysis, as an SOS-induced process that allows completion of the complementary strand synthesis on UV-irradiated single-stranded phiX174 phage DNA following infection of irradiated *E. coli* host cells (Caillet-Fauquet, et al. 1977). This is the condition *sine qua non* to obtain UV induced mutagenesis of this single-stranded phage. Later on, DNA sequencing of UV-induced mutations and construction of phage DNA with single defined UV lesion, provided further evidence that the UV-induced mutations indeed occur opposite lesions, such as pyrimidine dimers [review, Friedberg, 1995]. However, even after the identification of inducible genes, and their products, which are specifically required for induced mutagenesis, the mechanism of the mutagenic TLS in bacteria remained elusive until 1999. Early attempts (1975-78) by one of us (M.R.) with S. Spadari, G. Villani and S. Boiteux to detect *in vitro* bacterial and mammalian radiation-induced error-prone DNA polymerases activities were unsuccessful and left us only the observation that the AMV reverse transcriptase is capable of copying accross UV lesions in phiX174 DNA which normally block *E. coli* DNA polymerases I and III.

The latter blockage was accompanied by the unproductive turnover of dNTPs to dNMPs by the successive DNA polymerase and proof-reading exonuclease activities – a process we have called polymerase idling (Villani, et al. 1978). Thus, at that time, we concluded that the TLS required at least the absence or inhibition of the 3' to 5' proof-reading exonuclease (Villani, et al. 1978).

9. POLYMERASES ACT AS DNA "BAND-AIDS" CAPABLE OF AVOIDING AND CREATING MUTATIONS

Isolation from bacteria and eukaryotes of numerous members of a superfamily of new mutagenic DNA polymerases, performing TLS in the test tube, was such a surprise that in year 1999 more reviews were published than the original papers. However, the SOS hypothesis that guided this research was there for nearly three decades and the pioneering work of C. Lawrence and colleagues on yeast TLS polymerases Rev1 and Rev3/7 (DNA polymerase zeta) was published since 1996 (Nelson, et al. 1996). The discoveries that the properties of *E. coli* DNA polymerases V (UmuD'C) and IV (DinB) can account adequately for genetic experiments was of fundamental importance: the former is responsible for mutations targeted to the lesion and the latter for the "untargeted" mutations [reviews, (Radman M. 1999; Friedberg, et al. 2000)]. Phage lambda clear plaque mutation assays allowed even a morphological distinction of targeted and untargeted mutations (Caillet-Fauquet, et al. 1984). DNA polymerases IV and V add nucleotides one-by-one in an error-prone fashion: PolV produces base substitution and frameshift mutations opposite the damaged site and PolIV produces predominantly frameshift mutations explained by its capacity to extend synthesis on unpaired termini (Wagner, et al. 1999). Thus, it appeared as if PolV allows for a mutation-generating TLS, whereas PolIV allows for a general transient mutator effect, as deduced from the observation that only the UV induced untargeted mutations are subject to mismatch repair (Caillet-Fauquet, et al. 1984) suggesting that the template bears no lesion (or an unstable lesion or structure). The former acitivity helps survival and both appear to increase genetic variability [review, (Radman M. 1999)]

10. RING DIFFERENT EVOLUTION-ACCELERATING STRATEGIES

Given the presented interpretation of different strategies for the accelerated adaptive evolution, it is puzzling that several percent of rather strong genetic mutators are found among natural isolates of enterobacteria collected from all over the world and from most diverse environments (LeClerc, et al. 1996; Matic, et al. 1997). Why have bacteria been tricked into the costly strategy of second order selection of strong genetic mutators when the inducible mutator activity is an alternative? It may be, however, that the "smart" systems of inducible mutators and hypermutable genes are not efficient enough for adaptive evolution. For instance, the hypermutable genes with microsatellite runs are useful only for adaptation to a single specific and recurrent selective pressure, e.g., the immune system (Field, et al. 1999). Although the inducible mutator systems act on the entire genome, they will not be producing diversity during the periods of "easy life" but will do so during non-lethal selection. Therefore, when a lethal selective pressure hits an isolated small population, the inducible mutators may not have had the opportunity to produce adaptive mutations because all bacteria would be instantly killed. Strong genetic mutators are particularly favored when adaptation requires several genomic mutations, which may be the case for most adaptations to complex environmental changes (Ninio 1991).

In anthropomorphic terms, we may consider genetic mutators as genetic altruists: they pay a high cost (due to accumulation of deleterious mutations) in order to provide the population with mutations/alleles that can save life under new and most diverse selective pressures. An extensive analysis of 787 natural isolates of Escherichia coli shows the co-existence of the two mutational strategies among natural isolates of *Escherichia* and *Shigella* species: strains with a strong inducible mutator capacity seem to accumulate less constitutive mutators than strains with weak inducible mutator activity (Bjedov et al. 2003). Undergoing research in our laboratory, both modeling and experimental, already shows some compelling evidence that genes which have evolved in the mutationally damaged genomes of strong mutators can be rescued by non-mutators, via sex and recombination (Tenaillon, et al. 2000). Thus, the mutator subpopulations may arise and live as ephemeral genetic engineers which are themselves evolutionary dead ends, but which are often the source of new adaptive alleles that are transmitted horizontally to the non-mutators. Alternatively, mutator bacteria that have generated adaptation without having the time to accumulate deleterious mutations may restore lower mutation rate by suppressor mutations, reversion or horizontal transfer of functional anti-mutator genes (Denamur, et al. 2000).

Over evolutionary times, horizontal gene transfer has repeatedly allowed transition from mismatch repair deficient to mismatch repair proficient status (Denamur, et al. 2000). This is evidenced by the incongruencies between phylogenies of these genes and housekeeping genes. Furthermore, there is about 90% correlation between the sequence mosaicisms of the different mismatch repair genes and their hyper-recombination phenotypes (Denamur, et al. 2000).

As the hyper-rec phenotype experienced by mismatch repair deficient bacteria is not limited to their own loci, they may also have generated a high degree of genomic mosaicism. Thus, the global mosaic structure of enterobacterial genomes (Milkman and McCane 1995) may have resulted from the described evolutionary scenarios, and the periods of mismatch repair deficiency may have been the periods of the most intense genetic engineering of bacterial genomes in the course of their adaptive evolution.

ACKNOWLEDGEMENTS

I acknowledge the collaboration and generous exchange of ideas with I. Matic, F. Taddei and E. C. Friedberg. Our studies are supported by the "Programme de Recherche Fondamentale en Microbiologie et Maladies Infectieuses et Parasitaires - MENRT", the Ligue contre le Cancer and the Association pour la Recherche sur le Cancer

REFERENCES

1. Arjan J.A., Visser M., Zeyl C.W., Gerrish P.J., Blanchard J.L. and Lenski R.E. 1999. Diminishing returns from mutation supply rate in asexual populations. *Science* **283**: 404-6.
2. Boe L., Danielsen M., Knudsen S., Petersen J.B., Maymann J. and Jensen P.R. 2000. The frequency of mutators in populations of Escherichia coli. *Mutat Res* **448**: 47-55.
3. Bregeon D., Matic I., Radman M. and Taddei F. 1999. Inefficient mismatch repair: genetic defects and down regulation. *Journal of Genetics* **78**: 21-28.
4. Caillet-Fauquet P., Defais M. and Radman M. 1977. Molecular mechanism of induced mutagenesis. 1. *In vivo* replication of the single-stranded ultraviolet-irradiated fX174 phage DNA in irradiated host cells. *J. Molec. Biol.* **117**: 95-112.
5. Caillet-Fauquet P., Maenhaut-Michel G. and Radman M. 1984. SOS mutator effect in *E. coli* mutants deficient in mismatch correction. *EMBO J.* **3**: 707-712.
6. Cairns J., Overbaugh J. and Miller S. 1988. The origins of mutants. *Nature* **335**: 142-145.
7. Cairns J. and Foster P.L. 1991. Adaptive reversion of a frameshift mutation in *Escherichia coli. Genetics* **128**: 695-701.
8. Chao L. and Cox E.C. 1983. Competition between high and low mutating strains of *Escherichia coli. Evolution* **37**: 125-134.
9. Defais M., Fauquet P., Radman M. and Errera M. 1971. Ultraviolet reactivation and ultraviolet mutagenesis of lambda in different genetic systems. *Virology* **43**: 495-503.

10. Denamur E., Lecointre G., Darlu P., Acquaviva C., Sayada C., Sunjevaric I., Rothstein R., Elion J., Taddei F., Radman M. and Matic I. 2000. Evolutionary implications of the frequent horizontal transfer of mismatch repair genes. *Cell*
11. Drake J.W., Charlesworth B., Charlesworth D. and Crow J.F. 1998. Rates of spontaneous mutation. *Genetics* **148**: 1667-1686.
12. Echols H. 1981. SOS functions, cancer and inducible evolution. *Cell* 25: 1-2.
13. Field D., Magnasco M.O., Moxon E.R., Metzgar D., Tanaka M.M., Wills C. and Thaler D.S. 1999. Contingency loci, mutator alleles, and their interactions. Synergistic strategies for microbial evolution and adaptation in pathogenesis. *Ann N Y Acad Sci* **870**: 378-82.
14. Fijalkowska I.J., Dunn R.L. and Schaaper R.M. 1993. Mutants of Escherichia coli with increased fidelity of DNA replication. *Genetics* **134**: 1023-30.
15. Foster P.L. 1999. Mechanisms of stationary phase mutation: a decade of adaptive mutation. *Annu Rev Genet* **33**: 57-88.
16. Friedberg E.C., Walker G.C. and Siede W. 1995.DNA repair and mutagenesis.
17. Friedberg E.C., Feaver W.J. and Gerlach V.L. 2000. The many faces of DNA polymerases: strategies for mutagenesis and for mutational avoidance [comment]. *Proc Natl Acad Sci U S A* **97**: 5681-3.
18. Funchain P., Yeung A., Stewart J.L., Lin R., Slupska M.M. and Miller J.H. 2000. The consequences of growth of a mutator strain of escherichia coli as measured by loss of function among multiple gene targets and loss of fitness]. *Genetics* **154**: 959-70.
19. Glickman B.W. and Radman M. 1980. *Escherichia coli* mutator mutants deficient in methylation-instructed DNA mismatch correction. *Proc. Natl. Acad. Sci. USA* **77**: 1063-1067.
20. Gross M.D. and Siegel E.C. 1981. Incidence of mutator strains in *Escherichia coli* and coliforms in nature. *Mutat. Res.* **91**: 107-110.
21. Harris R.S., Longerich S. and Rosenberg S.M. 1994. Recombination in adaptive mutation. *Science* **264**: 258-260.
22. Jyssum K. 1960. Observation of two types of genetic instability in *Escherichia coli. Acta Pathol. Microbiol. Immunol. Scand.* **48**: 113-120.
23. LeClerc J.E., Li B., Payne W.L. and Cebula T.A. 1996. High mutation frequencies among *Escherichia coli* and *Salmonella* pathogens. *Science* **274**: 1208-1211.
24. Mao E.F., Lane L., Lee J. and Miller J.H. 1997. Proliferation of mutators in a cell population. *J. Bacteriol.* **179**: 417-422.
25. Matic I., Radman M., Taddei F., Picard B., Doit C., Bingen E., Denamur E. and Elion J. 1997. Highly variable mutation rates in commensal and pathogenic *E. coli*. *Science* **277**: 1833-1834.
26. McKenzie G.J., Harris R.S., Lee P.L. and Rosenberg S.M. 2000. The SOS response regulates adaptive mutation. *Proc Natl Acad Sci U S A* **97**: 6646-51.
27. Milkman R. and McCane M. 1995.DNA sequence variation and recombination in *E. coli*.. In: *Population Genetics of Bacteria*, (S. Baumberg, J.P.W. Young, M.H. Wellington & J.R. Saunders, eds) pp 127-142, Cambridge University Press, Cambridge.
28. Modrich P. and Lahue R. 1996. Mismatch repair in replication fidelity, genetic recombination, and cancer biology. *Annu. Rev. Biochem.* **65**: 101-133.
29. Napolitano R., Janel-Bintz R., Wagner J. and Fuchs R.P.P. 2000. All three SOS-inducible DNA polymerases (Pol II, Pol IV and Pol V) are involved in induced mutagenesis. *EMBO J.* in press:
30. Nelson J.R., Lawrence C.W. and Hinkle D.C. 1996. Thymine-thymine dimer bypass by yeast DNA polymerase zeta. *Science* **272**: 1646-9.
31. Ninio J. 1991. Transient mutators: a semiquantitative analysis of the influence of translation and transcription errors on mutation rates. *Genetics* **129**: 957-962.

32. Oliver A., Canton R., Campo P., Baquero F. and Blazquez J. 2000. High frequency of hypermutable Pseudomonas aeruginosa in cystic fibrosis lung infection. *Science* **288**: 1251-1253.
33. Radman M. 1974.Phenomenology of an inducible mutagenic DNA repair pathway in *Escherichia coli*: SOS repair hypothesis.
34. Radman M., Dohet C., Bourguignon M.-F., Doubleday O.P. and Lecomte P. 1981.High fidelity devices in the reproduction of DNA.
35. Radman M., Matic I., Halliday J.A. and Taddei F. 1995. Editing DNA replication and recombination by mismatch repair: from bacterial genetics to mechanisms of predisposition to cancer in humans. *Phil. Trans. R. Soc. Lond. B* **347**: 97-103.
36. Radman M. 1999. Enzymes of evolutionary change. *Nature* **401**: 866-7, 869.
37. Radman M., Matic I. and Taddei F. 1999. Evolution of evolvability. *Ann. N. Y. Acad. Sci.* **870**: 146-155.
38. Radman M., Taddei F. and Matic I. 2000. Evolution-driving genes. *Res. Microbiol.* **151**: 91-95.
39. Rayssiguier C., Thaler D.S. and Radman M. 1989. The barrier to recombination between *Escherichia coli* and *Salmonella typhimurium* is disrupted in mismatch-repair mutants. *Nature* **342**: 396-401.
40. Sniegowski P.D., Gerrish P.J. and Lenski R.E. 1997. Evolution of high mutation rates in experimental populations of *E. coli*. *Nature* **387**: 703-705.
41. Stambuk S. and Radman M. 1998. Mechanism and control of interspecies recombination in *Escherichia coli*. I. Mismatch repair, methylation, recombination and replication functions. *Genetics* **150**: 533-542.
42. Taddei F., Matic I. and Radman M. 1995. Cyclic AMP-dependent SOS induction and mutagenesis in resting bacterial populations. *Proc. Natl. Acad. Sci. USA* **92**: 11736-11740.
43. Taddei F., Radman M., Maynard-Smith J., Toupance B., Gouyon P.H. and Godelle B. 1997. Role of mutator alleles in adaptive evolution. *Nature* **387**: 700-702.
44. Tenaillon O., Toupance B., Le Nagard H., Taddei F. and Godelle B. 1999. Mutators, population size, adaptive landscape and the adaptation of asexual populations of bacteria. *Genetics* **152**: 485-93.
45. Tenaillon O., Le Nagard H., Godelle B. and Taddei F. 2000. Mutator and Sex in Bacteria: Conflict between Adaptive Strategies. *PNAS in press*
46. Treffers H.P., Spinelli V. and Belser N.O. 1954. A factor (or mutator gene) influencing mutation rates in *E. coli*. *Proc. Natl. Acad. Sci (USA)* **40**: 1064-1071.
47. Tröbner W. and Piechocki R. 1984. Selection against hypermutability in *Escherichia coli* during long term evolution. *Mol. Gen. Genet.* **198**: 177-178.
48. Varlet I., Dohet C., Petranovic M., Radman M. and Brooks P. 1991. Mismatch repair strand discrimination in *Xenopus* is directed by strand termini and not by hemimethylation.
49. Villani G., Boiteux S. and Radman M. 1978. Mechanisms of ultraviolet induced mutagenesis: extent and fidelity of *in vitro* DNA synthesis on irradiated templates. *Proc. Natl. Acad. Sci* **75**: 3037.
50. Vulic M., Dionisio F., Taddei F. and Radman M. 1997. Molecular Keys to Speciation: DNA Polymorphism and the Control of Genetic Exchange in Enterobacteria. *Proc. Natl. Acad. Sci. USA* **94**: 9763-9767.
51. Wagner J., Gruz P., Kim S.R., Yamada M., Matsui K., Fuchs R.P. and Nohmi T. 1999. The dinB gene encodes a novel E. coli DNA polymerase, DNA pol IV, involved in mutagenesis. *Mol Cell* **4**: 281-6.

52. Wang Y.C., Maher V.M. and McCormick J.J. 1991. Xeroderma pigmentosum variant cells are less likely than normal cells to incorporate dAMP opposite photoproducts during replication of UV- irradiated plasmids. *Proc Natl Acad Sci U S A* **88**: 7810-4.
53. Weigle J.J. 1953. Induction of mutation in a bacterial virus. *Proc. Natl. Acad. Sci. USA* **39**: 628-636.
54. Witkin E.M. 1967. The radiation sensitivity of *Escherichia coli* B: a hypothesis relating filament formation and prophage induction. *Proc. Natl. Acad. Sci. USA* **57**: 1275-1279.
55. Zhang Y., Yuan F., Wu X. and Wang Z. 2000. Preferential incorporation of G opposite template T by the low-fidelity human DNA polymerase iota. *Mol Cell Biol* **20**: 7099-108.

Index